DISCARDED
JENKS LRC
GORDON COLLEGE

Fundamentals in computer vision

Fundamentals in computer vision

An advanced course

Edited by

O.D. FAUGERAS

Institut National de Recherche en Informatique
et en Automatique (INRIA)

CAMBRIDGE UNIVERSITY PRESS
Cambridge
London   New York   New Rochelle
Melbourne   Sydney

Published by the Press Syndicate of the University of Cambridge
The Pitt Building, Trumpington Street, Cambridge CB2 1RP
32 East 57th Street, New York, NY 10022, USA
296 Beaconsfield Parade, Middle Park, Melbourne 3206, Australia

© Cambridge University Press 1983

First Published 1983

Printed in Great Britain at the University Press, Cambridge

Library of Congress catalogue card number: 82—14624

British Library Cataloguing in Publication Data

Computer vision
1. Image processing
I. Faugeras, O.D.
621 . 3819'598   TA1632

ISBN 0 521 25099 4

CONTENTS

Preface   vii
Contributors   ix

I Preprocessing
Kunt   *Acquisition and Visualization*   1
Abramatic   *Two-Dimensional Digital Signal Processing*   27
Granlund, Knuttsson, Wilson   *Image Enhancement*   57
Abramatic   *Digital Image Restoration*   69
Kunt   *Image Coding*   91

II Feature Extraction
Levialdi   *Edge Extraction Techniques*   117
Haralick   *Image Texture Survey*   145
Rosenfeld   *Motion: Analysis of Time-Varying Imagery*   173
Rosenfeld   *"Intrinsic Images": Deriving Three-Dimensional Information About a Scene From Single Images*   185

III Segmentation
Rosenfeld   *Digital Geometry: Geometric Properties of Subsets of Digital Images*   197
Haralick   *Image Segmentation Survey*   209
Rosenfeld   *Segmentation: Pixel-Based Methods*   225
Levialdi   *Basic Ideas for Image Segmentation*   239

IV Shape
Henderson   *Feature-Based 2-D Shape Models*   263
Henderson   *Syntactic and Structural Methods I*   273
Henderson   *Syntactic and Structural Methods II*   283
Faugeras   *3-D Shape Representation*   293
Faugeras   *Conversion Algorithms Between 3-D Shape Representations*   305
Rosenfeld   *Hierarchical Representation: Computer Representations of Digital Images and Objects*   315

V Control and Knowledge Representation
Latombe, Lux   *Basic Notions in Knowledge Representation and Control for Computer Vision*   325

Rosenfeld	*Relaxation: Pixel-Based Methods*	373
Faugeras	*Relational Structure Matching and Relaxation Labeling*	385

VI Hardware and Software

Kruse	*Algorithms and Hardware*	401
Kruse	*Hardware Structures for Parallel Picture Processing*	417
Kruse	*State-Of-The-Art Systems For Pictorial Information Processing*	425
Granlund, Arvidsson	*The GOP Image Computer*	443
Levialdi	*Languages for Image Processing*	459
Krusemark, Haralick	*Achieving Portability in Image Processing Software Packages*	479

PREFACE

This book is intended to cover some of the major areas in Computer Vision. It contains six main Sections concerned with Preprocessing, Feature Extraction, Image Segmentation, Shape Representation, Knowledge Representation and Control, Hardware and Software. These Sections follow more or less logically from an analysis of the various building blocks of a possible Computer Vision System. Their order does not reflect any hierarchy and is simply a pedagogical compromise.

The Section on Preprocessing is concerned with various ways of transforming an image into another image : converting an analog image into a digital one and vice versa (I.1) convolving a digital image with a given impulse response (I.2), getting a better looking or easier to analyse image (I.3), eliminating a degredation (I.4), compressing the number of bits without loosing quality (I.5).

Section II on Feature Extraction is concerned with various ways of extracting more symbolic information from an image : Edges (II.1), Textures (II.2), Motion (II.3), Distance (II.4).

Section III on Segmentation first discusses the problems related to the use of a discrete grid for representing continuous surfaces (III.1). A general survey of Segmentation techniques is given in III.2, Pixel based techniques are described in III.3 and Region based techniques are introduced in III.4.

Section IV introduces the problems of Shape representation. For planar shapes feature based techniques are first discussed (IV.1). Syntactic and Structural methods are then presented (IV.2, IV.3). The three-dimensional case is discussed in IV.4 and IV.5. Hierarchical representations are introduced in IV.6

Section V deals with the problem of the control of the various elements involved in the Vision process as well as with the related question of representing knowledge. Basic Notions are introduced in V.1. A special type of control structure, relaxation labeling is described at the pixel level (V.2) and at a higher symbolic level (V.3).

Hardware and Software implementations of some of the algorithms and techniques described in Sections I to V are described in Section VI. Hardware issues are dealt with in VI.1 to VI.4, Software issues such as languages for Image Processing and Portability are discussed in VI.5 and VI.6.

CONTRIBUTORS

Fundamentals in Computer Vision
An advanced course
Paris, June 1982

Course sponsored by The Commission of the European Community
Organized by INRIA (Institut National de Recherche en
Information et en Automatique)

Course Director
O.D. Faugeras,
INRIA, Domaine de Voluceau—Rocquencourt,
78153 Le Chesnay Cedex,
FRANCE.
Scientific Secretary
M. Amirchahy,
INRIA, Domaine de Voluceau—Rocquencourt,
78153 Le Chesnay Cedex,
FRANCE.

Contributors
J.F. Abramatic,
Domaine de Voluceau—Rocquencourt,
78153 Le Chesnay Cedex,
FRANCE.

J. Arvidsson,
Linkoeping University,
Valla, 581 83 Linkoeping,
SWEDEN.

O.D. Faugeras,
Domaine de Voluceau—Rocquencourt,
78153 Le Chesnay Cedex,
FRANCE.

G.H. Granlund,
Picture Processing Laboratory,
Linkoeping University,
Linkoeping,
SWEDEN.

R.M. Haralick,
Virginia Polytechnic Institute and State University,
Blacksburg, Virginia 24061,
U.S.A.

Contributors

T.C. Henderson,
Department of Computer Science,
The University of Utah,
Salt Lake City, Utah 84112,
U.S.A.

H. Knutsson,
Picture Processing Laboratory,
Linkoeping University,
Linkoeping,
SWEDEN.

B. Kruse,
IMTEC,
Image Technology AB,
Box 5047,
S-58005 Linkoeping
SWEDEN.

S.W. Krusemark,
Department of Electrical Engineering,
Virginia Polytechnic Institute and State University,
Blacksburg, VA24061,
U.S.A.

M. Kunt,
Laboratoire de Traitement des Signaux de l'EPF-Lausanne,
16, Chemin de Bellerive,
CH-1007 Lausanne,
SUISSE.

J.C. Latombe,
Laboratoire IMAG,
BP 53 X,
38041 Grenoble Cedex,
FRANCE.

S. Levialdi,
Institute of Information Sciences University of Bari,
ITALIE.

A. Lux,
Laboratoire IMAG,
BP 53 X,
38041 Grenoble Cedex,
FRANCE.

A. Rosenfeld,
Computer Vision Laboratory,
Computer Science Center,
University of Maryland,
College Park, MD 20742,
U.S.A.

R. Wilson,
Department of Electrical and Electonical Engineering.
The University of Aston in Birmingham,
19 Coleshill Street,
Birmingham B47BP,
ENGLAND.

ACQUISITION AND VISUALIZATION

Murat KUNT
Laboratoire de
Traitement des Signaux
de l'EPF-Lausanne
16, chemin de Bellerive
CH-1007 Lausanne

1. Introduction

Image processing in general and computer vision in particular, require the acquisition and the visualization of images. These operations are carried out using image sensors and displays. From the technological point of view, an image can be interpreted as an energy distribution over a surface. This energy, electromagnetic in nature, is radiated by different sources covering an important part of the spectrum.

Image sensors generate usually a time varying analog signal which represent the spatial energy variations in the image. Digital processing of images requires the sampling and the quantization of this signal. When the digital processing is completed, the digital image should very often be transformed into an analog signal for display. Throughout the whole chain, the human observer may intervene at any given level to observe the image.

In this chapter, we first review the most commonly used image sensors and displays. Because of the important role played by the human observer, the human visual system is discussed from neurobiological and engineering point of view. The sampling and the quantization of image signals are described. Finally, important pre and post processing operations are summarized.

2. Major imaging and display devices [7]-[8]

2.1 *General remarks*

Image sensors should scan a two dimensional area. Usually a spot of light is directed to a sub-area of the surface to be scanned. Either the transmitted (transparency film) or the reflected (photographic paper or opaque surface) light is analysed to determine the energy transmitted or reflected by this sub-area. The spot size is usually made small enough so that the numerical value extracted from the analog signal corresponds to a very small area of the image. Depending on how the scanning is made, sensors may be classified in three cathegories :

1. Electronic scanning
2. Electromechanical scanning
3. Mechanical scanning.

An image sensor is characterized by a number of features. The main ones are :
- signal to noise ratio : a measure of the useful information extracted from the sensor's signal;
- dynamic range : possible values representing the variation range of the light energy;
- resolution : a measure of the smallest detail in the image which can be retained by the sensor;
- transfer function : relationship between the incoming light energy and the output signal;
- integration time : time interval in which the sensor accumulates charges generated by the incoming light;
- reading speed : determines the scanning time for a given resolution and picture size.
- spectral sensitivity : indicates the portion of the electromagnetic spectrum for which the sensor can be used.

2.2 *Electronic scanning devices*

Electronic scanning image sensors are the fastest. They are generally based on electronic tubes, matrices of photodiodes or charge transfer devices. The main one are briefly described hereafter :

- image dissector camera : the image dissector camera is an electronic tube which has a photo-emissive cathode at one end converting incoming photons into electrons (fig. 1). The scene to be examined is projected on the photocathode by means of a lens. A deflection system (frequently magnetic) focuses the electron beam on an aperture plate. Only those electrons emitted by an elementary area of the photocathode pass through the aperture and reach the electron multiplier to produce the signal current. Since there is no integration, the image dissector camera is less sensitive than other tubes.

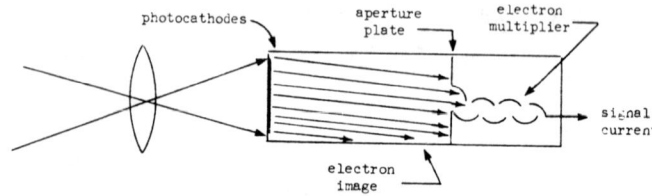

Fig. 1 : The image dissector camera

Its resolution may reach 8000 points across its diameter. Its spectral sensitivity covers the range from ultraviolet to infrared.

Acquisition and visualization

- <u>Image orthicon camera</u> : the image orthicon camera (fig. 2) is made up by three sections : imaging, scanning and multiplication. The incoming light hits the photocathode which emits electrons proportional to the intensity of the light. These electrons are focused on a target where a secondary emission takes place. A mesh placed in front of the target collects the secondary emission electrons which can be viewed as a gain factor.

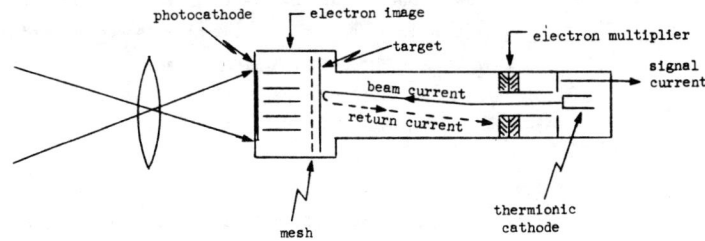

Fig. 2 : The image orthicon camera

The back side of the target (scanning section) is scanned periodically by an electron beam which neutralizes the accumulated charge on an element-by-element basis. The surplus beam current goes back to the vicinity of the electron gun where it enters an electron multiplier. The output current is the difference between the beam current and the accumulated charges on the target. Its main disadvantage is the high noise level for dark areas due to the reflected beam.

- <u>Vidicon camera</u> : the vidicon camera converts an optical image into an electron flow by means of a photoconductive target. The target (see fig. 3) is in contact with a transparent and conductive signal plate at the front of the camera. The increase of illumination decreases the resistance of the photoconductive plate. An electron beam is swept periodically on the back

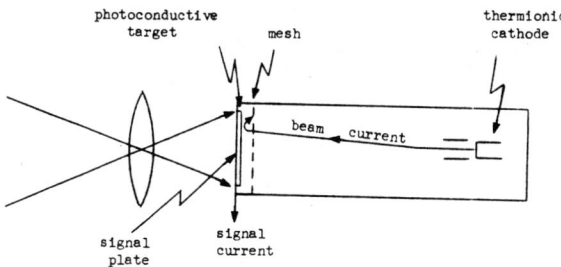

Fig. 3 : The vidicon camera

of the target depositing charges retained by capacitance across the target. Between two scanning each local capacitors is discharged proportionally to the local incoming light. The output signal is obtained by coupling capa-

ditively the signal plate to the element charging current. In the vidicon
camera, the photoconductive plate is simultaneously the transducer and the
integrator. This limits some of its performances. Its signal to noise ratio
is not very high. It exibits luminance shading and has a limited resolution

- <u>Other tubes</u> : there is a large variety of other electron tubes used as camera.
  A non exhaustive list is the following : Isocon, phimbicon, Leddicon,
  Vistacon, Saticon, chalnion, Silicon-target vidicon, Ultraviolet and infrared
  vidicon, pyricon vidicon, x-ray vidicon. They are not used as widely as the
  ones described above.

- <u>Solid state arrays</u> : present day technology offers three types of solid state
  two-dimensional arrays : photodiodes matrices, charge coupled device matrices
  (CCD) and charge injection device matrices (CID). Photodiode matrices can be
  as large as 100 x 100 with two dimensional addressing. CCD matrices are
  obtained by assembling linear CCD arrays. In CID matrices, reading and
  writing are independant and separated. This makes them suitable for frame
  averaging. More details on these devices are given in the next section for
  linear solid state arrays.

2.3 *Electromechanical scanning devices*

Electromechanical scanning devices are slower than fully electronic scanning
devices because one of the dimensions of the image field is scanned mechanically.
At each step of the mechanical scanning an entire picture line is scanned
electronically. This can be achieved by photodiode arrays, by charge coupled
devices or by charge coupled photodiodes.

- <u>Photodiode arrays</u> : these arrays are scanned by multiplexing. Each element
  is made of a photodiode cascaded with a MOS transistor. The scanning generator
  is a digital shift register with parallel output. Each one of these outputs
  is connected to the gate of a transistor. The scanning is started with a
  pulse loading the first bit of the shift register. While a bit (0 or 1) is
  shifted in the register, it opens or closes each of the MOS switches. At the
  sampling of each diode, the charge accumulated is discharged in the output
  line. Arrays as large as 2048 elements are available with present day
  technology.

- <u>Charged coupled devices</u> (CCD) : the principle of charge coupled devices
  results from the charge transfer devices, i.e., analog shift registers.
  In addition, the MOS transistor used are produced specially by covering
  a semiconductor with an isolator and a metallic electrode to make them
  photosensitive. The charges accumulated proportionally to the incoming

Acquisition and visualization

light are shifted in the analog register to produce the output signal. Such devices are available today with arrays as large as 1800.

- <u>Charge coupled photodiode arrays</u> (CCPD) : these arrays are obtained by coupling elementary photodiodes with an analog CCD shift register. It is a synthesis of two previously described devices. Arrays up to 1800 elements are available.

2.4 *Mechanical scanning devices*

The mechanical scanning devices are the slowest imaging devices. Both dimensions are scanned mechanically. The main advantage of these systems is the very high resolution they allow. Three major alternatives are the drum scanner, the flat bed scanner and the laser scanner.

- <u>Drum scanner</u> : the image to be digitized is fitted on a drum, rotating at a relatively high speed. At each revolution one picture line is scanned with a unique photodiode or photomultiplier system. When such a line is fully scanned, the support of the photodiode is shifted mechanically by a sampling step and the following line is scanned.

- <u>Flat bed scanner</u> : the image to be digitized is fixed to a flat surface. Either the photodiode assembly or the bed is moved in a x-y fashion to perform the scanning .

- <u>Laser scanner</u> : a laser beam is projected on the image to be analysed through a mirror. The mirror is fixed on a galvanometer which is driven with a sawtooth signal. This signal permits the scanning in one dimension whereas the scanning in the other dimension is obtained by the step ba step advance of the image support.

2.5 *Display devices*

The most commonly used display device is the TV monitor driven with a video signal. This signal is generated by reading digital data from a refresh memory and by converting it to an analog signal through an D/A converter and putting it in the appropriate format. The memory refreshes the display at the regular frame rate (50 readings/sec). In addition, all the mechanical scanning devices described previously can also be used for hard copy reproduction of images on photographic paper or film. In this case, the image to be digitized is replaced with an unexposed film or paper and the electrical signal representing the digital image is converted into light to expose the film.

3. Human observer

3.1 *Description and definitions*

The human visual system uses parts of the nervous system. The latter is doubtlessly the most complicated communication network. It is managed by the most powerfull computer : the brain. The communication in this network is carried out through nervous cells called neurons. The brain contains about 10^{11} neurons, roughly the same number as that of stars in our galaxy.

A neuron has a body of size varying between 5 and 100 µm. A main fibre called axon and a number of fiber branches called dendrites are attached to this body. See fig. 4 for some typical neurons.

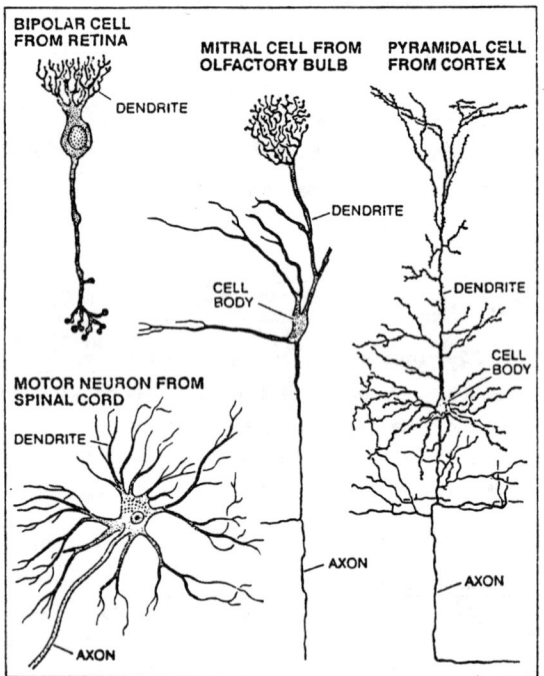

Fig. 4 : Some typical neurons (after [1])

The information transfer from one neuron to another is made electrochemically. The junction between two neurons is called synapse. The transmitting and the receiving neurons are called respectively presynaptic and postsynaptic. The information generated in a presynaptic neuron travels its axon like an electrical signal in a cable. Terminal branches of the axon transmit this signal to the dendrites ot the postsynaptic neuron. During this transmission, the electrical signal generates chemicals at the end of the axon which are deposited on the

Acquisition and visualization

postsynaptic neuron. In turn, these chemicals generate a new electrical signal on the postsynaptic neuron. A neuron can receive signals from thousands of presynaptic neurons and can transmit to thousands postsynaptic neurons. A given neuron can handle up to 200'000 synapses. The action of a neuron on another can be of two types : **excitatory** or inhibitory. The first one generates pulses in the postsynaptic neuron whereas the second inhibits the existing pulses.

The nervous system can be studied neuron by neuron with three basic tools of the neurobiologist : the microscope, the selective stain and the microelectrode. The signal observed at a given neuron is a pulse train. Its frequency is proportional to the intensity of the stimulus. In the brain the distinction between two identical signals representing two different phenomena is made on the basis of the wiring and not the signal. Only some specialized neurons react to a given exitation.

3.2 The eye

The eye is the receiver for visual signals. It focuses them to form the image on the retina. This latter analyse the image and sends the message to the brain through the optical nerve and optical paths in the head. Very roughly speaking, the eye can be viewed as a photo camera (see fig.5).

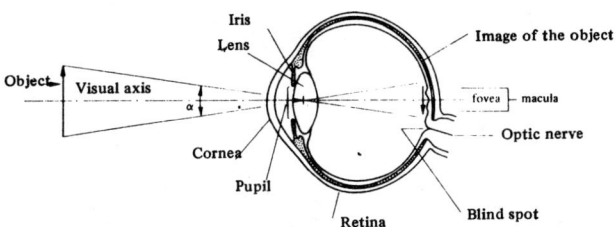

Fig. 5 : Section of the eye.

The light passes through the cornea and aqueous humor and enters the inner eye through the pupil. The amount of light allowed to enter is regulated by the pupil. The lens focuses the light on the sense cells of the retina. The volume of eye is small (6,5 cm^3). The diameter is about 24 mm and the weight 7 gr. The lens of the eye is not perfect. This imperfection is the source of the spherical aberration which appears as a blur on the focal plane. Such a blur can be modeled as a two-dimensional low-pass filter. The pupil's diameter varies between 2 and 9 mm. The highest cut-off frequency corresponds to 2 mm. continuous enlargement of the pupil's diameter decreases the cut-off frequency.

3.3 The retina

The retina is the neurosensorial layer of the eye and its area is about 12,5 cm². It transforms the light it receives into electrical signals which are transmitted to the visual cortex through the optic nerve. Fig. 6 shows the optical path in the visual system.

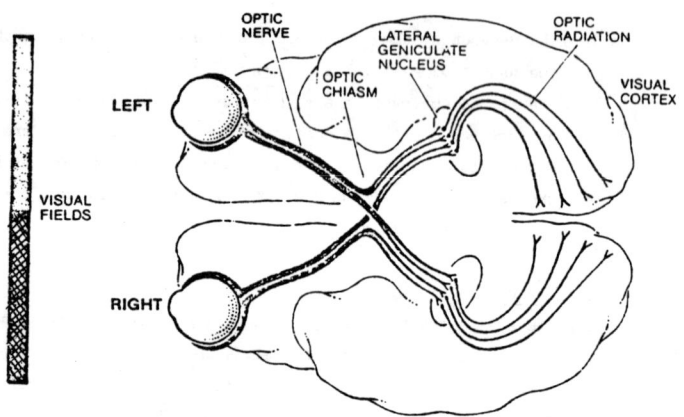

Fig. 6 : The optical path in the visual system.

It should be noticed that at the optic chiasm, the output of each eye is divided into two. Consequently, the information content of the left half of the visual field is processed by the right side of the brain and the information content of the right half of the visual field is processed by the left side of the brain. The retina contains five types of cells organized in layers (fig. 7). The layer of photocells is the farmost layer from the incoming light. There are two types of photocells : rods and cones. A normal eye contains about 130 millions of rods and 6,5 millions of cones. In serial connections, photocells are connected to bipolar cells which in turn are connected to ganglion cells, whose axons make up the optical nerve. In parallel connections, the horizontal cells receive synapses from the photocells and may act on bipolar cells and photocells. Amacrin cells receive synapses from bipolar cells and may act on ganglion cells and bipolar cells. This indicates feed-back loops in parallel connections.

Crossing the retina in its thickness, the number of cells decreases progressively. The information is concentrated more and more at each level.

3.4 The ganglion cells

To study specialized cells, such as ganglion cells, and the other cells of the

Acquisition and visualization

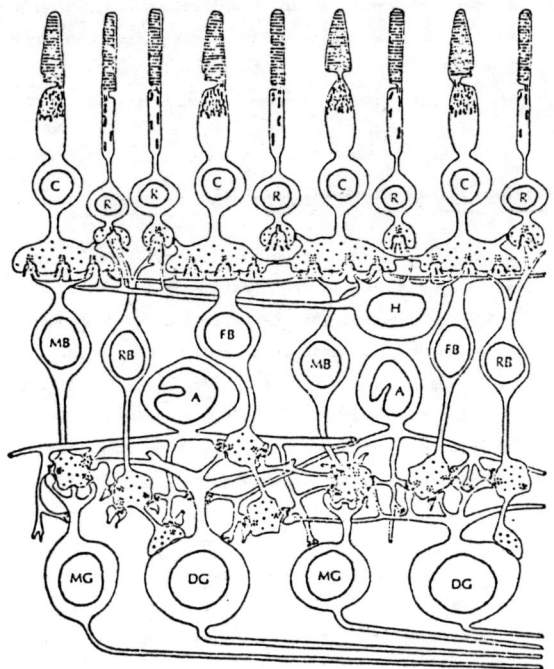

Fig. 7 : Retinal cells (after [2])

visual cortex, two notions need to be defined. The first one is that of the receptive field associated with a cell. It is simply the area of the retina which can influence the behaviour of that cell. The second is that of the effective excitation. It is the stimulus which produces a pulse train of highest frequency.

Because of the parallel and feed-back connections, a receptive field is divided in regions. Some of them produce excitation whereas some others inhibition. They are often called "on area" and "off area".

The effective excitation of a ganglion cell is a circular spot whose diameter is about 0,2 mm. The receptive fields are also circular and there are two types : those which are on in the center and off at the border and those which are off in the center and on at the border. At this level, the information is processed independently from spatial orientation. It should also be noted that

ganglion cells ignore the absolute light intensity and measure only differences
in their receptive field. Simultaneous contrast effects and edge detection are
mainly due to ganglion cells. Their effect can be modeled as a two-dimensional
high-pass filter. Lateral inhibitions and excitations are also the sources of
the so-called Mach phenomena which is the subjective enhancement of sharp
brightness changes.

3.5 The lateral geniculate nucleous

The information processed by the retina reaches the lateral geniculate nucleous
after being divided at optic chiasma. The cellular analysis of this nucleous
shows also a layered organization of the cells. Each layer receives information
only from one eye. This is called ocular dominance. Crossing these layers, one
observes an alternance in the ocular dominance. In addition, here also, neurons
receiving information from a given area of the retina are grouped independently
from the specific layer. The functionning of the cells in the lateral geniculate
nucleous are almost identical to those of ganglion cells. The independency
from the orientation is maintained.

3.6 The visual cortex

The visual cortex is a folded layer of neurons of about 2 mm in thickness. It
is located at the back of the brain. The information transmitted from the lateral
geniculate nucleous is received at the "area 17" of the cortex which is connected
to areas 18 and 19. Cellular analysis of the cortex indicates also that 10^{10}
neurons contained in it are hierarchicly organized in layers with only a few
types of neurons. These are classified as simple cells, complex cells,
hypercomplex cells and higher order hypercomplex cells.

The receptive field of a simple cell is elliptical. The center is bar shaped
and surrounded with two opposed areas. The effective excitation is not a
circular spot but a slit oriented in the same direction as the central bar of
the receptive field. If the slit is rotated other simple cells start reacting
depending on the angle. It is at this level of the visual system that the
dependency on the orientation is introduced. For a fixed orientation, if
the excitation is moved in the receptive field the pulses vanish progressively.
Fig. 8 shows the receptive field of a simple cell.

The complex cells are also sensitive to the orientation of the excitation.
In contrast with simple cells however, they are not sensitive to the position
of excitation in the receptive field.

Complex cells indicate orientation independently of position.
Fig. 9 shows the receptive field of a vertical complex cell.

Acquisition and visualization

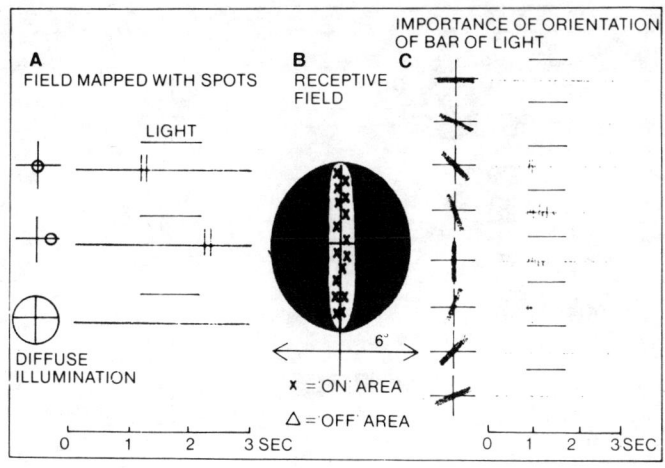

Fig. 8 : Receptive field of a simple cell (after [1])

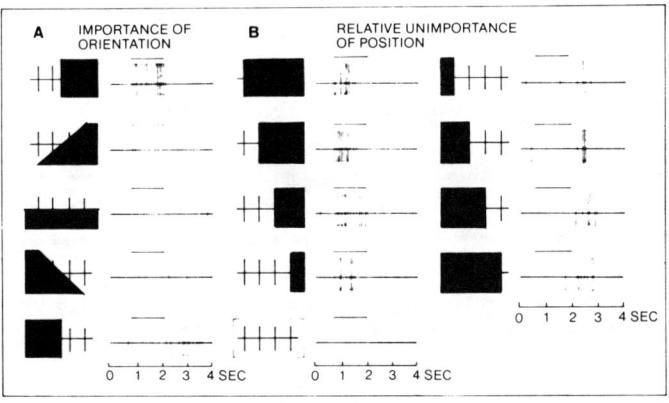

Fig. 9 : Receptive field of a complex cell (after [1])

The effective excitation of hypercomplex cells needs to have also a specific orientation, but in addition, it requires a discontinuity such as a corner or the end of a line.

Fig. 10 summarizes the functions of these various cells in the perception of a white rectangle on a dark background.

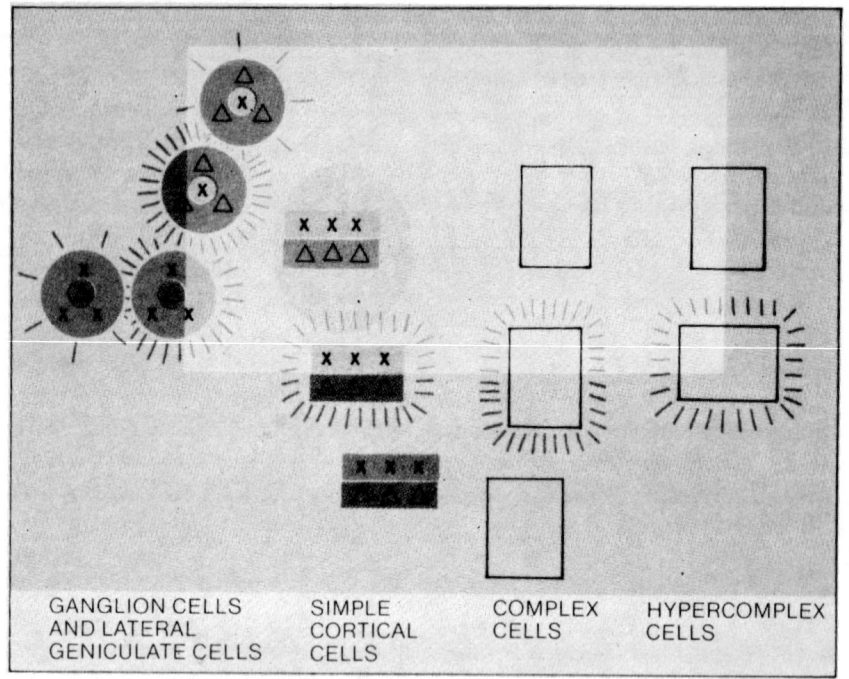

Fig. 10 : Perception of a rectangle (after [1])

3.1 *Columnar organization of the cortex and a global model*

Hubel and Wiesel [3] carried out a very complete cell by cell analysis of the cortex. They observed a columnar organization in both ocular dominance and direction dominance. Penetrating the cortex perpendicularly with a micro-electrode, one notices that all the cells encountered respond to excitation coming from one given eye only. A neighbouring penetration shows cells responding to the other eye. Finally if the penetration is slanted, from one layer to the next the ocular dominance alternates from one eye to the other.

If the same experiment is repeated by looking for not the ocular dominance but the response to a given orientation, the same columnar organization can be observed. A perpendicular penetration shows that all the cells encountred respond to a given orientation. In the neighbouring column, the orientation is slightly different. There are roughly 30 quantized directions. Combining these results, one obtains the columnar model of the cortex as shown in Fig. 11 where the bars in the columns indicate the preferred directions.

All the properties described in this section can be summarized in a block diagram (see fig. 12) where parts related to the lens, to the retina and to the cortex are indicated.

Bars in the boxes indicate the directional filters followed by another filter bank for detecting the intensity of the stimulus.

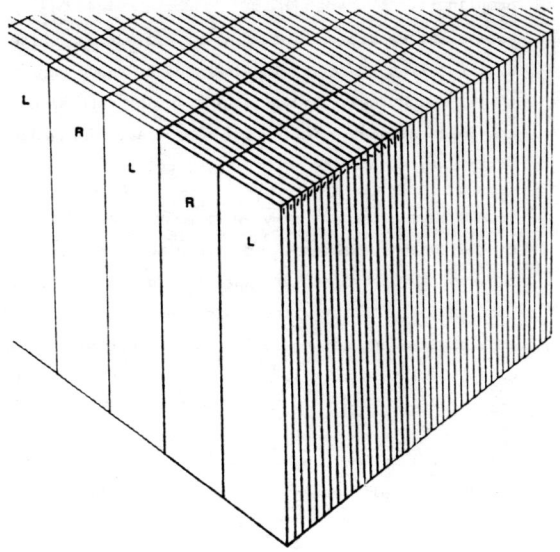

Fig. 11 : Columnar organization of the cortex (after [3])

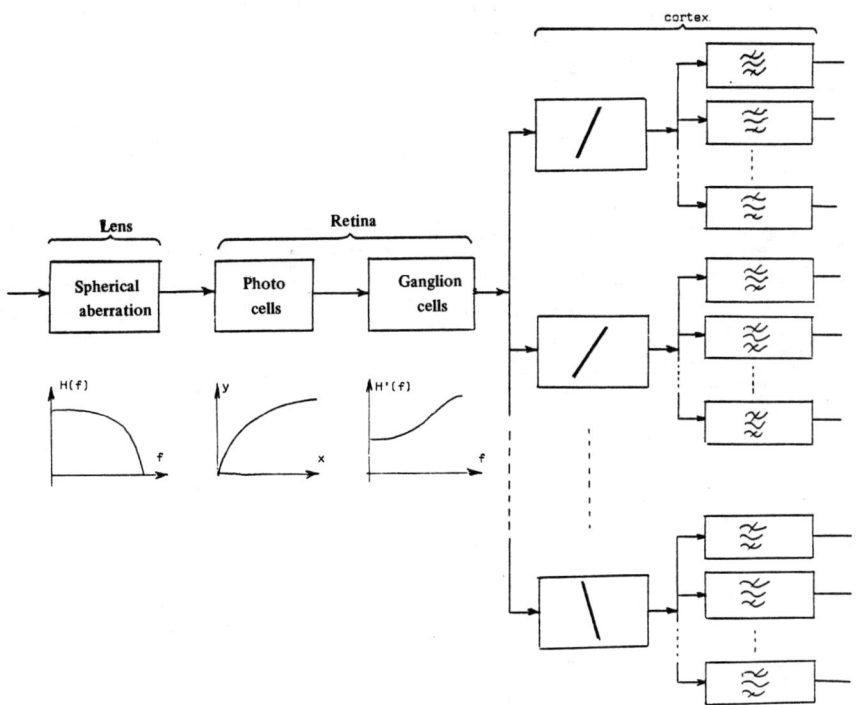

Fig. 12 : Block-diagram model of the visual system.

4. Sampling

From a mathematical point of view, a black-and-white image will be represented by a function x(u,v) of two independant space variables u and v. The value of this function x(u,v) at a given point corresponds to the brightness of the image at that point. Color images are represented by three functions of this type, one for each primary color.

From a physical point of view, the brightness x(u,v) is a continuous function of the space variables u and v. Furthermore, although its dynamic range, i.e. the range of permissible values of x(u,v) is finite, it can take any value from an infinite set within this range. To be processed in a computer or in a digital processing system, this function should be sampled and quantized. We shall discuss in this section the conditions required for sampling and its effects.

Let $x_a(u,v)$ be the analog (continuous) brightness signal and $X_a(f,g)$ its two-dimensional Fourier transform defined by

$$X_a(f,g) = \int_{-\infty}^{+\infty} \int_{-\infty}^{+\infty} x_a(u,v) \, e^{-j2\pi(fu + gv)} \, du \, dv \quad (1)$$

The most common way to sample $x_a(u,v)$ is the periodical sampling in both u and v axes, with increments or periods Δu and Δv. The sampled signal can then be expressed as :

$$x(k\Delta u, l\Delta v) = x(k,l) = x_a(u,v) \Big|_{\substack{u = k\Delta u \\ v = l\Delta v}} \quad (2)$$

The main problem of sampling is the determination of the sampling periods Δu and Δv. The requirement is to be able to reconstruct exactly $x_a(u,v)$ by interpolation between the samples x(k,l). Let us restrict ourselves to the so-called idealized sampling. In this case the sampled signal $x_s(u,v)$ can be expressed as :

$$x_s(u,v) = x_a(u,v) \cdot p(u,v) \quad (3)$$

where

$$p(u,v) = \sum_{k=-\infty}^{+\infty} \sum_{l=-\infty}^{+\infty} \delta(u - k\Delta u, v - l\Delta v) \quad (4)$$

and where $\delta(u,v)$ is the two-dimensional Dirac's delta function. Fig. 13 shows this multiplication.

In the Fourier or frequency domain, the relationship which corresponds to (3) is the convolution integral :

Acquisition and visualization

$$X_e(f,g) = X_a(f,g) ** P(f,g)$$

$$= \int_{-\infty}^{+\infty}\int_{-\infty}^{+\infty} X_a(f',g') P(f-f', g-g') \, df' \, dg' \qquad (5)$$

By generalizing one-dimensional definitions and properties of Dirac's delta function, it can be shown that

$$P(f,g) = \frac{1}{\Delta u \Delta v} \sum_{m=-\infty}^{+\infty} \sum_{n=-\infty}^{+\infty} \delta(f - \frac{m}{\Delta u}, g - \frac{m}{\Delta v}) \qquad (6)$$

This equation shows that the Fourier transform of a two-dimensional pulse train of periods Δu and Δv is also a pulse train of periods $1/\Delta u$ and $1/\Delta v$. A second important property is the following

$$Y(\alpha,\beta) ** \delta(\alpha-a, \beta-b) = \int_{-\infty}^{+\infty}\int_{-\infty}^{+\infty} Y(\alpha',\beta') \delta(\alpha-\alpha'-a, \beta-\beta'-b) \, d\alpha' \, d\beta'$$

$$= Y(\alpha-a, \beta-b) \qquad (7)$$

Substituting (6) in (5) and taking into account (7), we obtain

$$X_e(f,g) = \frac{1}{\Delta u \Delta v} \sum_{m=-\infty}^{+\infty} \sum_{n=-\infty}^{+\infty} X_a(f - \frac{m}{\Delta u}, g - \frac{n}{\Delta v}) \qquad (8)$$

According to (8), the Fourier transform of the sampled signal is obtained by repeating periodically with periods $1/\Delta u$ and $1/\Delta v$ in both directions f and g the Fourier transform $X_a(f,g)$ of the analog signal (see fig.13).

To reconstruct entirely the function $x_a(u,v)$ it is necessary to isolate $X_a(f,g)$ from its periodical replicas. This operation can, in principle, be done by filtering $X_e(f,g)$ with a low-pass filter. However, even with an ideal low-pass filter, $X_a(f,g)$ can be isolated if and only if it is limited in frequencies to $F = 1/2\Delta u$ and to $G = 1/2\Delta v$. In other words, it is necessary to have

$$X_a(f,g) = 0 \quad \text{for } f > F = \frac{1}{2\Delta u}$$

$$\text{and } g > G = \frac{1}{2\Delta v} \qquad (9)$$

This condition, known as the two-dimensional sampling theorem, gives the appropriate values for the sampling periods Δu and Δv provided that the bandwidths F and G of $x_a(u,v)$ are known.

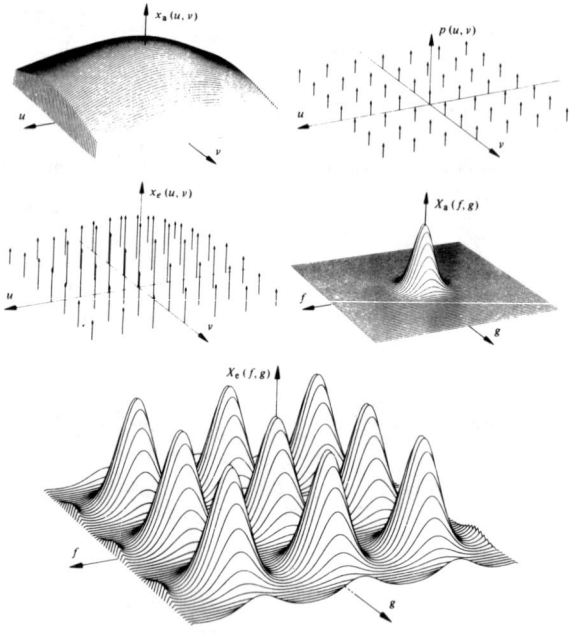

Fig. 13 Two-dimensional idealized sampling

$$\Delta u = 1/2F \quad \text{and} \quad \Delta v = 1/2G \tag{10}$$

The sampling theorem can be stated as follows : an analog signal $x_a(u,v)$ with bounded frequency extent F and G (cycles/unit distance) can be reconstructed exactly from its samples $x_a(k\Delta u, l\Delta v)$ if and only if the sampling periods Δu and Δv are less than or equal to 1/2F and 1/2G respectively.

If the conditions (9) are not satisfied, $X_a(f,g)$ overlaps with its periodical replicas. In this case, even if the main part of $X_a(f,g)$ is isolated with a low-pass filter, the output of the filter will be a disturbed version of $x_a(u,v)$. Fig. 14 shows the overlapping case.

Overlapping may lead to Moire figures if the undersampled image contains a set of lines. If, for example, a set of parallel lines of frequency f_r in the direction θ is undersampled, in the reconstructed image after sampling, another set of lines with another frequency f'_r in another direction θ' appears. Fig. 15 illustrates this effect.

Acquisition and visualization

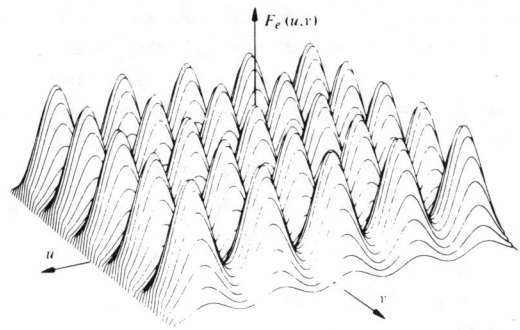

Fig. 14 : Overlapping Fourier transforms.

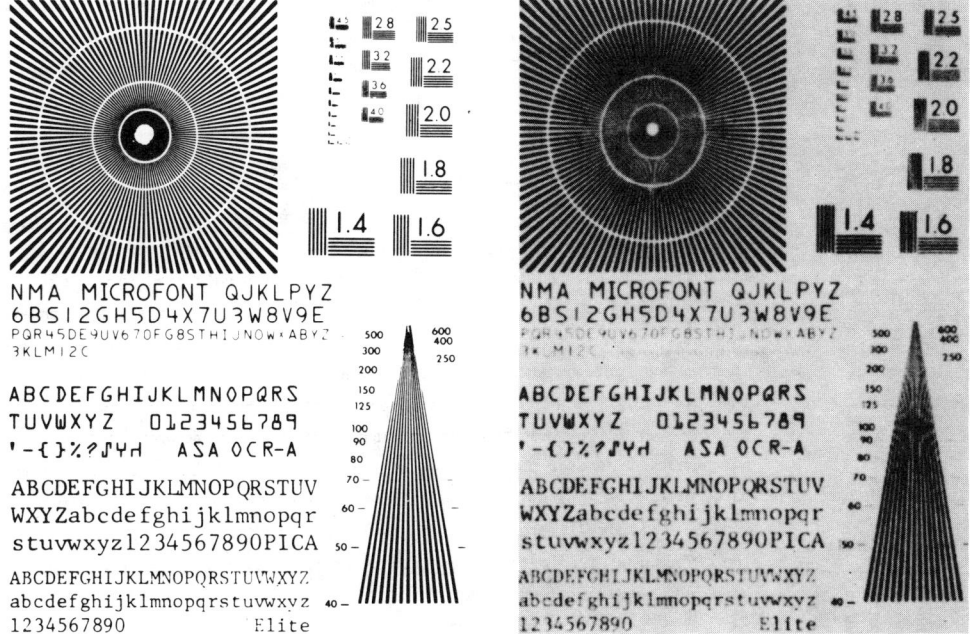

Fig.15 : Moiré figures resulting from undersampling.

The sampling theorem is purely theoretical. It is difficult to apply it in practice. One problem is related to the measurement of F and G for a given function $x_a(u,v)$. With lacking methods and instruments, the frequency **extent** of $x_a(u,v)$ can only be **coarsely** estimated by using as much as possible all the a priori knowledge available on the signal $x_a(u,v)$. **Luxurious** solutions which consist **of choosing** very small values for Δu and Δv are to be prohibited because they lead to an enormous number of samples, difficult or even impossible to handle in digital **processors.** Another problem is related to the destiny of the sampled image. In the context of a given **application,** the practical principle is to select Δu and Δv as large as possible, while guaranteeing that all the information needed **is** carried by the samples.

If the sampled image is to be processed to become another image observed by humans, the properties of the human visual system should also be considered for selecting Δu and Δv. It is well known that the eye behaves like an integrator (low-pass filter) for spatial radial frequencies higher than roughly 10 cycles/degree. The attenuation in the frequency response becomes very important around 60 cycles/degree. Very often, the eye is used as low-pass filter to isolate $X_a(f,g)$ in (8). In such cases, the sampling periods should be slightly less than the separating power of the eye (1 minute of arc). For an observer at 30 cm of an image, this corresponds to about Δu = Δv = 1/10 mm. Very often, larger sampling periods can be used if the physical support of the reconstructed image performs some sort of integration.

A comment is in order about ideal and practical sampling. Practical hardware systems which scan the picture cannot perform the ideal sampling as described. The spot size cannot be made infinitely small. Accordingly, the measured brightness is, in practice, an average brightness over the area covered by the aperture of the spot. The detailed analysis of this situation will be discussed in section 6.

Practical sampling systems use very often a square sampling grid with Δu = Δv. This may not be quite true for some TV cameras for example, where technological constraints require Δu slightly different than Δv. In a square Cartesian sampling grid, each picture point - called also picture element or pixel - is surrounded with 8 neighbours. Half of them are along diagonals and are at a distance of Δu√2 from the center whereas the other half are at a distance Δu. In some applications where geometrical measurements on the images are involved, the existence of two different distances creates complications. They can be greatly overcome by weighting the contribution of each pixel according to its neighborhood. Another possibility is to use a hexagonal sampling grid. In this case, each pixel has six neighbours at the same distance. Such a grid is used in biomedical image processing [4]. An approximation to hexagonal grid can be obtained from a cartesian grid by shifting even numbered (or odd numbered) sampling lines by Δu/2.

5. Quantization

To complete the digitization of the brightness function $x_a(u,v)$, the sampling, as previously discussed, should be followed by the quantization of samples. The principle of quantization is to divide the brightness dynamic range into a finite number or intervals and to assign the same value at every brightness value within a given interval. The problem which will be discussed in this section is to find the number of intervals and the variation of their width as a function of the brightness.

Acquisition and visualization

In areas of an image where the brightness varies very little from one pixel to the next (low spatial frequency areas), the quantization will create a constant level. Between two such areas, there will be a sudden jump from one level to another, whereas the original brightness has a continuous variation. These jumps create generally false contours which do not correspond to any real object. They are known as quantization noise.

There are two important factors which intervene in the quantization : the first one is the human visual system (subjective factor) and the second one is the physical support of the image where it will be reproduced in quantized form (objective factor).

In order to satisfy the requirements of the human visual system, the gray-level difference between two adjacent intervals should be less than the so-called "just noticeable difference". This difference is determined by the Weber-Fechner experience. In this experiment, the visual field is divided into two regions of uniform brightness B and B + ΔB respectivelly. The observer is asked to increase ΔB untill he sees a difference between the two brightnesses. If this measurement is repeated for several values of B, it is observed that the ratio $\Delta B/B$ is constant.

$$\frac{\Delta B}{B} = C_W \qquad (11)$$

This relationship is known as Weber-Fechner law. It is valid over a large brightness interval ranging roughly from 1.to 1000 nits (candela per square meter).

The physical support of the image determines the available dynamic range by its photometric properties. Generally, it is characterised either by its transparency (film) or by its reflectance (photographic paper or opaque supports). Let us assume a film with transparencies τ_1 and τ_2 when it is exposed to saturation and when it is not exposed at all respectively. The dynamic range of this film, measured in optical density unit is given by

$$D = d_{\tau_1} - d_{\tau_2} = -(\log_{10} \tau_1 - \log_{10} \tau_2) \qquad (12)$$

The just noticeable difference coefficient C_W in eq. (11) is given in per cent. It varies between 0,01 and 0,02 depending on the observer. In order to express it in optical density unit, one needs to consider the term $(1+C_W)$ to translate the origin of the log function. The number of quantization intervals is given by

$$N_q = \frac{D}{\log_{10}(1+C_W)} = \frac{\log_{10} \tau_2 - \log_{10} \tau_1}{\log_{10}(1+C_W)} \qquad (13)$$

The number of quantization levels used according to (13) does not produce false contours and the levels are linearly distributed.

Such a quantization scheme does not take into account the distribution of the brightness to adjust the width of the intervals to the probability of occurence of the corresponding brightness value. A quantitative quantization scheme can be derived by optimizing the width of the intervals according to a given criterion (minimum mean square quantization error for example [5]).

Fig. 16 shows the input-output characteristics of a quantizer where the decision levels x_i in the input signal and the reconstruction levels y_i on the output signal are defined. As shown in this figure, we assume that all input values between x_i and x_{i+1} are represented as y_i. If the range of the input brightness is bounded by x_{min} and x_{max}, the quantization error is given by :

$$\varepsilon = E[(x-y)^2] = \int_{x_{min}}^{x_{max}} (x-y)^2 p_x(x)dx = \sum_{i=0}^{N-1} \int_{x_i}^{x_{i+1}} (x-y_i)^2 p_x(x)dx$$

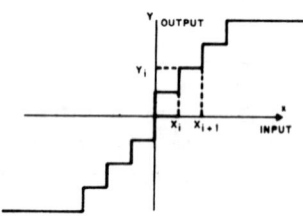

Fig. 16 : Characteristics of a quantizer. x is the input and y is the output. $\{X_i\}$ and $\{Y_i\}$ are decision and representative levels, respectively. Inputs between X_i and X_{i+1} are represented by Y_i.

where N is the total number of quantization levels between x_{min} and x_{max}. For N sufficiently large, it can be assumed that the distribution p(x) is approximately constant over each quantization band. Accordingly (14) is reduced to

$$\varepsilon \simeq \sum_{i=0}^{N-1} p_x(x_i) \int_{x_i}^{x_{i+1}} (x-y_i)^2 dx$$

$$= \frac{1}{3} \sum_{i=0}^{N-1} p_x(x_i) [(x_{i+1} - y_i)^3 - (x_i - y_i)^3] \qquad (15)$$

The optimum assignment of the reconstruction levels y_i can be found by letting

Acquisition and visualization

$$\frac{\partial \varepsilon}{\partial y_i} = 0 \qquad (16)$$

which yields the solution :

$$y_i = \frac{x_{i+1} + x_i}{2} \qquad (17)$$

Accordingly, the reconstruction levels should be at the midpoint between each pair of decision levels. Substituting (17) in (15) gives :

$$\varepsilon = \frac{1}{2} \sum_{i=0}^{N-1} p_x(x_i) [x_{i+1} - x_i]^3 \qquad (18)$$

Minimizing (18) with Lagrange multipliers leads to

$$x_i = \frac{\int_{x_i}^{x_{i+1}} x p_x(x) \, dx}{\int_{x_i}^{x_{i+1}} p_x(x) \, dx} \qquad (19)$$

Which gives the optimum assignment of the decision levels. Numerical tabulations of (17) and (19) are given by Max [6]. For non-uniform probability densities, the spacing of decision levels is narrow for likely brightness and wide for less likely brightness. This may result in small amount of granular noise but poor representation of detailed areas and edges. For subjective reasons, the design of an image quantizer should not only be based on the mean square error criterion. The combination of both quantitative and subjective design procedures can be made in a number of ways. Two most commonly used approaches are the design which maintains the quantization error at the threshold of visibility while minimizing the number of quantizer levels and the use of a weighted mean square error where subjective constraints are introduced as weights.

It should be noted that nonlinear quantization can be performed by companding, i.e. by transforming nonlinearly the brightness input so that it has a uniform density, by quantizing this variable uniformly and by transforming back with the inverse nonlinear transformation (fig. 17).

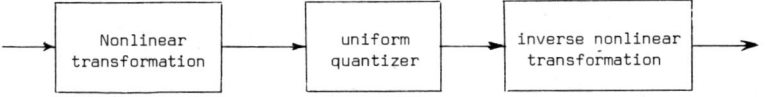

Fig. 17 : Companding quantizer.

6. Preprocessing and post-processing [9]

Any type of processing which is standard for all images and which can be included on sampling and quantizing, can be called preprocessing. The purpose here is to alleviate the task of the ad-hoc subsequent processing by, for example, correcting the imperfections of the imaging system.

The most common preprocessings and imaging imperfections can be analysed through the two-dimensional linear system theory. For example, imperfections in the lens system introduce a spatial low-pass filtering. In the Fourier domain, this can be described by

$$Y(f,g) = G(f,g) \cdot X(f,g) \qquad (20)$$

where $X(f,g)$, $G(f,g)$ and $Y(f,g)$ are respectively the Fourier transform of the ideal input image $x(k,l)$, the frequency response of the lens imperfection and the Fourier transform of the spatially low-pass filtered image $y(k,l)$.

In the image domain, the equivalent relationship to (20) is

$$y(k,l) = \sum_{k'} \sum_{l'} g(k',l') \, x(k-k', l-l') \qquad (21)$$

for the discrete case, or

$$y_a(\alpha, \beta) = \iint g_a(u,v) \, x_a(\alpha-u, \beta-v) \, du \, dv \qquad (22)$$

for the analog case.

The main source of spatial bandlimitation is more likely to be the spot size than optical imperfections. In practical scanning systems the spot size is not infinitely small. The brightness measured at a given point (u_0, v_0) is in fact a local average brightness in a small area S surrounding this point. If $h_a(u,v)$ represent the form of the spot, the measured brightness is given by

$$y_a(u_0, v_0) = \iint_S h_a(u,v) \, x_a(u_0-u, v_0-v) \, du \, dv \qquad (23)$$

which has the same form as (22). In Fourier domain we have

$$Y_a(f,g) = H_a(f,g) \, X_a(f,g) \qquad (24)$$

The real brightness is thus filtered with a filter whose frequency characteristic is $H_a(f,g)$. For currently used spots, $H_a(f,g)$ is of the low-pass type. If the spot size is large, the extent S of $h_a(u,v)$ is large and $H_a(f,g)$ is a severe low-pass filter. Fig. 18 shows typical spot sizes and their Fourier transforms.

Acquisition and visualization

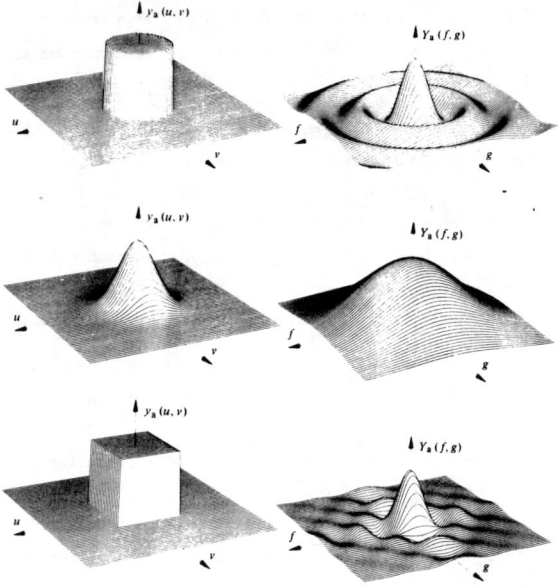

Fig. 18 : Typical spot shapes and their Fourier transforms.

If the shape $h_a(u,v)$ is known and if the filtering (24) is disturbing, it is possible to correct for it by an approximative inverse filtering by carefully avoiding the zeros of $H_a(f,g)$. The frequency response of the ideal inverse filter is simply $1/H_a(f,g)$. It can be used provided that the zeros of $H_a(f,g)$ are outside the range where this effect should be eliminated.

Another example of preprocessing is the measure and the correction for non uniform scene illumination. In such a case, the recorded brightness $y(k,l)$ can be modeled as

$$y(k,l) = p(k,l) \, x(k,l) \qquad (31)$$

where $x(k,l)$ is the ideal image and $p(k,l)$ a two dimensional function representing the non uniform illumination. This latter can be measured if a really uniform illumination is recorded through the same imaging system. The ideal image is then given simply by division

$$x(k,l) = y(k,l) / p(k,l) \qquad (32)$$

A priori known, point by point gray scale modifications, simple linear filtering or non linear filtering such as edge extraction, frame averaging, correction for geometric distortions can also be viewed as proprocessing.

At the other end of processing chain, just prior to visualization certain type of processing might be necessary to prepare the digital image for final display. The set of these processings is called postprocessing. It might include compensations for distortions in the display device, interpolation or decimation for adapting the resolution of the processed image to that of its final support or any transformation which has the inverse effect of those applied as preprocessing. A typical example of this type is logarithmic and exponential transformations applied to pixel brightness.

Another example is that of the exact reconstruction of the ideal image from its samples. Assuming that the sampling is made according to the sampling theorem, this can be achieved by spatial filtering the samples with an appropriate filter. The frequency characteristics of this filter should be such that, in eq. (8) the main spectrum $X_a(f,g)$ can be isolated from its periodical replicas. The frequency response of an ideal low-pass filter which can be used for this purpose is given by

$$H(f,g) = \begin{cases} \Delta u \Delta v & \text{if } |f| < F \text{ and } |g| < G \\ 0 & \text{otherwise} \end{cases}$$ (2)

The impulse response of this filter is

$$h(u,v) = \int_{-\infty}^{+\infty} \int_{-\infty}^{+\infty} H(f,g) \, e^{j\,2\pi(fu + gv)} \, df \, dg$$

$$= \Delta u \Delta v \int_{-F}^{F} \int_{-G}^{G} e^{j\,2\pi(fu + gv)} \, df \, dg = \frac{\sin(\pi u/\Delta u)}{\pi u/\Delta u} \cdot \frac{\sin(\pi v/\Delta v)}{\pi v/\Delta v}$$

Filtering $X_e(f,g)$ by $H(f,g)$ is expressed as

$$X_a(f,g) = X_e(f,g) \cdot H(f,g)$$ (2)

In the image domain, this is equivalent to

$$x_a(u,v) = \sum_{k=-\infty}^{+\infty} \sum_{l=-\infty}^{+\infty} x(k,l) \frac{\sin \frac{\pi}{\Delta u}(u-k\Delta u)}{\frac{\pi}{\Delta u}(u-k\Delta u)} \cdot \frac{\sin \frac{\pi}{\Delta v}(v-l\Delta v)}{\frac{\pi}{\Delta v}(v-l\Delta v)}$$ (2)

Acquisition and visualization

This relation is an infinite sum interpolation formulae to reconstruct the ideal image. Its practical use requires an approximation in which the number of sinx/x functions is limited. If, in eq. (25) the square area 2Fx2G is changed into a circular area, we obtain

$$H(f,g) = \begin{cases} \Delta u \Delta v & \text{if } \sqrt{f^2 + g^2} < r_0 \\ 0 & \text{otherwise} \end{cases} \quad (29)$$

The impulse response of this filter is

$$h(u,v) = r_0 \frac{J_1(2\pi r_0 \sqrt{u^2 + v^2})}{\sqrt{u^2 + v^2}} \quad (30)$$

where J_1 is the first order Bessel function.

These ideal reconstruction filters are often difficult to implement in practice. Other reconstruction filters, more suitable for practical implementation, can also be used, such as zero order or first-order (linear) interpolation filters.

7. Conclusions

We have discussed in this chapter major acquisition and visualization techniques used in computer vision. The imaging and display devices briefly described constitute a small part in a wide range of equipments covering the electromagnetic spectrum from acoustic waves to gamma rays.

Since the human observer is an important "component" of a computer vision chain, the properties of the human visual system have been described from neurobiological and engineering point of views yielding a relatively detailed block diagram for the whole system.

A large majority of imaging and display devices uses analog signals. For digital processing it is mandatory to convert them to digital signals and then back to analog for display. Sampling and quantization are two fundamental operations in this respect. They have been discussed in details. Finally, simple pre and post processing are discussed viewing them as possible modular parts of the acquisition and/or visualization.

References

[1] S.W. Kuffler and J.G. Nichols, "From neuron to brain : A cellular approach to the functions of the nervous system", Sinauer Ass. Inc., Sunderland, Massachusetts, USA, 1976.

[2] T.M. Cornsweet, "Visual perception", Academic Press Inc, New-York, 1974.

[3] D.H. Hubel and T.N. Wiesel, "Brain mechanism of vision", Scientific American, Special issue on the brain, Vol. 241, No 3, September 1979, pp 130-146.

[4] M. Ingram, K. Preston Jr, "Automatic analysis of blood cells", Scientific American, pp 72-82, November 1970.

[5] B. Smith, "Instantaneous companding of quantized signals", Bell Syst. Techn. Journal, Vol. 36, No 3, Jan. 1957, pp 653-709.

[6] J. Max, "Quantization for minimum distortion", IEEE Trans. Information Theory, Vol. IT-6, pp 7-12, March 1960.

[7] L.M. Biberman and S. Nudelman (Eds.), "Photoelectronic imaging devices" Vol. 1 and Vol. 2, Plenum Press, New York, 1971.

[8] B. Kazan, "Advances in Image Pickup and Display", Academic Press, New York, Vol. 1, 1974 and Vol. 3, 1977.

[9] W.K. Pratt, "Digital Image Processing", John Wiley and Sons, New York, 1978.

TWO-DIMENSIONAL DIGITAL SIGNAL PROCESSING

J.F. ABRAMATIC
I.N.R.I.A. Domaine de Voluceau - Rocquencourt
B.P. 105, 78153 LE CHESNAY Cédex FRANCE

INTRODUCTION

Two-Dimensional (2-D) digital signal processing techniques provided the first breakthroughs that allowed to perform image processing using digital computers. This started in the late sixties. Since then, new developments took place in hardware technology. Furthermore research efforts succeeded in generalizing 1-D signal processing techniques to 2-D signals. These results led to efficient image processing techniques that solve coding, enhancement and restoration problems. These are image processing tasks where inputs and outputs are digital images. Some of these techniques have also been successfully used for feature extraction problems such as edge detection for example where the result is no longer a digital image.

More than ten years the first research works, theory of 2-D signal processing has emerged. References [1-3] give an idea of the progress that has occured. This chapter after introducing the basic notions tries to give an outline of the main results.

Among the very numerous techniques available for digital image processing, it is worthwhile to pick out the linear space invariant (LSI) filtering operations. Although the overall process involved is very often globally non linear and sometimes space variant, it is very common that at least one step of the whole processing consists in a 2-D LSI filtering operation. This is the reason why we shall focus our attention to this problem.

PROBLEMS

When şomeone is led to perform 2-D LSI filtering, he has very often used the following procedure.

1. <u>Define an ideal 2-D LSI filter</u> that solves his problem or a part of his problem. This step is very much application dependent. It will thus be described in other chapters such as image enhancement or image restoration. We shall suppose in the course of this chapter that this ideal filter is known. Its representation can be of various forms (spatial or spectral characteristics for example).

2. <u>Define the implementation procedure</u> that is to be used. This step takes into account quantitative characteristics of the problem (size of the images, size of the convolution,..) and also parameters derived from the hardware available (general purpose versus special purpose operators, real time versus batch operations...).

3. <u>Design</u> a 2-D LSI filter that is "close" to the ideal one given by step 1 and is implementable using the constraints derived from step 2.

These implementation and design problems are the core of this chapter and are proceded by an introduction to 2-D digital signal processing that presents the various representations of 2-D signals and systems.

I. REPRESENTATIONS OF 2-D SIGNALS AND SYSTEMS

a) <u>Definitions</u> :

SIGNAL

A 2-D signal $u(.,.)$ (often called image in the following of this text) is a real-valued(*) function of two integer variables. In most cases, this function is non-zero on a finite set (e.g. a rectangle).

(*) We leave out of the range of this text problems related to quantification.

LSI FILTER

A 2-D LSI filter is an application of the set of images into itself defined by :

$$y(i,j) = \sum_{-\infty}^{+\infty} \sum_{-\infty}^{+\infty} h(k,\ell) u(i-k, j-\ell) \qquad (1)$$

where $u(.,.)$ and $y(.,.)$ are input and output images respectively. $h(.,.)$ is called the impulse response of the LSI filter. The right-hand side of equation (1) is called the convolution of $x(.,.)$ and $h(.,.)$.

b) <u>Spectral Representations</u> :

Signals and systems can be represented either in the spatial domain as in equation (1) or in the spectral domain.

CONTINUOUS FOURIER TRANSFORM (CFT)

Definition :

The continuous Fourier transform is defined by :

$$X(e^{j\omega_1}, e^{j\omega_2}) = \sum_{-\infty}^{+\infty} \sum_{-\infty}^{+\infty} x(k,\ell) e^{-j(\omega_1 k + \omega_2 \ell)} \qquad (2)$$

Convolution Property :

The CFT of a convolution product is a simple product. If $Y(.,.)$, $H(.,.)$ and $U(.,.)$ denote the CFT's of $y(.,.)$, $h(.,.)$ and $u(.,.)$ respectively, we derive from equation (1) :

$$Y(e^{j\omega_1}, e^{j\omega_2}) = H(e^{j\omega_1}, e^{j\omega_2}) U(e^{j\omega_1}, e^{j\omega_2}) \qquad (3)$$

DISCRETE FOURIER TRANSFORM (DFT)

Definition :

The Discrete Fourier Transform of a 2-D signal of <u>finite extent</u>

is given by :

$$X(m,n) = \sum_{k=0}^{N-1} \sum_{\ell=0}^{N-1} x(k,\ell) \, e^{-\frac{2j\pi}{N}(km+\ell n)} \quad (*) \tag{4}$$

for $m=0,\ldots,N-1$; $n=0,\ldots,N-1$

Remark :

The DFT of a signal is not unique but depends upon the size N of the interval over which the DFT is calculated.

INVERSE DISCRETE FOURIER TRANSFORM (IDFT)

The IDFT is given by :

$$x(k,\ell) = \frac{1}{N^2} \sum_{m=0}^{N-1} \sum_{n=0}^{N-1} X(m,n) \, e^{\frac{2j\pi}{N}(mk+n\ell)} \tag{5}$$

for $k=0,\ldots,N-1$; $n=0,\ldots,N-1$.

Convolution Property :

The IDFT of the product of two DFT's is the order N <u>circular</u> convolution of the original signals. The circular convolution of two signals is given by the same expression of equation (1) with the constraint that indices are calculated modulo N.

$Z_1 Z_2$ TRANSFORM

The $Z_1 Z_2$ Transform of a 2-D signal is given by :

$$X(z_1,z_2) = \sum_{-\infty}^{+\infty} \sum_{-\infty}^{+\infty} x(m,n) \, z_1^{-m} z_2^{-n} \tag{6}$$

The $Z_1 Z_2$ Transform of a 2-D impulse response is called the transfer function of the 2-D LSI filter.

(*) We restrict the developments to "square" signals for clarity purposes only.

c) <u>2-D LSI Filters. Properties</u> :

FINITE and INFINITE IMPULSE RESPONSE FILTERS

A 2-D LSI Filter is said to be an FIR (Finite Impule Response) filter iff :

$$h(k,\ell) \neq 0 \quad K_1 \leq k \leq K_2 \quad ; \quad L_1 \leq \ell \leq L_2 \tag{7}$$

If condition (7) does not hold, the filter is said to be an IIR (Infinite Impulse Response) filter.

BIBO STABILITY

A 2-D filter is said to be stable in the BIBO (Bounded Input-Bounded Output) sense iff :

$\forall u(.,.)$ such that $|u(i,j)| < U$ for all (i,j) then $|y(i,j)| < Y$ for all (i,j) when $y(i,j)$ is given by equation (1).

Theorem :

A 2-D LSI filter is BIBO stable iff

$$\sum_{-\infty}^{+\infty} \sum_{-\infty}^{+\infty} |h(k,\ell)| < +\infty \tag{8}$$

RECURSIVE FILTERS

Implementation of IIR filters is impossible in general. More precisely, an infinite number of arithmetic operations is required to calculate one output value using equation (1). On the other hand, in some cases, the use of FIR filters is too restrictive for the application. This is the reason why a lot of efforts have been devoted to study IIR filters that would be implementable, i.e., that would require a finite number of arithmetic operations to calculate one output point. Recursive filters fulfill this property. One output point is calculated using a finite set of input values and a finite set of "<u>previously</u>" calculated output values. Defining more precisely the word "previously" is not straightforward when one deals with 2-D signals. One has to define the notion of <u>causality</u>.

Causality :

As opposed to the 1-D case, where a natural notion of causality is suggested by the total order relationship over integers, there are various notions of causality in the 2-D case. Two of them are most popular.

Quarter-Plane Causality

A 2-D filter is said to be quarter-plane causal iff :

$$h(k,\ell) \neq 0 \qquad k \geq 0 \; ; \; \ell \geq 0 \qquad (9)$$

Non-Symmetric Half Plane Causality (NSHP) :

A 2-D LSI filter is said to be NSHP causal iff :

$$h(k,\ell) \neq 0 \quad \begin{array}{l} 1) \; k > 0 \quad \forall \ell \\ 2) \; k = 0 \quad \ell \geq 0 \end{array} \qquad (10)$$

Quarter-Plane Recursive

Quarter-Plane recursive filter are defined by their transfer function :

$$H(z_1,z_2) = \frac{N(z_1,z_2)}{D(z_1,z_2)} = \frac{\sum_{r=0}^{N_1} \sum_{s=0}^{N_2} b_{r,s} z_1^{-r} z_2^{-s}}{\sum_{r=0}^{N_3} \sum_{s=0}^{N_4} a_{r,s} z_1^{-r} z_2^{-s}} \qquad (11)$$

$a_{0,0}$ is supposed normalized to unity.

Thus $y(i,j)$ can be computed through :

$$y(i,j) = \sum_{r=0}^{N_1} \sum_{s=0}^{N_2} b_{r,s} u(i-r,j-s) + \sum_{\substack{r=0 \\ (r,s) \neq (0,0)}}^{N_3} \sum_{s=0}^{N_4} a_{r,s} y(i-r,j-s) \qquad (12)$$

Equation (12) shows that a finite number of calculations is necessary. Equation (11) shows that such a filter is IIR.

TYPICAL TWO-DIMENSIONAL PROBLEMS

Most of the notions, we just introduced, have their counter part in the 1-D signal processing area. However typical problems arise due to the multidimensional situation. We have already seen that the notion of causality was not easy to define in the 2-D case. Another typical problem is due to the lack of generalization for the fundamental theorem of algebra.

In general a polynomial of degree n has an infinite number of zeroes in the multidimensional case. Thus a transfer function of a 2-D FIR filter cannot be factorized into a product of elementary filters. We shall see later how this result influences 2-D FIR filter implementation. Another consequence can be found in the stability testing problem for recursive filters. As in the 1-D case, the stability of 2-D recursive filters is related to the poles of the transfer function of the filter. As there is an infinity of such poles, the stability test becomes difficult to handle. A lot of efforts have been devoted to this issue of stability testing [1-3] .The main result is due to Huang :

A 2-D LSI quater-plane recursive filter is stable iff :

$D(z_1,0) \neq 0 \qquad |z_1| \geq 1$

$D(z_1,z_2) \neq 0 \qquad |z_1| = 1 \text{ and } |z_2| \geq 1$

where $D(Z_1,Z_2)$ is the denominator of the transfer function.

After this short introduction, we can get into the two main pratical problems of 2-D signal processing, implementation and design.

II. IMPLEMENTATION OF 2-D LSI FILTERS

An implementation problem is characterized by two types of parameters.

i) <u>Parameters defining the problem</u> :

- Size of the input image.
- Size of the impulse response (if it is finite).
- Order of the recursive filter.

ii) <u>Parameters defining the available system</u> :

- Size of the primary memory.
- Arithmetic operations execution times.
- Access times to the secondary memory.
- Availability of special purpose hardware operators.

A very important issue to be considered also is whether or not the application has some real-time constraints. A first part of this paragraph is devoted to batch operations using general purpose computers. A second part presents techniques allowing real-time implementation.

a) <u>Batch operations</u> :

This category of implementation techniques can be split into three classes.

- Direct method.
- Fourier methods.
- Recursive methods.

The first two suppose the filter is FIR, the third supposes it is not. For all of them, we try to assess the computational burden and the input-output requirements.

DIRECT METHOD

This method trivially exploits the definition of the convolution operation.

$$y(i,j) = \sum_{k=-L_1}^{L_2} \sum_{\ell=-L_1}^{L_2} h(k,\ell)u(i-k,j-\ell) \qquad (13)$$

It is easy to see that this method requires in general $(L_1+L_2+1)^2$ arithmetic operations(*) per output point. The data management problem from secondary to primary memory is also simple. If the data are supposed to be stored on a line by line basis, a maximum number of lines is read at a time, computations take place, output results are stored and these steps are repeated until the whole image has been processed. Each point of the input image is read only once.

In the case where the impulse response has some symmetries, the number of multiplications can be reduced accordingly. This is often possible when processing images.

FOURIER METHODS

The global technique

We have seen in the previous paragraph that multipying DFT's of signals and taking the IDFT of the result, one gets the circular convolution of the original signals. However it is easy to see that if the DFT 's are computed on a large enough domain, circular and linear convolutions are equivalent. More precisely if one convolves an N x N signal with an L x L impulse response, the output signal is (N+L-1)x(N+L-1). If the original DFT's are computed over an (N+L-1)x(N+L-1) domain, the IDFT of their product is the linear convolution. This implementation is described by Figure 1.

(*) An arithmetic operation will be equivalent to one addition and one multiplication.

```
u(i,j) → [2-D FFT (N+K-1)×(M+L-1)] → U(n,m) ⊗ Y(n,m) → [2-D FFT⁻¹ (N+K-1)×(M+L-1)] → y(i,j)
```

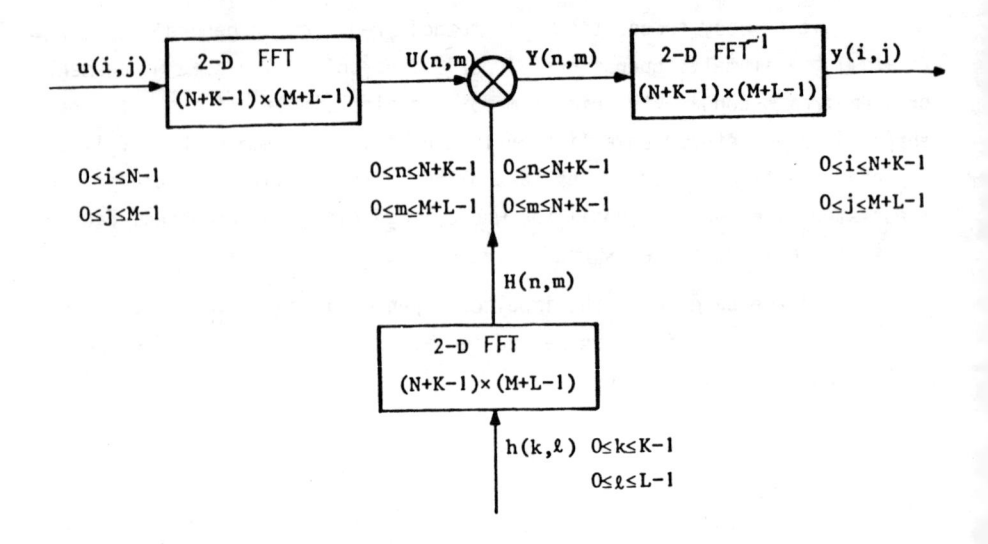

0≤i≤N-1   0≤n≤N+K-1   0≤n≤N+K-1   0≤i≤N+K-1
0≤j≤M-1   0≤m≤M+L-1   0≤m≤N+K-1   0≤j≤M+L-1

H(n,m)

[2-D FFT (N+K-1)×(M+L-1)]

h(k,ℓ) 0≤k≤K-1
        0≤ℓ≤L-1

Figure 1. Implementation of FIR filter using the Fourier method.

This implementation requires to compute the DFT of a 2-D signal efficiently. A lot of efforts [4] have been devoted to this problem of generalizing to 2-D signals the Fast Fourier Transforms (FFT) algorithms available for 1-D signals. The main difficulty occurs in the data management part of the algorithm.

The separability property of the DFT definition allows to split the computations into two majors steps :

$$\xi(k,n) = \sum_{\ell=0}^{N-1} x(k,\ell) e^{-\frac{2j\pi}{N} n\ell} \qquad (14a)$$

$$X(m,n) = \sum_{k=0}^{N-1} \xi(k,n) e^{-\frac{2j\pi}{N} mk} \qquad (14b)$$

The total number of arithmetic operations to compute the size NxN DFT of a 2-D signal is then on the order of $N^2 \log_2 N$ if one uses 1-D FFT algorithm to compute $\xi(.,.)$ and $X(.,.)$ as in equations (14a-b). This result allows to predict a good efficiency for the Fourier methods when considering the computational burden criterion.

Unfortunately, difficulties arise when data management is taken into account. The 2-D FFT requires to transpose an image that is supposed too large to be stored at once in the primary memory. Various authors have studied this problem [4]. They propose algorithms that basically allow to perform image with $N\log_2 N$ disk accesses.

Other methods do not take advantage of the separability used in equation(14a-b). They propose to compute the 2-D DFT on an iterative basis generalizing more directly the principles of the 1-D FFT algorithm. Their performances are comparable to the ones just mentioned that optimize the data management criterion.

The main progress can be made when one remarks that in most cases the size of the impulse response in much lower than the size of the image. In this case, the block convolution has to be used.

Block convolution :

The image is decomposed into blocks small enough to be stored in primary memory. Each block is convolved with the impulse response using the Fourier method. However no data management problem occurs at that level. Edge effects have to be taken care of to obtain the correct global results. Again various authors have studied this method (HUNT [5], TWOGOOD et al.[6], DERICHE-ABRAMATIC [7]).The main problem here is to optimize the size of the blocks once the characteristics of the problem are given.

Table 1 compares the three major techniques we just presented in the case of a typical processing convolution problem.

FILTER SIZE	DIRECT CONVOLUTION Sec.	BLOCK CONVOLUTION Sec.	FOURIER CONVOLUTION Sec.
5x5	90	92	495
10x10	310	110	495
15x15	650	138	495
20x20	1150	143	495
25x25	1900	156	495
30x30	2570	196	495
35x35	3590	232	495
40x40	4800	267	495

Table 1. Execution times for FIR filtering on a 32-bit minicomputer for a 512x512 image.

RECURSIVE METHODS

Recursive techniques are another way of improving efficiency of 2-D filtering implementation. We emphasize implementation of quarter-plane recursive filters that are widely used. Same conclusions may be drawn for NSHP filters.

A general quarter-plane filter of order N is implemented by the following recursive equation :

$$y(i,j) = \sum_{r=0}^{N-1} \sum_{s=0}^{N-1} b_{r,s} u(i-r,j-s) + \sum_{\substack{r=0 \\ (r,s) \neq (0,0)}}^{N} \sum_{s=0}^{N} a_{r,s} y(i-r,j-s) \quad (15)$$

One quarter-plane recursive filter requires $2N^2$ arithmetic operations. In most cases, the prototype filter lies in the four quadrants of the impulse response plane. Four quarter-plane filters have to be used as described in Figure 2.

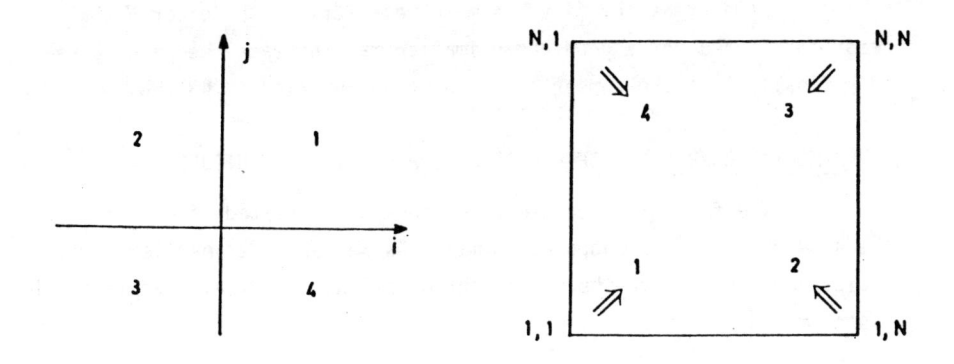

Figure 2a. Impulse Response Domain.   Figure 2b. Quarter-Plane Filters Implementation.

A total number of $8N^2$ arithmetic operations is thus involved in the filtering operations. As for the data management problem, when images are stored on a line by line basis, filters 1 and 2 of Figure 2b can be implemented simultaneous then output of filters 3 and 4 are computed and added to the previous results. The input image has to be read twice. An intermediary file has to be written and read once.

Separable Denominator Recursive (SDR) filters :

We have seen that stability testing is a very difficult problem. One way to eliminate this problem is to restrict the class of recursive filters to SDR filters. Their transfer function is defined as :

$$H(z_1,z_2) = \frac{N(z_1,z_2)}{D_1(z_1)D_2(z_2)} \tag{16}$$

Stability testing reduces to two 1-D stability tests and is thus quite simple. Moreover such filters will be seen in the next paragraph to have nice design techniques.

However as the class is more restrictive, the order N that needs to be used for a good approximation may increase. We have to remark finally that the number of computation involved is $4N^2+8N$.

EXPERIMENTAL COMPARISON OF FOURIER and RECURSIVE TECHNIQUES

The final user of 2-D filtering is interested of determining the good choice he is supposed to make. As we mentioned earlier, this choice depends upon the hardware configuration. A typical example is illustrated in Figure 3.

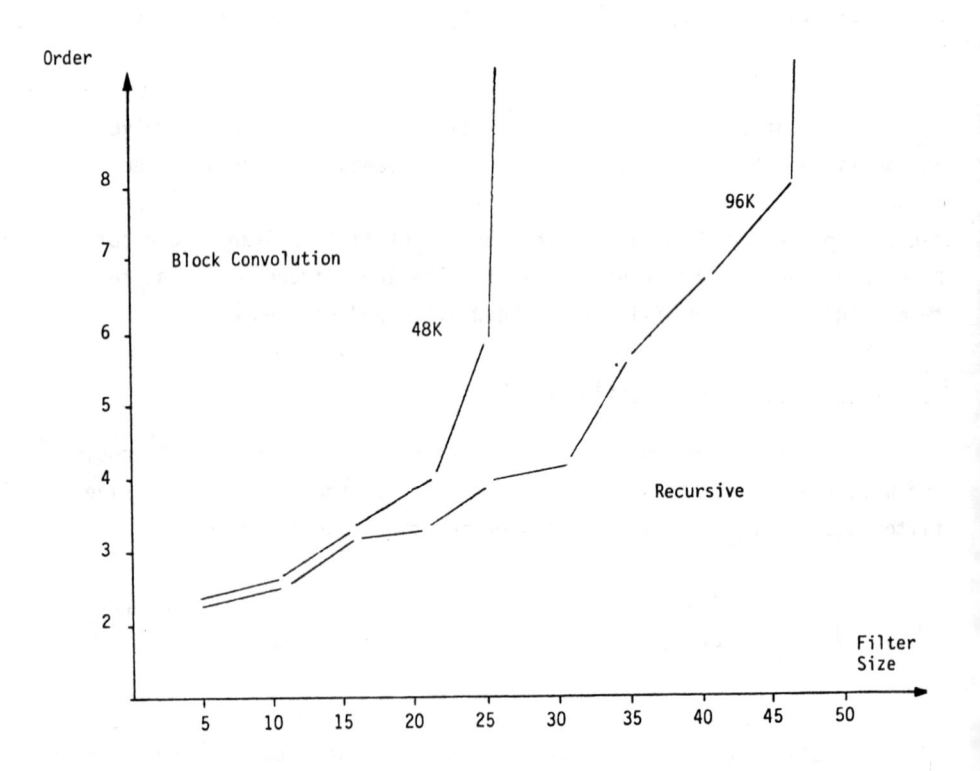

Figure 3. Optimal choices for convolution techniques for two primary memory sizes.

b) **Real time operations** :

State of the art hardware components can be used to perform the convolution of a 512 x 512 image stored in a refresh memory with 3x3 kernels in less than one thirtieth of a second. Special purpose operators can thus be included in image display devices to perform 2-D filtering at TV rate. Unfortunately it is very common that applications require to implement larger convolutions. This paragraph is concerned by providing algorithms that permit to use the 3x3 convolver as a basic element in sequential or parallel processeses that perform large convolutions.

SGK Filters

The first idea consists in sequentially using "small" size convolvers. Such implementation is described in Figure 4, where $P_i(z_1,z_2)$ are the z_1,z_2 transforms of "small" size impulse responses. The filter is called basic "Small" Generating Kernel filter (SGK).

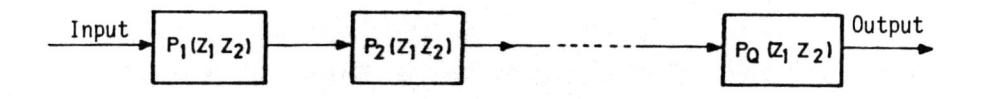

Input → $P_1(Z_1 Z_2)$ → $P_2(Z_1 Z_2)$ → ------ → $P_Q(Z_1 Z_2)$ → Output

Figure 4 : Implementation of basic SGK filter

If the size of the arithmeticKernels is $(2P+1)^2$ and Q of them are used, the overall filter is a $(2PQ+1)^2$ FIR filter. One has to remark that the lack of fundamental theorem of algebra for 2-D polynomials implies that, in general, a $(2PQ+1)^2$ filter is not decomposable in a cascade of Q elementary $(2P+1)^2$ FIR filters. The basic SGK filters are thus a class of FIR filters that have the decomposition property. Design procedures are thus needed to find the SGK that best approximates a desired prototype filter.

On the other hand implementation of $(2PQ+1)^2$ FIR filters requires $(2PQ+1)^2$ arithmetic operations as we saw earlier. SGK filters only require $Q(2P+1)^2$ arithmetic operations. The loss of generality is somewhat compensated by a gain in implementation efficiency.

In some cases, the class of basic SGK filters is too small to obtain good design results. The class of general SGK filters is thus introduced that contains the class of basic SGK filters and preserves most of the advantages of the sequential convolution process. Such filters are described by figure 5.

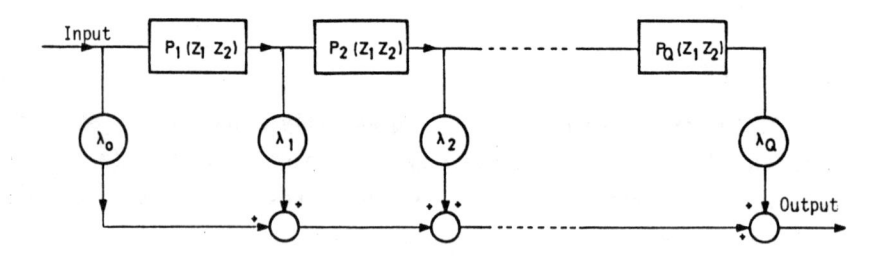

Figure 5. Implementation of general SGK filters.

The number of arithmetic operations required is $Q(2P+1)^2+Q$, but the sequential convolution property remains. Moreover it is possible to implement such filters when two image memories are available (see detailed explanations in ABRAMATIC-FAUGERAS [5]). Such filters require special purpose design procedures described in the next section.

In the case where the P_i's are constrained to be equal, an efficient design procedure due to Mc Clellan [6] is available.

SVD/SGK filters

Another idea consists in decomposing a given FIR prototype filter in a sum of separable 2-D filters. This can be efficiently done if one makes use of the Singular Value Decomposition (SVD) theorem. If one looks at an LxL FIR impulse response $h(k,\ell)$ as a matric $\underline{H}$. The SVD theorem says that such a matrix can be decomposed :

$$\underline{H} = \sum_{r=1}^{R} \lambda_r \underline{U}_r \underline{V}_r^t \qquad (17)$$

where R is the rank of H, $\{\lambda_r\}$ are the square roots of the eigenvalues of $\underline{HH}^T$, $\{U_r\}$ and $\{V_r\}$ are the eigenvectors of $\underline{HH}^T$ and $\underline{H}^T\underline{H}$ respectively. Equation (17) can be rewritten as :

$$h(k,\ell) = \sum_{r=1}^{R} \lambda_r u_r(k) v_r(\ell) \qquad (18)$$

Moreover the SVD theorem says that if one truncates the decomposition to K terms, the mean square error on the impulse response is given by :

$$\varepsilon_2 = \sum_{r=K+1}^{R} \lambda_k \qquad (19)$$

All these results lead to propose the so-called SVD/SGK implementation of an FIR filter described in figure 6.

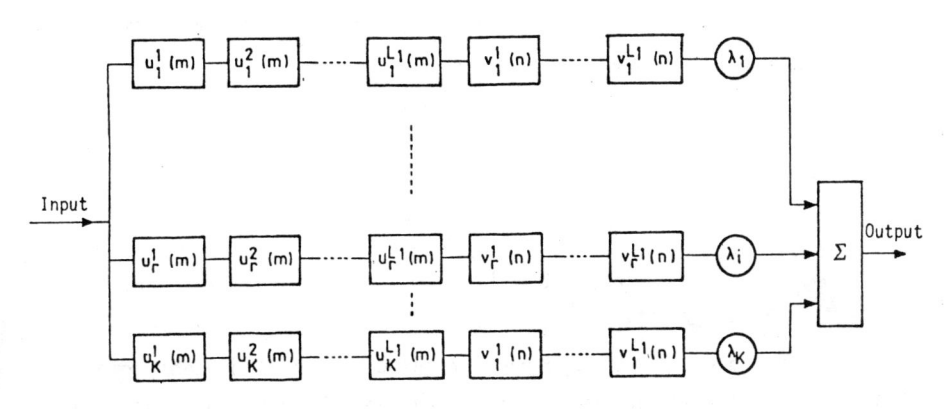

Figure 6 . Implementation of SVD/SGK filter.

Each 1-D filter has been decomposed in a cascade of 1x3 or 3x1 filters. This is always possible thanks to the fundamental theorem of algebra. As we can see, the design procedure of an SVD/SGK filter is trivial and reduces to picking the first K terms of the summation. Unfortunately implementation of such filters is more costly than the one used for SGK filters. The main drawback is the fact that three image memories (instead of two) are required. ABRAMATIC [7] gives a complete comparison of the two implementation procedures including fixed-point computations issues.

III. DESIGN OF 2-D LSI FILTERS

Again problems and techniques differ if one looks for a recursive (IIR) or non-recursive (FIR) filter.

a) <u>Recursive filters</u> :

We concentrate on the design of quarter-plane recursive filters. For these filters, the design procedure can be set as follows :

Given :

1) A prototype filter described by its impulse response $\{h(k,\ell)$; $0 \leq k \leq K$; $0 \leq \ell \leq L\}$ or its transfer function $H(z_1,z_2)$.

2) The order of the recursive filter N.

Find :

The coefficients $\{a_{r,s}\}$ et $\{b_{r,s}\}$ that minimize an error criterion between $H(z_1,z_2)$ and $A(z_1,z_2)$ where :

$$A(z_1,z_2) = \frac{\sum_{r=0}^{N-1} \sum_{s=0}^{N-1} b_{r,s} z_1^{-r} z_2^{-s}}{\sum_{r=0}^{N} \sum_{s=0}^{N} a_{r,s} z_1^{-r} z_2^{-s}}$$

Various methods have been proposed. We first present some of them that illustrate the difficulties involved. We give them the name of their first user in the 1-D case as they are direct generalizations of 1-D methods to the 2-D case.

PRONY METHOD

This method consists in solving a linear problem with a mean square error criterion ε_p

$$\varepsilon_p = \int_{\Gamma_1} \int_{\Gamma_2} |H(z_1,z_2)D(z_1,z_2) - N(z_1,z_2)|^2 W(z_1,z_2) \frac{dz_1}{2i\pi z_1} \frac{dz_2}{2i\pi z_2} \quad (*) \qquad (20)$$

where $W(z_1,z_2)$ is a weighting positive function. Unknown $\{a_{r,s}\}$ and $\{b_{r,s}\}$ appear linearly in the mse criterion. Optimization is thus done by solving a linear system.

APLEVITCH METHOD

Strictly speaking Prony's method does not minimize a mse criterion between the prototype and the recursive filter. The most natural error is ε_A.

$$\varepsilon_A = \int_{\Gamma_1} \int_{\Gamma_2} \left| H(z_1,z_2) - \frac{N(z_1,z_2)}{D(z_1,z_2)} \right|^2 W(z_1,z_2) \frac{dz_1}{2i\pi z_1} \frac{dz_2}{2i\pi z_2}$$

Unknowns $\{a_{r,s}\}$ now appear in a non-linear fashion. The most popular method to solve this optimization problem is to use a gradient method.

MARIA & FAHMY [8] utilized such a technique. They generalized it to an L^p criterion and restricted it to the case where the denominator is built out of second order 2-D polynomials. This constraint simplifies the stability test.

STEIGLITZ METHOD

STEIGLITZ [9] proposed an alternative method in between the two previous ones. SHAW & MERSEREAU [10] first used it in the 2-D case. This technique uses an iterative algorithm. At each iteration, the criterion is :

$$\varepsilon_S^{(k)} = \int_{\Gamma_1} \int_{\Gamma_2} \frac{W(z_1,z_2)}{|D_{k-1}(z_1,z_2)|} \left| H(z_1,z_2) D^k(z_1,z_2) - N^k(z_1,z_2) \right| \frac{dz_1}{2i\pi z_1} \frac{dz_2}{2i\pi z_2}$$

Each optimization problem is thus done by solving a linear system. SHAW & MERSEREAU [10] also studied the design of Separable Denominator Recursive (SDR) filters using this method. Each iteration is split into two sub-iterations for preserving the linear dependance with respect to denominator coefficients.

(*) Γ_1, Γ_2 are the unit circles in the z_1 and z_2 planes respectively.

SEPARABLE DENOMINATOR RECURSIVE FILTERS

When poles are simple, SDR transfer function can be decomposed :

$$A(z_1,z_2) = \frac{N(z_1,z_2)}{D_1(z_1)D_2(z_2)} = \sum_{r=1}^{N}\sum_{s=1}^{N} \frac{\lambda_{r,s}}{(1-p_r z_1^{-1})(1-q_s z_2^{-1})} \qquad (21)$$

This decomposition allowed ROSENCHER [11] and ABRAMATIC et al [12] to propose a specific design algorithm that minimizes the ε_A criterion. Taking the derivatives of ε_A with respect to $\{\lambda_{r,s}\}$, $\{p_r\}$, and $\{q_s\}$ and setting them to zero gives N^2+2N non linear equations of N^2+2N unknowns. An iterative approach similar to Steiglitz technique allows to solve these equations for the poles $\{p_r\},\{q_s\}$ only. $\{\lambda_{r,s}\}$ are then given by solving an N^2 linear system. ABRAMATIC [7] also proposed a way of adjusting the numerator coefficients to minimize a Chebyshev criterion. Comparative results are given for a deconvolution filter shown in Figure 6. Figure 7 shows the original image, the prototype output and two SDR of order 3 outputs designed with L^2 and L^∞ criteria respectively

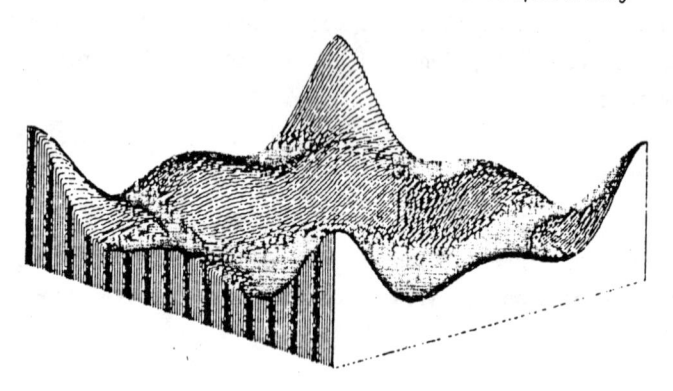

Figure 6. Deconvolution prototype.

SINGULAR VALUE DECOMPOSITION DESIGN APPROACH

TREITEL & SHANKS [13] first proposed to perform the SVD of FIR impulse responses (equation (17)). Using 1-D recursive designs for each of the $u_r(.)$ and $v_r(.)$, they obtained SDR filters of order KN if K is the number of retained eigenvalues and N is the order of the 1-D recursive filters.

DERICHE & ABRAMATIC [14] proposed to perform the K 1-D filter designs in one step. They constrain the 1-D filters to have the same denominator. This gives an order N SDR filters. Figure 8 shows example of prototype and SDR filter designed with this approach.

Blurred Image

Image filtered by the prototype.

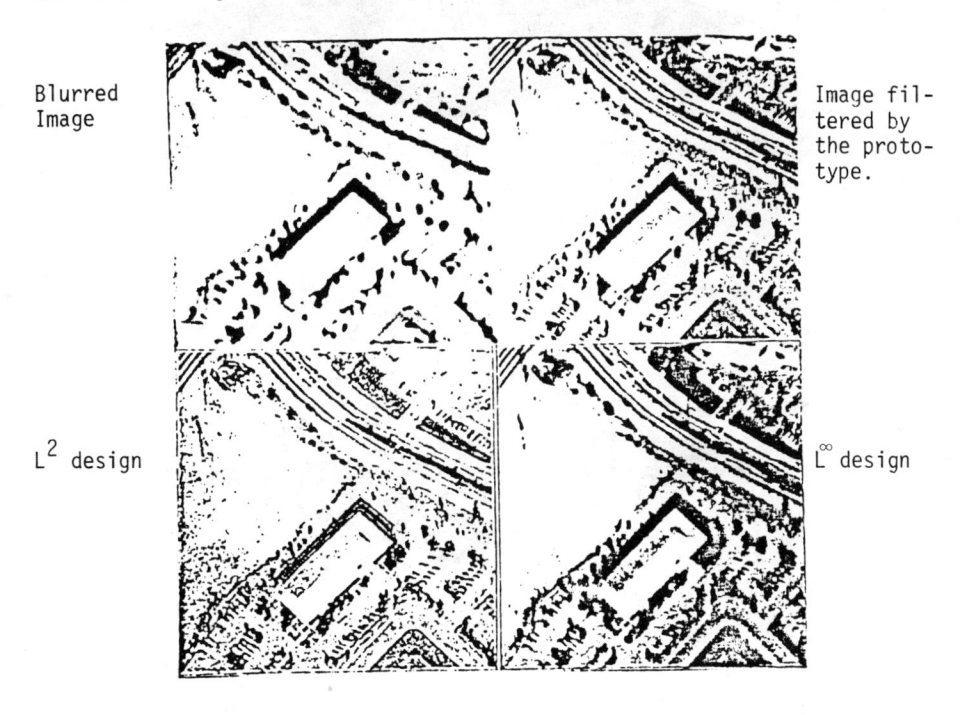

L^2 design

L^∞ design

Figure 7. Deconvolution example.

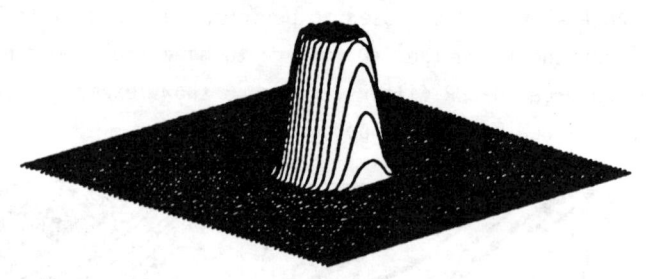

Prototype low pass filter.

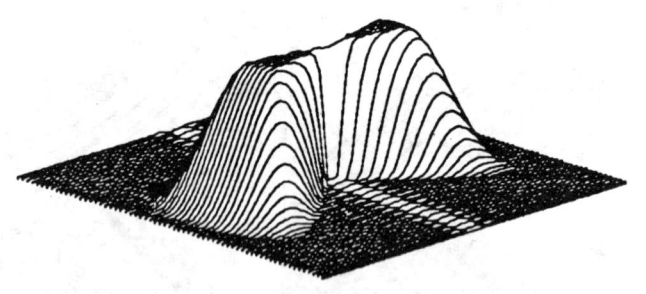

Prototype filter.

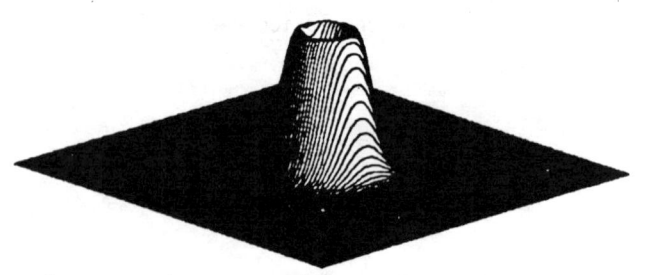

Prototype bandpass filter.

Figure 8. Design Examples.

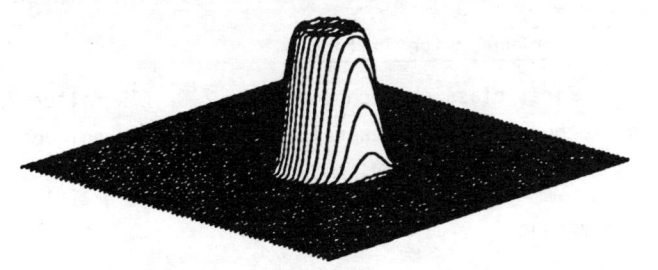

SDR of 6th order. NMSE = .84E-2.

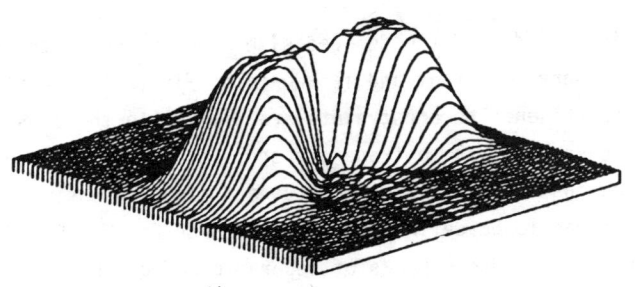

SDR filter of 6th order NMSE = .32E-2.

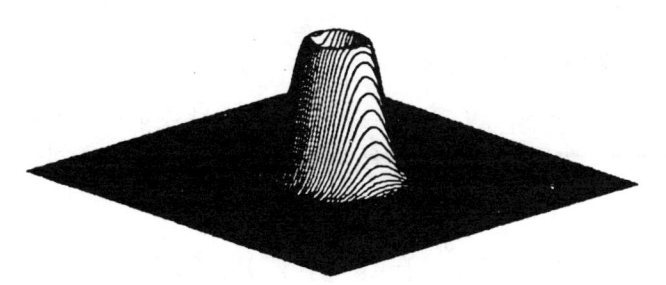

SDR bandpass filter of 6th order NMSE = .19E-2.

Figure 8. Design Examples. (Continued).

b) <u>Non-Recursive Filters</u> :

A first class of techniques for 2-D FIR filter design is composed of methods directly generalized from 1-D techniques. Another class contains more specific algorithms.

WINDOWING METHOD

The most simple and thus popular method is the windowing method. Starting from a large impulse response h(.,.), the FIR filter of desired size is obtained through :

$$a(k,\ell) = h(k,\ell) \cdot w(k,\ell) \qquad (22)$$

w(.,.) is a window of the desired size. The choice of the shape of this window is very important and is two-fold. First of all, its smoothness has to be chosen. The constraints are similar to the ones appearing in the 1-D case. One has to trade off sharp transition in the frequency domain against small ripples in pass and stop regions. Second of all the 2-D shape has to be defined. HUANG [15], SPEAKE & MERSEREAU [16] compared the advantages and drawbacks of separable and circular symmetric shapes.

Windowing techniques are very simple to implement. On the other hand, it is hard to evaluate quantitatively the performance of such methods. This is the reason why efforts have been devoted to optimization techniques.

OPTIMIZATION METHODS

Following 1-D signal processing works, a lot of efforts was concentrated on optimization techniques for Chebyshev problems. The criterion to be minimized is :

$$\varepsilon_\infty = \max_{\omega_1 \omega_2 \in D} |H(e^{j\omega_1}, e^{j\omega_2}) - A(e^{j\omega_1}, e^{j\omega_2})|$$

HU & RABINER [17] first studied the case where the domain D is restricted to the transition band. The small size of the optimization area justified the use of linear programming techniques. When applications require to choose $[0,2\pi] \times [0,2\pi]$ as the D domain, computational problems arise. FIASCONARO [18], KAMP & THIRAN [19], HARRIS & MERSEREAU [20] presented various attemps to alleviate these problems. For this aim, generalization of exchange algorithms are desirable. These algorithms are iterative. They consist in solving the optimization problem on a subset of the domain D and modify this subset iteratively until reaching the solution of the global optimization problem. Exchange strategies are made difficult by the multidimensionality of the problem (see RICE [21]). In any case, these methods are restricted in practice to 15x15 FIR filters.

Mc CLELLAN TRANFORMATION and SGK FILTERS

A very original idea due to Mc CLELLAN [6] consists in designing a 2-D FIR filter with a quadrilateral symmetry through a geometric transformation on a 1-D FIR linear phase filter.

$$G(e^{j\omega}) = \sum_{k=0}^{N} a(k) \cos \frac{2\pi}{N} \omega k = \sum_{k=0}^{N} \hat{a}(k) (\cos \frac{2\pi}{N} \omega)^k$$

The Mc Clellan transformation is defined by

$$\cos \omega = A \cos \omega_1 + B \cos \omega_2 + C \cos \omega_1 \cos \omega_2 + D$$

We thus obtain after a trivial rearrangement :

$$H(e^{j\omega_1}, e^{j\omega_2}) = \sum_{k=0}^{N} \sum_{\ell=0}^{N} h(k,\ell) \cos \frac{2\pi}{N} k\omega_1 \cos \frac{2\pi}{N} \ell\omega_2$$

which is a 2-D FIR filter quadrilateral symmetry. Mc Clellan remarked that some interesting choices could be made for $\{A,B,C,D\}$. As an example :

$$A = B = C = -D = \frac{1}{2}$$

gives "almost" circular curves in the $\{\omega_1,\omega_2\}$ plane when ω is fixed. This choice allows to design "almost" circular symmetric FIR filters through transformation.

Furthemore MECKLENBRAUKER, MERSEREAU [22] remarked that 2-D filters designed through Mc Clellan transformations have the SGK property under the constraints that the kernels are 3x3 and equal. This is another reason for designing FIR filters using this very efficient technique.

MERSEREAU & al [23] generalized the transformation to the case where kernels are larger than 3x3, but the design procedure becomes more complicated.

When applications require the use of general SGK filters with different kernels, special purpose design algorithms have to be used. ABRAMATIC & FAUGERAS [5] proposed to use a block relaxation approach. This iterative method consists in fixing the values of Q-1 kernels and optimize the Qth remaining kernel at each iteration.

IV. CONCLUSION : Trends in 2-D filtering digital systems

At the early beginning, during the sixties, digital image processing used to require the most advanced computer centers. The costs involved were imposing to share the facilities with other applications. Results were obtained through batch operations. Emphasis was put on the techniques we presented in the first part of the chapter although the main delay for the user was caused by the fact that the computer was shared among a lot of users. In the early seventies computing costs decreasing, it became affordable for an image processing laboratory to set up its own computing facilities (minicomputer, image display device). Batch methods found their full domain of application and their efficiency was crucial as it became possible to build interactive image processing systems. Time response of the systems were directly related to algorithms efficiency.

Nowadays man-machine communication can still be enhanced by means of real time computations performed on intelligent display devices. Hardware evolution permits to transfer from the host computer to the display device an increasing number of tasks. It is also possible to perform convolutions right after data acquisition before storing the digitized data. Devices performing such operations are called "smart" sensors. They deliver processed data such as edge maps or texture features. Again this lighten the computational burden assigned to the minicomputer and save its capabilities for higher level tasks.

Tomorrow problems are characterized by the introduction of a third dimension. After having dealt with fixed images, users are looking forward

to process sequences of images. Other areas require to process volumic
signals such as seismic signals for example.If hardware improvements
allow to proceed in this direction, algorithmic researches are needed to
adapt the processing techniques to available facilities. As an example,
the SGK concept can be generalized to 3-D signals. Another class of
investigations is devoted to parallel machines and receive a great deal of
interest. The chapter devoted to architectures for image processing describes the first attempts in this direction.

REFERENCES

[1] T.S. HUANG "Picture Processing and Image Filtering". Topics in Applied Physics, Springer Verlag. (1979).

[2] M.P. EKSTROM, S.K. MITRA. "Two-Dimensional Digital Signal Processing". Benchmark Papers in Electrical Engineering and Computer Science (V.20) Dowden, Hutchinson & Ross. Strondsburg Pa (1978).

[3] T.S. HUANG. "Two-Dimensional Digital Signal Processing". Linear Filters. Topics in Applied Physics, Vol.42, Springer Verlag. (1981).

[4] T.S. HUANG. "Two-Dimensional Digital Signal Processing". Transforms and Median Filters. Topics in Applied Physics, Vol.43, Springer Verlag.(1981)

[5] J.F. ABRAMATIC, O.D. FAUGERAS. "Sequential Convolution Techniques for Image Filtering". IEEE Trans. on ASSP. Vol.30. N°1 (February 1982). (Corrections in April 1982 issue).

[6] J.H. Mc CLELLAN "The Design of Two-Dimensional Digital Filters by Transformation". Seventh Annual Princeton Conf. Inf. Sci. & Syst. : 247-251. (1973).

[7] J.F. ABRAMATIC. "Approximation de Filtres Biindiciels et Traitement Numérique des Images". Thèse de Doctorat d'Etat. Université de Paris VI. (1980).

[8] G.A. MARIA, M.M. FAHMY. "An LP-Design Technique for Two-Dimensional Digital Recursive Filters". IEEE Trans. Acous. Speech & Signal Processing 22(1) : 15-21. (1974).

[9] K.STEIGLITZ, L.E. Mc BRIDE. "A Technique for the Identification of Systems". IEEE Trans. Aut. Cont. 10 ( ) : 461-464. (1965).

[10] G.A. SHAW, R.M. MERSEREAU. "Space-Domain Design of Two-Dimensional Recursive Digital Filters". Proc. IEEE Int. Conf. on Acoust. Speech and Signal Processing. Washington DC. (1979).

[11] Emm ROSENCHER. "Approximation Rationnelle des Filtres à un et deux indices : Une Approche Hibertienne". Thèse d'Ing. Doc. Univ. Paris IX. (1978).

[12] J.F. ABRAMATIC, F.GERMAIN, Emm. ROSENCHER. 'Design of 2-D Separable Denominator Recursive Filters". IEEE Trans. ASSP-27. pp.445-453 (1979).

[13] S. TREITEL, J.L. SHANKS. "The Design of Multistage Separable Planar Filters". IEEE Trans. Geos. Electron. 9(1) : 10-27. (1971).

[14] R. DERICHE, J.F. ABRAMATIC. 'Design of 2-D Recursive Filters Using Singular Value Decomposition Techniques". Proc. ICASSP-82, Paris.

[15] T.S. HUANG . "Two-Dimensional Windows". IEEE Trans. Audio-Electroacoust. 20(1) : 88-89. (1972).

[16] T.S. SPEAKE, R.M. MERSEREAU. "A comparison of Different Window Formulation for Two-Dimensional FIR Filter Design". Proc. ICASSP-79.Washington DC. (1979).

[17] J.V. HU, L.R. RABINER. "Design Techniques for Two-Dimensional Digital Filters". IEEE Trans. Audio-Electroacoust. 20(4) : 249-257. (1972).

[18] J.G. FIASCONARO. "Two-Dimensional Non Recursive Filters". In Topics in Applied Physics. Vol.6. Springer Verlag. (1979).

[19] Y. KAMP, J.P. THIRAN. "Chebyshev Approximation for Two-Dimensional Non Recursive Digital Filters". IEEE Trans. Circ. & Syst. 22(3) ; 208-218. (1975).

[20] D.B. HARRIS, R.M. MERSEREAU. "A comparison of Algorithms for Minimax Design of 2-D Linear Phase FIR Digital Filters". IEEE Trans. ASSP-25.pp.492-500. (1977).

[21] J.R. RICE "The Approximation of Functions". Vol.1 : Linear Theory. Addison-Wesley, Reading, Mass. (1969).

[22] W.F.G. MECKLENBRAUKER, R.M. MERSEREAU. "Mc Clellan Transformations for Two-Dimensional Digital Filtering : II-Implementation". IEEE Trans. Circ. & Syst. 23(7) : 414-422. (1976).

[23] R.M. MERSEREAU, W.F.G. MECKLENBRAUKER, J.F. QUATIERI. "Mc Clellan Transformations for Two-Dimensional Digital Filtering : I-Design" IEEE Trans. Circ. & Syst. 23(7) : 405-414. (1976).

IMAGE ENHANCEMENT

G.H. Granlund and H. Knutsson
Picture Processing Laboratory, Linkoeping University,
Linkoeping, Sweden

R. Wilson, Department of Electrical and Electronical Engineering, The University of Aston in Birmingham, 19 Coleshill Street, Birmingham B4 7BP, England

Abstract. Enhancement is an important preprocessing operation for automated analysis of images, or for making images more appealing visually. Improvements can be made through pixelwise modification of the gray scale transfer function, or through processing of image segments within a certain size neighborhood. For effective enhancement it is important to take into account the image content, and procedures have to involve a low level analysis of the image. This allows the use of isotropic filters which can suppress noise without blurring edges and lines.

INTRODUCTION

Image enhancement involves operations upon an image with the intention of producing another image which is improved in some respect. The related problems of enhancing and restoring noisy images have received a considerable amount of attention in recent years. Restoration methods have generally been based on minimum mean-squared error operations, such as Wiener filtering or recursive filtering (See Helstrom (1967), Hunt (1973) and Pratt (1972)). The rather vague title of enhancement has been given to a wide variety of more or less ad-hoc methods, such as median filtering, (Justusson 1978), which have nonetheless been found useful. In most cases, however, the aim is the same: an improvement of the subjective quality of the image.

It should be noted that enhancement is an information lossy operation, which means that we can never obtain more information in the transformed image than in the original one. Enhancement can however improve the image in some particular respect by imposing restrictions upon its low level structure.

Image enhancement is an important step for preprocessing of images for analysis, or to make images more appealing visually. A reduction of noise may greatly improve the performance of analysis procedures such as edge and line detection. The effects for visual interpretation on the other hand should not be overestimated. No experiments so far have been able to unambiguously demonstrate improvements of visual interpretation, e.g. of X-ray images, after enhancement. The visual system is itself able to perform a very efficient content dependent filtering, which can hardly be surpassed with present technical methods.

The main effects of enhancement are the following:

1 Enhancement of certain properties of the image
2 Reduction of noise
3 Correction of distortions

The requirements implicit in these operations are by no means simple. A fundamental problem is the distinction between image features and noise. No distinction between a relevant image feature and noise can be made based upon the knowledge of a single pixel. An effective distinction requires knowledge about the processed image as well as rules about how image features are expected to behave. This requires that a partial analysis of the image content is performed, the result of which is controlling the enhancement operation.

In the following sections we will look at different operations useful for enhancement.

POINT OPERATIONS

Point operations are used to modify the gray scale of an image. Each output pixel corresponds directly to a pixel having the same coordinates in the input image. These operations are often characterized as contrast enhancement operations.

A point operation produces an output image $G(x,y)$ from an input image $F(x,y)$, where the transformation can be expressed as:

$$G(x,y) = g(F(x,y)) \qquad (1)$$

The point operation is completely specified by the function g, which gives the mapping of input gray level to output gray level.

These functions may be of different types, such as linear, piece-wise linear, logarithmic, exponential, etc, to accomplish the required task. They can for example be used for canceling of effects of image sensor non-linearities, or adaptation of image data to display characteristics.

An interesting type of enhancement is obtained through so called histogram equalization. This implies that the distribution of gray scale values is changed from that in the original image, to produce an image having a flat or rectangular histogram.

NEIGHBORHOOD OPERATIONS

The most common concept related to image enhancement is spatial filtering. This requires access to neighborhoods of a certain size. The simplest operations are linear and position invariant, characterized by a convolution relationship. The input image is convolved with a point-spread function to produce an ouput image which is hopefully improved by the operation.

This type of image enhancement can be performed in the spatial domain or the frequency domain dependent upon the characteristics of the problem.

An input image $F(x,y)$ is operated on by a filter $H(x,y)$ to produce the output image $G(x,y)$

$$G(x,y) = F(x,y) * H(x,y) \tag{2}$$

where the two-dimensional convolution is defined as:

$$F(x,y) * H(x,y) = \int\int_{-\infty}^{\infty} F(\xi,\eta) H(x-\xi,y-\eta) \, d\xi d\eta = \tag{3}$$
$$= \int\int_{-\infty}^{\infty} F(x-\xi,y-\eta) H(\xi,\eta) \, d\xi d\eta$$

The frequency domain relationships are given by:

$$G(u,v) = F(u,v) \cdot H(u,v) \tag{4}$$

where the frequency domain functions constitute the two-dimensional Fourier transforms of the spatial domain functions. For example:

$$F(u,v) = \int\int_{-\infty}^{\infty} F(x,y) \exp[-j2\pi(ux+vy)] \, dxdy \tag{5}$$

Given the availability of fast Fourier transforms, linear shift invariant filtering operations are commonly performed in the frequency domain. These methods can be generalized to certain classes of non-linear operations characterizing what is called homomorphic systems. This allows a treatment of functions which are related by multiplication or convolution. By implementing the inverse of the characteristic system function, it is possible to reduce the non-linear operation (e.g. multiplication) to addition, which reduces the system to a linear one.

Procedures have been proposed for noise supression in images using linear shift-invariant filters. It is in fact possible to prescribe an optimal filter function in the least squares sense, given information about the frequency spectrum of the image as well as of the noise. However, such attempts have not appeared too useful, due to a strong lack of correspondence between model and reality. Some of the reasons for this are that images are not modeled well by stationary stochastic processes and effects of noise depend on local image content. The validity of the least squares quality criterion applied at either the pixel level or simply weighted in the frequency domain is questionable.

IMAGE CONTENT DEPENDENT OPERATIONS

It appears that what constitutes image detail and what constitutes noise has to be determined in a partial analysis of the image content. This means that an effective filtering procedure will have to include at least a low level analysis of the image content. Various attempts have been made to find an approach to take into account the non-stationary properties of image information (See Anderson & Netravali (1976), Abramatic & Silverman (1979) and Ingle & Woods (1979)). One way to attack these problem is using the "general-operator" model of description of image features suggested by Granlund (1978), Granlund and Knutsson (1982). This model allows the description at the images as a hierarchical structure, in terms of different level primitives.

An enhancement operation derived from the above model is a two-stage process. First, the image is convolved with a set of line and edge extraction filters of several orientations to produce a control or "bias" image. This image is complex valued, each point having a magnitude and direction representing the local edge or line information. For computational reasons the number of orientations of the filters is restricted to 4 in the range (o, π). The magnitude, $B(x,y)$, and direction, $\theta(x,y)$, of the bias image at the point (x,y) are equivalent to the ordinary GOP operations for line and edge detection. They are derived from the original image F(x,y) in the following way. First the image F(x,y) is convolved with the set of 4 line and edge filters, $L_i(x,y)$ and $E_i(x,y)$ respectively, to give the magnitude in the i:th direction, $B_i(x,y)$:

$$B_i(x,y) = \sqrt{S_i^2(x,y) + C_i^2(x,y)} / V(x,y) \qquad i=1,2,3,4 \qquad (6)$$

where

$$S_i(x,y) = F(x,y) * E_i(x,y) \qquad (7)$$
$$C_i(x,y) = F(x,y) * L_i(x,y)$$

and

$$V(x,y) = [\sum_{i=1} \sqrt{S_i^2(x,y) + C_i^2(x,y)}]^\beta \qquad 0<\beta<1 \qquad (8)$$

In eqn (7), * denotes convolution. In eqn (8) the exponent β is used to control the degree to which absolute (β=o) or relative (β=1) magnitude is important. The overall magnitude, B(x,y) is then estimated by adding the squared differences of the components in orthogonal directions:

$$B(x,y) = \sqrt{[B_1(x,y) - B_3(x,y)]^2 + [B_2(x,y) - B_4(x,y)]^2} \qquad (9)$$

and the direction $\theta(x,y)$ is estimated using

$$\sin 2\theta(x,y) = [B_2(x,y) - B_4(x,y)]/B(x,y) \qquad (10)$$
$$\cos 2\theta(x,y) = [B_1(x,y) - B_3(x,y)]/B(x,y) \qquad (11)$$

Note that the estimation of $\theta(x,y)$ is expressed in the form of eqns. (10) and (11) to avoid the degeneracy associated with the inverse trigonometric functions.

The design of the line and edge extraction filters can be carried out using a least-squares approach in the frequency domain (See Knutsson & Granlund (1980)). The function chosen should have good interpolation properties, be separable in radius and angle and have smooth variation (to allow a good finite impulse response approximation).

E_i and L_i denote edge and line filters in the i:th direction. ρ and θ are Fourier domain radius and angle respectively.

$$E_i(\rho,\theta) = e_i(\rho) \cdot e_i(\theta); \quad L_i(\rho,\theta) = \ell_i(\rho) \cdot \ell_i(\theta) \qquad (12),(13)$$

where

$$e_i(\rho) = \ell_i(\rho) = \exp - [\frac{4 \ln 2}{\ln^2 B} \ln^2(\rho/\rho_c)] \qquad (14)$$

$$\ell_i(\theta) = \cos^2[\theta - \theta_i] \qquad (15)$$

$$e_i(\theta) = \ell_i(\theta) \operatorname{sgn}[\cos(\theta - \theta_i)] \qquad (16)$$

For implementation of the filters (kernels), square windows of 15x15 pixels have been used.

The finite impulse response approximation to the filter functions can be derived by minimization of the squared error between its transform and those of eqns. (12)-(16) above. Fig. 1. Having estimated the edge magnitude and direction at each point, it is possible to construct an anisotropic filter for the enhancement operation. This filter is the sum of two components: an isotropic low-pass smoothing function (a squared cosine) and a line extraction filter oriented in the direction given by the bias image. The relative proportions of two components are controlled by the bias magnitude. Thus in "flat" regions of the image, the filter is isotropic, but as an edge is approached it becomes increasingly anisotropic, with a bandwith in the direction parallel to the edge which is much lower than that perpendicular to the edge (Figs. 2 and 3).

The isotropic smoothing filter function $H(\rho,\theta)$ can be expressed:

$$H(\rho,\theta) = h(\rho) = \begin{cases} \cos^2(\frac{\pi\rho}{1.8}) & \rho < 0.9 \\ 0 & \text{else} \end{cases} \quad (17)$$

The line extraction filter $M(\rho,\theta)$ was chosen to give a reasonably flat overall response:

$$M(\rho,\theta) = m(\rho) \cdot m(\theta) \quad (18)$$

with

$$m(\theta) = \cos^2\theta \qquad |\theta| < \frac{\pi}{2}, \quad |\theta-\pi| < \frac{\pi}{2} \quad (19)$$

and

$$m(\rho) = \begin{cases} 1 - H(\rho) & \rho < 0.9 \\ 1 & 0.9 \leq \rho < \pi \\ \cos^2[\frac{\pi}{1.8}(\rho-\pi+0.9)] & \pi-0.9 \leq \rho < \pi \end{cases} \quad (20)$$

The processed image, $G(x,y)$ can therefore be expressed as:

$$G(x,y) = \alpha_s F(x,y)*H(x,y) + \alpha_e B(x,y)[F(x,y)*M(x,y,\theta(x,y))] \quad (21)$$

where $H(x,y)$ is the smoothing function and $M(x,y,\theta)$ is the line extraction filter. The constant α_s is chosen to maintain the mean gray level.

In practice, $M(x,y,\theta)$ is obtained by interpolation of the filter responses in the 4 fixed directions. Once the bias image is obtained, it is possible to process the noisy image iteratively, by repetition of the operation expressed by eqn. (21). (Figs. 2c and 3c).

$$G_i(x,y) = \alpha_s G_{i-1}(x,y)*H(x,y) + \alpha_e B(x,y)[G_{i-1}(x,y)*M(x,y,\theta(x,y))] \quad (22)$$
$$i = 1,2,3,\ldots$$

with

$$G_0(x,y) = F(x,y) \qquad (23)$$

EXAMPLES OF ENHANCEMENT

In figures 4 and 5 examples are given of context-sensitive enhancement. The original image in figure 4a contains noise at a level of 10 dB S/N. Figure 4b shows the result after two iterations. Figure 5 shows the use of enhancement as a preprocessing step for analysis. A noisy fingerprint is subjected to enhancement with 5b showing the result after one iteration and 5c the result after two iterations. The processing has been performed using the GOP Image Processor.

REFERENCES

Abramatic, J.F., Silverman, L.M. Non Stationary Linear Restoration of Noisy Images. Proc. IEEE Conf. on Decision and Control, pp 92-99, Ft. Lauderdale, 1979.

Anderson, G.L., Netravali, A.N. Image Restoration Based on a Subjective Criterion. IEEE Trans. Syst. Man and Cyber., SMC-6, 12, pp 845-853, 1976.

Granlund, G.H. In Search of A General Picture Processing Operator. Comput. Graph. and Image Proc., 8, 2, pp 155-173, 1978.

Granlund, G.H., Knutsson, H. Hierarchical Processing of Structural Information in Artificial Intelligence. Proc. of 1982 IEEE Int. Conf. on Acoustics, Speech and Signal Processing, pp 11-16, Paris, 1982.

Helstrom, C.W. Image Restoration by the Methods of Least Squares. J. Opt. Soc. Am., 57, 3, pp 297-303, 1967.

Hunt, B.R. The Application of Constrained Least Squares Estimation to Image Restoration by Digital Computer. IEEE Trans. Comput., C-22, pp 805-812, 1973.

Ingle, V.K., Woods, J.W. Multiple Model Recursive Estimation of Images. Proc. IEEE Conf. on Acoustics, Speech and Signal Processing, pp 642-645, Washington, 1979.

Justusson, B. Noise Reduction by Median Filtering. Proc. 4th Int. Joint Conf. on Pattern Rec., pp 502-504, Kyoto, Japan, 1978.

Knutsson, H., Granlund, G.H. Fourier Domain Design of Line and Edge
        Detectors. Proc. IEEE Conf. on Pattern Rec., pp 45-48, Miami,
        1980.
Pratt, W.K. Generalised Wiener Filtering Computation Techniques. IEEE
        Trans. Comput., C-21, 7, pp 297-303, 1972.
Wilson, R., Knutsson, H., Granlund, G.H. Image Coding Using a Predictor
        Controlled by Image Content. Proceedings of IEEE Int. Conf. on
        Acoustics, Speech and Signal Processing, Paris, France, pp
        432-435, 1982.

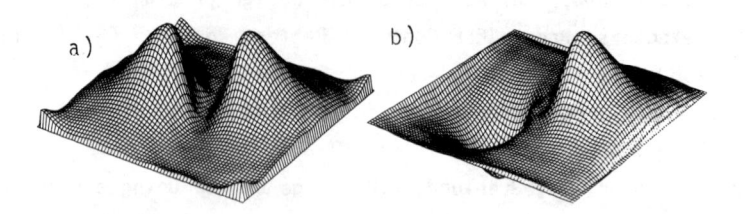

Figure 1 Fourier domain responce of (a) line mask (b) edge mask. Parameters here are B = 4 $\rho_c = 1.11$ $\theta_i = 22.5$

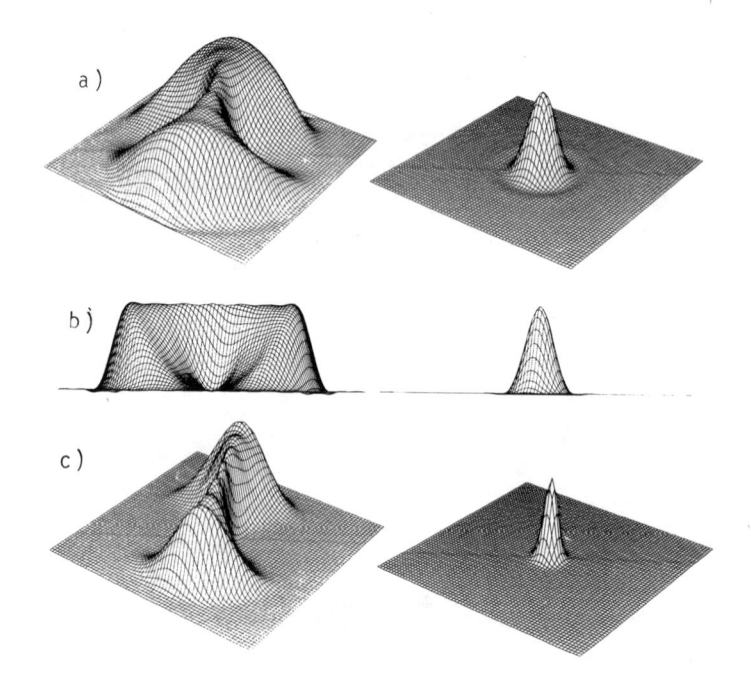

Figure 2 a) Line extraction filter
b) Side view
c) Resulting line extraction filter after 4 iterations

Figure 3 a) Smoothing filter
b) Side view
c) Resulting smoothing filter after 4 iterations

Image enhancement

Figure 4 Context sensitive enhancement of portrait
a) Original image with 10 dB S/N
b) Enhanced image after two iterations

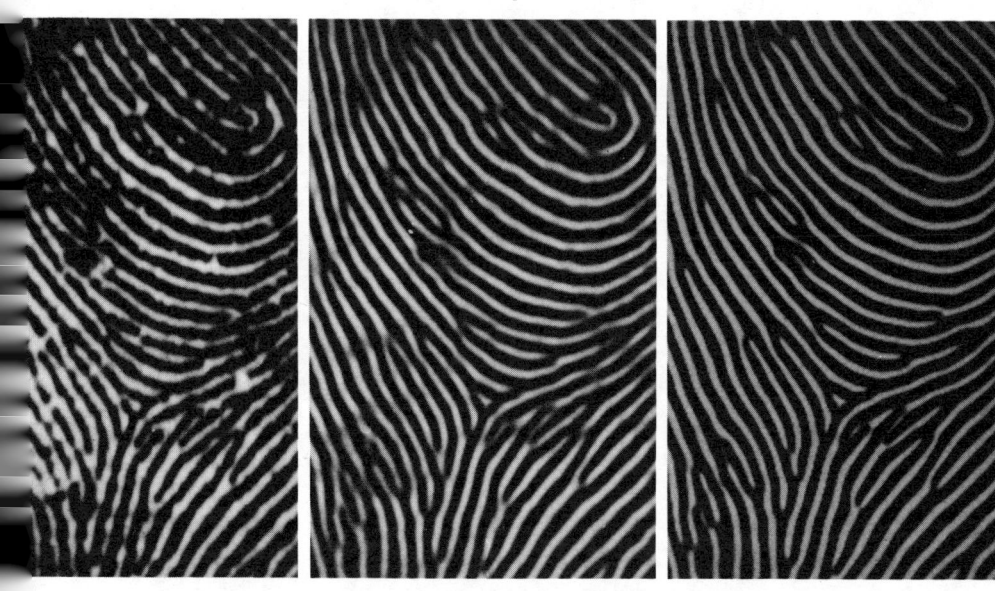

Figure 5 Context sensitive enhancement of fingerprint
a) Original image
b) Result after one iteration
c) Result after two iterations

DIGITAL IMAGE RESTORATION

J.F. ABRAMATIC

Domaine de Voluceau - ROCQUENCOURT

B.P. 105 - 78153 - LE CHESNAY CEDEX - FRANCE

INTRODUCTION

Chapters devoted to image data acquisition have already indicated that image formation processes could never be perfect. Three major types of degradation can arise.

- non linearity
- bandwidth reduction
- noise distorsion

We do not reexamine here these problems in details. We just recall first that non linearity occurs in every recording process. Photochemical and photoelectronic techniques lead to different situations. However, in both cases, saturation and logarithmic correspondance occur. It is also clear that any physical recording system is band limited. Thus a loss of resolution cannot be avoided. Finally the measurement process introduces noise. Poisson and Gaussian noises are popular models and have been extensively studied and evaluated. The overall recording process can be described with the two following operations.

$$g(x,y) = s\{ \int_{-\infty}^{+\infty} \int_{-\infty}^{+\infty} h(x,y,r,s) f(r,s) \, drds \} + n(x,y) \qquad (1)$$

$s(.)$ represents the non linearity that is modeled as a point transformation. $h(.,.,.,.)$ is a linear spatial operator that takes into account the bandwitdth limitation. $n(.,.)$ is the noise supposed to be additive. $f(.,.)$ is the ideal image, $g(.,.)$ is the measured one. Digital processing requires to use a discretized version of equation (1).

$$g(k,l) = s\{ \sum\sum h(k,l,m,n) f(m,n) \} + n(k,l) \qquad (2)$$

Approximations are necessary to get equation (2) from equation (1). They have been properly described in ANDREWS-HUNT [1]. Equation (2) is the one we would like to invert. More precisely digital image restoration is the digital process that tries to recover $f(.,.)$ from $g(.,.)$. This problem is, in general, impossible for various reasons, which we try to classify.

1. - <u>Deterministic degradations</u>

a) $s(.j, h(.,.,.,.)$ are often unknown. They have to be calculated from the measurement data $g(.,.)$. In any case, this identification process is never perfect and induces errors.

b) If $s(.)$ and $h(.,.,.,.)$ are known from some a priori information (characteristics of the film, calibration of the camera,...), the mathematical process of inversion may not lead to a unique solution.

c) If the solution can be proved to be unique, the inversion process can be ill-conditioned. In this case, computations are difficult to perform.

2. - <u>Stochastic degradations</u>

a) Values taken by the noise process are always unknown. In some cases even statistics of the noise are unknown and need to be computed from the data.

b) In a stochastic framework, it is very often that the image itself is modeled as a stochastic process. Statistics of this process also have to be computed from the measured image.

For these computations to be possible, assumptions such as stationnarity and ergodicity need to be made. It is very hard to check whether or not such assumptions are valid.

In most cases, image restoration appears to be an estimation problem. One looks for an estimation $\hat{f}(.,.)$ of $f(.,.)$ given the information $g(.,.)$. A criterion has to be introduced that one tries to minimize between the ideal image and the proposed estimate. So as to deal with a mathematically tractable problem, the mean square error (mse) criterion is very often used. If needed, some constraints are introduced to take into account a priori informations.

We shall put emphasis on estimation techniques with mse criterion.

It has to be pointed out that, in general, there is no physical reason to use such criterion. We shall also have the opportunity to criticize this criterion and propose various ways to tackle this situation.

As we have already mentioned image restoration was among the early topics digital image processing dealt with. A large variety of techniques has thus been proposed. Books (ANDREWS-HUNT[1]),chapters of books (PRATT [2], HUANG [3], GONZALES-WINTZ [4], ROSENFELD [5])have been devoted to this topic. It is not possible to present now all of these techniques in such a chapter and we shall emphasize the major methods in the case of a Linear Shift Invariant (LSI) degradation where efficient techniques developed in the chapter related to 2-D Digital Processing can be used. Moreover these methods have gathered a large part of the image restoration works. We shall then extend the results in various directions to illustrate the amount of work that has been performed. Space-variant degradations will be covered. Non linear restoration techniques will be presented. Some of them use criteria derived from human vision considerations.

I. - LSI DECONVOLUTION TECHNIQUES

Two major assumptions have to be made to reduce a restoration problem to a deconvolution problem.

i) non-linearity

The easiest case occurs when the non-linearity is supposed negligible. This can be justified in some cases when the processing uses density data rather than intensity data. Another way of eliminating the non linearity is to suppose that it distributes over addition. Equation (2) can then be transformed as

$$s^{-1}\{g(k,l)\} \simeq \Sigma\Sigma\ h(k,l,m,n)\ f(m,n) + s^{-1}\{n(k,l)\} \tag{3}$$

ii) non-stationnarity

To get a deconvolution, we need to further assume that $h(.,..,...)$ is space-invariant

$$h(k,l,m,n) = h(k-m,l-n,0,0) \qquad \forall\ k,l,m,n \tag{4}$$

With these assumptions, the distorsion process is modelled as

$$g(k,l) = \sum\sum h(k-m,l-n) f(m,n) + n(k,l) \qquad (5)$$

2-D filtering techniques tell us that equation (5) can be rewritten in the Fourier Domain as

$$G(u,v) = H(u,v) F(u,v) + N(u,v) \qquad (6)$$

The major linear techniques then propose to obtain $\hat{F}(u,v)$ the 2-D Fourier transform of the estimated image as the output of a restoring filter $R(u,v)$ whose input is the measured image.

$$\hat{F}(u,v) = R(u,v) G(u,v) \qquad (7)$$

To finish with notations, when spectra are required to compute the restoring filter, they are denoted $S_{gg}(u,v), S_{ff}(u,v), S_{nn}(u,v)$ for the measured image, the ideal image, the noise respectively. σ_n^2 will be the variance of the noise. The first class of techniques is related to the case where all the a priori information is available.

a) <u>deconvolution with a priori information available</u>

The techniques presented here vary with the criterion that one uses to define an error between the ideal and estimated images.

1. <u>Inverse filter</u>

When the criterion is a mse error between the measured image and the estimated image filtered by $h(.,.)$:

$$\varepsilon_2 = E[g(k,l) - h * \hat{f}(k,l)]^2 \qquad (8)$$

the restoring filter is

$$R(u,v) = \frac{H^*(u,v)}{|H(u,v)|^2} \qquad (9)$$

where $H^*(u,v)$ denotes the conjugate of $H(u,v)$. This filter is the most trivial one could think of. It does not take into account the noise distorsion. It is thus very inefficient near the zeroes of $H(u,v)$ where $R(u,v)$ goes to infinity. If the noise is non-zero for these frequencies, the estimated image will be poor

$$\hat{F}(u,v) = F(u,v) + \frac{N(u,v)\, H^*(u,v)}{|H(u,v)|^2} \qquad (10)$$

2. Constrained mse

To take into account the noise influence, one can choose to minimize a linear function of the estimated image under a constraint related to the noise

$$\varepsilon^2 = E[L * \hat{f}(k,l)]^2 \qquad (11)$$

with

$$E[g(k,l) - h * \hat{f}(k,l)]^2 = \sigma_n^2 \qquad (12)$$

The restoring filter is then

$$R(u,v) = \frac{H^*(u,v)}{|H(u,v)|^2 + \gamma |L(u,v)|^2} \qquad (13)$$

where γ is ajusted to satisfy (12). Various choices have been proposed for $L(u,v)$.

i) $L(u,v) = 1 \qquad (14)$

$$R(u,v) = \frac{H^*(u,v)}{|H(u,v)|^2 + \gamma}$$

This filter is called the pseudo-inverse filter. It alleviates the noise problem described in the previous paragraph.

ii) $L(u,v) = 1 - \dfrac{\cos u + \cos v}{2} \qquad (15)$

$$R(u,v) = \frac{H^*(u,v)}{|H(u,v)|^2 + \gamma(1 - \frac{\cos u + \cos v}{2})^2}$$

This filter tries to minimize the energy of the gradient of the estimated image. It thus limits again the effects of the noise.

iii) $L(u,v) = \left\{ \dfrac{S_{nn}(u,v)}{S_{ff}(u,v)} \right\}^{1/2} \qquad (16)$

$$R(u,v) = \frac{H^*(u,v)\, S_{ff}(u,v)}{|H(u,v)|^2 S_{ff}(u,v) + \gamma\, S_{nn}(u,v)}$$

The minimization of the energy of the restored image is here weighted by the confidence one can have in the restored image. This confidence is measured through the noise to signal ratio. This filter is called the parametric wiener filter. When γ varies from 0 to 1, this filter varies from the inverse filter to the wiener filter.

3. Wiener filter

This filter has been widely used in signal processing. It uses as a criterion

$$\varepsilon^2 = E[f(k,l) - \hat{f}(k,l)]^2 \qquad (17)$$

which is a real mse between the ideal and restored image considered as a realization of stochastic processes. The restoring filter is given by

$$R(u,v) = \frac{H^*(u,v) S_{ff}(u,v)}{|H(u,v)|^2 S_{ff}(u,v) + S_{nn}(u,v)} \qquad (18)$$

This filter is essentially a low-pass filter. The noise is generally supposed to lie in a larger frequency band than the signal.

4. Power Spectrum Equalization

Another approach to deconvolution was proposed by CANNON [6]. It consists in forcing the spectrum of the restored image to be equal to the spectrum of the ideal image.

The spectrum $S_{\hat{f}\hat{f}}(u,v)$ of the restored image is

$$S_{\hat{f}\hat{f}}(u,v) = |R(u,v)|^2 S_{gg}(u,v) \qquad (19)$$

Thus

$$S_{\hat{f}\hat{f}}(u,v) = |R(u,v)|^2 \{|H(u,v)|^2 S_{ff}(u,v) + S_{nn}(u,v)\} \qquad (20)$$

Enforcing this spectrum to be equal $S_{ff}(u,v)$ leads to the restoring filter

$$R(u,v) = \left\{ \frac{S_{ff}(u,v)}{|H(u,v)|^2 S_{ff}(u,v) + S_{nn}(u,v)} \right\}^{1/2} \qquad (21)$$

This filter will be seen to have an interesting property when a priori information is not available. This filter can also be considered as the geometric mean between the inverse filter and the Wiener filter.

5. Geometric Mean filter

Digital image restoration can be seen as a trade off between gain in resolution and sensitivity to noise. A good resolution can be achieved by means of inverse filtering but the sensitivity to noise was seen to be very high. On the other hand, Wiener filtering has good properties regarding noise effects but as a low-pass filter induces a loss in resolution. STOCKHAM proposed to use a filter in between inverse and Wiener filter, the geometric mean filter. More precisely, to be able to trade off between these two filters, he proposed the following filter.

$$R(u,v) = \{\frac{H^*(u,v)}{|H(u,v)|^2}\}^\alpha \{\frac{H^*(u,v) S_{ff}(u,v)}{|H(u,v)|^2 S_{ff}(u,v) + \gamma S_{nn}(u,v)}\}^{1-\alpha} \quad (22)$$

α and γ are two tuning parameters that can be chosen according to the specified problem at hand. One can notice that choosing $\alpha = \frac{1}{2}$, $\gamma = 1$ leads to the Power Spectrum Equalization (PSE) filter of CANNON, as we just mentioned. The PSE filter thus appears as one possible compromise in the resolution/noise problem. We shall also see that this formulation allows further developments when α and γ are no longer tuned for a whole image but are made dependent upon the local situation in the image.

b) blind deconvolution

Another result of CANNON'S work is related to deconvolution when a priori information is lacking. This situation is called blind deconvolution. CANNON'S approach is the following.

Having justified the PSE filter as a trade off between inverse and Wiener filtering, the problem is to calculate this filter with limited information.

Looking at equation (21), it is easy to see that the denominator of the right hand side is the spectrum $S_{gg}(u,v)$ and thus can be computed from the measured image which is available. Spectrum evaluation is performed by sectioning the measured image of size N x N into blocks of size M x M. M is chosen to be much larger than the supposed size of the unknown impulse response h(.,.) For each block i, the power spectrum $S_{gg}^i(u,v)$ is computed. $S_{gg}(u,v)$ is then computed by averaging the $S_{gg}^i(u,v)$.

To set up the restoring filter, one needs also to know its numerator $\{S_{ff}(u,v)\}^{1/2}$. One possibility is to compute it from similar unblurred images available. This amounts to having some a priori information. Another possibility is to recognize H(u,v) by its signature in the measured image spectrum. Evaluating the noise spectrum in a uniform part of the image allows then to estimate $S_{ff}(u,v)$ from

$$S_{gg}(u,v) = |H(u,v)|^2 S_{ff}(u,v) + S_{nn}(u,v) \qquad (23)$$

Examples of this blind deconvolution technique can be found in CANNON [6] and OPPENHEIM [7].

c) Comments on deconvolution techniques

All the methods we just presented find the restored image by LSI filtering the measured one. Two reasons justified this type of approach. First of all, the mathematical modelling allows to compute such filters. Second, a number of techniques are available for implementing 2-D LSI filtering as we have seen in another chapter.

Unfortunately, LSI filtering does not give satisfactory results in many instances. Except for the inverse filter, all of these filters are low-pass or band-pass filters. They attenuate high frequencies and thus lead to a loss of resolution which is seldom acceptable. When the restored image is judged by a human observer, details like edges appear blurred. On the other hand such filters do an excellent job in more uniform areas where no details are of interest and where the noise appears the most.

The idea thus arose that adapting the filter to local situations (edge vs uniform area) would give a better result as far as the eye is concerned. This approach is now developped.

II. ADAPTIVE RESTORATION TECHNIQUES

a) ANDERSON-NETRAVALI'S Approach

The first attempt to tune the restoring filter to the local characteristic of the image is due to ANDERSON-NETRAVALI [8]. They studied the noise only case and started from a restoring technique first presented by BACKUS-GILBERT [9] who were dealing with 1-D geophysical signals. FRIEDEN in HUANG [3] suggested that this technique could be employed for image restoration. Basically, Frieden's filter can be seen as another compromise between inverse and Wiener filter, comparable to the parametric Wiener filter. The contribution of ANDERSON-NETRAVALI consists in bringing out the idea that the compromise could vary inside the image. Furthermore they studied how this variation could be controlled. They introduced a so-called masking function M(k,l) which quantifies the edge busyness at point (k,l).

$$M(k,l) = \sum_{m,n} \sum c^{-\sqrt{(k-m)^2 + (l-n)^2}} (|m^H(m,n)| + |m^V(m,n)|) \tag{86}$$

where $m^H(.,.)$ and $m^V(.,.)$ are the gradients of the image in the horizontal and vertical directions respectively. c controls the decaying influence of the edge when the point (k,l) gets further away. The next step in ANDERSON-NETRAVALI'S work was to quantitatively associate the noise visibility and the masking function. They introduced a visibility function $\varphi(M)$ which for a given masking value M, measures the visibility of the noise. For a masking value of zero, no edge is present and the visibility is total

$$\varphi(0) = 1$$

When the masking value grows, the edge "hides" the noise and the visibility function goes to zero

$$\varphi(+\infty) = 0$$

Psychophysical experiments were set up to measure precisely the visibility function on various images.

b) Comments on ANDERSON-NETRAVALI'S Approach

The following remarks can be made :

i) In their conclusion, ANDERSON-NETRAVALI noticed that the visibility functions computed on various images were not widely different. They tried the experiments where they restored an image based upon a visibility function calculated on a different image. The results were not significantly worse. This leads to the conclusion that a regularly decreasing function from 1 to 0 would be a good guess for the visibility function. This prevents from estimating this function.

ii) Computation of the masking function involves estimation of horizontal and vertical derivatives of the original image. Two points have to be made here

- these derivatives may allow to set up non-isotropic filters. These filters may be more efficient to smooth the noise in edge regions.
- these derivatives need to be estimated from the distorted image which is the only one to be available.

iii) starting from these observations, ABRAMATIC-SILVERMAN [10] proposed an adaptive restoration technique that somewhat bridges the gap between the LSI techniques previously introduced and the adaptive method we just mentionned.

c) <u>ABRAMATIC-SILVERMAN's Approach</u>

This adaptive restoration technique can be seen as a non linear filtering method. It is also presented in the noise only case to put forward the comparison with the ANDERSON-NETRAVAILI's method. Generalization to the deconvolution case will be mentionned at the end of this paragraph. The non linear problem is set up as follows.

i) the measurement equation is modified as

$$g(k,l) = v(k,l) + n^*(k,l) \tag{25}$$

Where $v(.,.)$ is a noisy version of the ideal image which is supposed to look identical to this ideal image for a human observer. $n^*(.,.)$ is the residual noise that the restoring process should try to filter.

This residual noise should depend upon the local situation

$$n^*(k,l) = \gamma_{k,l}(f(.,.),n(k,l)) \tag{26}$$

ii)) the criterion is a modified mse criterion that depends upon the local situation

$$\varepsilon_2(k,l) = N_{k,l}(r_{k,l}(.,.),f(.,.)\sigma_n^2) + S_{k,l}(r_{k,l}(.,.),f(.,.)\Lambda_{ff}(.,.)) \quad (27)$$

where $N_{k,l}$ and $S_{k,l}$ are adaptive versions of the noise and resolution terms of the standard Wiener criterion :

$$\varepsilon_2 = N(r(.,.),\sigma_n^2) + S(r(.,.),\Lambda_{ff}(.,.)) \quad (28)$$

$$N(r(.,.),\sigma_n^2) = \sigma_n^2 \Sigma\Sigma \; r^2(k,l) \quad (29)$$

$$S(r(.,.),\Lambda_{ff}(.,.)) = \Sigma\Sigma\Sigma\Sigma([\delta(k,l)-h(k,l)]\Lambda_{ff}(k-k',l-l') \\ [\delta(k',l')-h(k',l')]) \quad (30)$$

$\Lambda_{ff}(.,.)$ is the autocorrelation function of the ideal image.

$\delta \;(.,.)$ is the 2-D Kronecker symbol

$r(.,.)$ is the impulse response of the restoring filter.

The adaptive filter is thus

$$\hat{f}(k,l) = \underset{mn}{\Sigma\Sigma} \; r_{k,l}(m,n) \; g(k-m,l-n) \quad (31)$$

where $r_{k,l}(.,.)$ minimizes $\varepsilon_2(k,l)$

The choices for $\lambda_{k,l}, S_{k,l}$ lead to various filters.

Two of them are of most interest.

i) If the adaptivity is just introduced through $N_{k,l}$ as

$$N_{k,l} = \varphi(M(k,l)) \; \sigma_n^2 \; \Sigma\Sigma \; r_{k,l}(m,n) \quad (32)$$

where $\varphi(.)$ is the visibility function, one obtains an adaptive version of the parametric Wiener filter.

$$R_{k,l}(u,v) = \frac{S_{ff}(u,v)}{S_{ff}(u,v) + \varphi_{k,l}\sigma_n^2} \quad (*) \tag{33}$$

ii) If the adaptivity is introduced through $\gamma_{k,l}$

$$\gamma_{k,l} = \varphi_{k,l} \cdot n(k,l) \tag{34}$$

one obtains an adaptive linear combination of the identity filter (which plays here the role of the inverse filter), and the Wiener filter

$$R_{k,l}(u,v) = (1-\varphi_{k,l}) + \varphi_{k,l} \frac{S_{ff}(u,v)}{S_{ff}(u,v) + \sigma_n^2} \tag{35}$$

This filter has a very efficient implementation.

More details about these filters can be found in ABRAMATIC-SILVERMAN [10].

One can find also in this paper various ways of handling the non isotropic nature of the restoration for points that are close to Another contribution of the ABRAMATIC-SILVERMAN [10] approach is to propose an estimation of the masking function which answers one question we arose at the end of the previous paragraph. Finally solutions to the deconvolution problem are given following the lines we just presented.

$$R_{k,l}(u,v) = \frac{H^*(u,v) S_{ff}(u,v)}{|H(u,v)|^2 S_{ff}(u,v) + \varphi_{k,l}\sigma_n^2} \tag{36}$$

$$R_{k,l}(u,v) = \varphi_{k,l} \frac{H^*(u,v) S_{ff}(u,v)}{|H(u,v)|^2 S_{ff}(u,v) + \sigma_n^2} + (1-\varphi_{k,l}) \frac{H^*(u,v)}{|H(u,v)|^2} \tag{37}$$

(*) $\varphi_{k,l}$ is a condensed notation for $\varphi(M(k,l))$

Digital image restoration

III.- RESTORATION OF LSV DEGRADATIONS

a) SAWCHUK's approach

SAWCHUK [15] showed that a number of space-variant blur such as motion and optical aberrations blurs could be decomposed into three operations. A geometric distortion is the first step followed by an LSI blur and a second geometric distortion. This space variant blur is described in Figure 1.

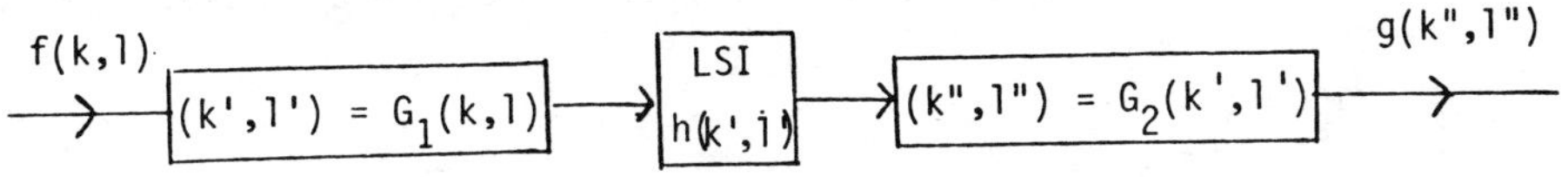

$$f(k,l) \rightarrow \boxed{(k',l') = G_1(k,l)} \rightarrow \boxed{\text{LSI } h(k',l')} \rightarrow \boxed{(k'',l'') = G_2(k',l')} \rightarrow g(k'',l'')$$

Figure 1 : A class of space-variant blur

SAWCHUK then showed easily that the deblurring process could be implemented by the scheme described in figure 2.

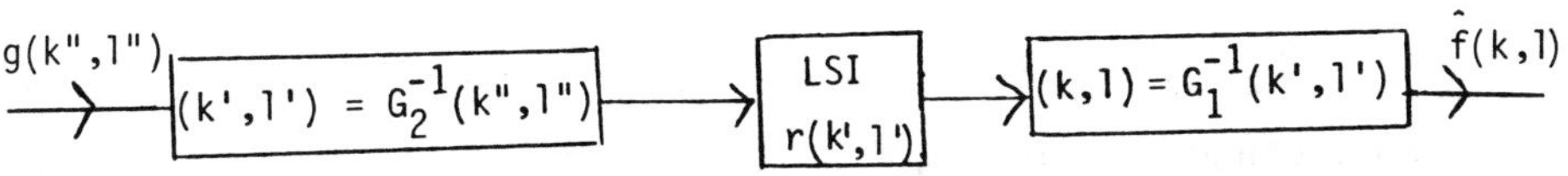

$$g(k'',l'') \rightarrow \boxed{(k',l') = G_2^{-1}(k'',l'')} \rightarrow \boxed{\text{LSI } r(k',l')} \rightarrow \boxed{(k,l) = G_1^{-1}(k',l')} \rightarrow \hat{f}(k,l)$$

Figure 2 : Restoration system

Spectacular illustrations of this work can be found in SAWCHUK [15].

b) General algebraic approach

Equation (2) can be rewritten in matrix formas

$$\underline{g} = s([H]\underline{f}) + \underline{n} \tag{38}$$

Where $\underline{g}$ and $\underline{f}$ are $N^2 \times 1$ vectors obtained by stacking columns of the measured and ideal images. [H] is an $N^2 \times N^2$ matrix built from the values of $h(.,.,.,.)$. If we suppose again that the non-linearity can be neglected. We are led to a problem of matrix inversion in presence of noise.

$$\underline{g} = [H]\underline{f} + \underline{n} \tag{39}$$

In practical image restoration problems, [H] is too large to have an inverse that could be computed. As in the LSI case, the restoration problem appears as an approximation problem. Again the choice of the criterion leads to various solutions. This approach thus generalizes to the LSV case the results previously obtained.

i) Inverse filter :

If the criterion $E(\hat{\underline{f}})$ is

$$E(\hat{\underline{f}}) = \| \underline{g} - [H]\underline{f} \|^2 \tag{40}$$

The restored image is given by

$$\hat{\underline{f}} = (H^{*t} H)^{-1} H^{*t} \underline{g} \tag{41}$$

ii) Constrained least-square approach :

Minimizing a linear function of $\hat{\underline{f}}$

$$E[\hat{\underline{f}}] = [L]\hat{\underline{f}} \tag{42}$$

under the constraint

$$\| \underline{g} - [H]\hat{\underline{f}} \|^2 = \| \underline{n} \|^2 \tag{43}$$

leads to the constrained least-square filter

$$\hat{\underline{f}} = (H^{*t}H + \gamma L^{*t})^{-1} H^{*t} \underline{g} \tag{44}$$

We leave to the reader as an exercise to derive the various filters one could obtain when picking different L's - Details can be found in ANDREWS-HUNT [1].

iii) **Wiener filter** :
If $E[\underline{f}]$ is chosen is be

$$E[\hat{\underline{f}}] = ||\phi_f^{-1/2} \phi_n^{1/2} \underline{f}||^2 + ||\underline{g}-[H]\underline{f}||^2 \qquad (45)$$

one obtains the LSV Wiener filter

$$\hat{\underline{f}} = (H^{*t} H + \phi_f^{-1} \phi_n)^{-1} H^{*t} \underline{g} \qquad (46)$$

where ϕ_f and ϕ_n are the covariance of the ideal and noise image respectively.

iv) **Singular Value Decomposition approach**

Typical to the algebraic approach is the use of the Singular Value Decomposition (SVD) to pseudo-inverse the matrix [H]. We suppose here that the noise is negligeable. It is known that various definition are available for pseudo-inversion of a singular matrix. The filter previously named pseudo-inverse filter is often called a least-square inverse filter as it minimizes a mse criterion between $\underline{g}$ and $H[\underline{f}]$.

Another pseudo-inverse can be defined by using the SVD theorem which says that any rank R matrix H can be decomposed as

$$[H] = \sum_{k=1}^{R} \lambda_k^{1/2} u_k v_k^t \qquad (47)$$

$$[H] = [U] [\Lambda^{1/2}] [V]^t \qquad (48)$$

where λ_k are the eigenvalues of $[H]^t [H]$ and $[U]$ and $[V]$ the eigenvectors of this matrix and its transposed.
A possible pseudo-inverse $[H]^+$ is given by

$$[H]^+ = [V] [\Lambda^{-1/2}] [U]^t \qquad (49)$$

the restored image is then given by

$$\hat{\underline{f}} = \sum_{k=1}^{R} \lambda_k^{1/2} (u_k^t \underline{g}) \underline{v}_k \tag{50}$$

ANDREWS-HUNT [1] show some examples of this restoration technique where they illustrate the convergence of the right-hand side of (50) when k varies from 1 to R. In the case of severe degradations where R is small, the last terms are hard to compute and noise is introduced.

c) <u>Comments on LSV restoration techniques</u>

When SAWCHUK's method is applicable to the problem, it should be used as it allows to efficiently implement the restoration process.

The algebraic approach is conceptually appealing as it generalizes previouly introduced techniques. Unfortunately it requires a huge amount of computations.

In general, LSV degradations are very hard to identify.

For these reasons, LSV restoration techniques are not commonly used nowadays.

IV. - <u>NON LINEAR ALGEBRAIC TECHNIQUES</u>

There are various reasons why one is led to use non-linear techniques. We have seen in the introduction that most realistic problems have a built-in non-linearity(e.g. in the sensing device). If this non linearity is neglected which is the assumption we have made so far, non linear algorithms may have to be used because the criterion defining the restoration introduces the non-linearity. We have gone through that issue in section II. It has to be mentioned here that non-linearity can also be introduced by taking into account constraints such as positivity or finite energy constraints. In this case, inequalities force to use nonlinear programming techniques the most popular of which is the gradient descent algorithm. In some cases, positivity constraints can be taken care of by using the homomorphic approach proposed by STOCKHAM [16] . The idea of homomorphic filtering applied to image restoration leads to processing density values rather than intensity values. Going back from the density domain to the intensity domain after the processing guarantees that the intensity data are positive. A number of other methods have been tried to tackle positivity constraints some of them are now quickly reviewed.

a) Positivity-constrained restoration techniques

A very comprehensive approach to positivity-constrained restoration problem was proposed by FRIEDEN [17]. It is based upon a random grain model and leads to a maximum entropy techniques. It deals with the LSI degradation

$$\underline{g} = [H]\underline{f} + \underline{n} \tag{51}$$

for an N x N image. Two assumptions are made :

i) A total number P of image grain information is given and has to be shared among the N^2 elements of $\underline{f}$.

ii) A total number N of noise grain information is given and has to be shared among the N^2 element of $\underline{n}$.

The idea is then to maximize the occurence of $\underline{f}$ and $\underline{n}$ under the constraints (51) (52).

$$\sum_{i=1}^{N^2} f_i = P \tag{52}$$

The occurences of $\underline{f}$ and $\underline{n}$ are given by

$$O(f) = \frac{P!}{\prod_{i=1}^{N^2} f_i} \tag{53}$$

$$O(n) = \frac{N!}{\prod_{i=1}^{N^2} n_i} \tag{54}$$

Recalling that, for large numbers, the Stirling formula gives

$$\log N! \simeq N \log N \tag{55}$$

The problem can be reformulated as

$$\text{minimize } N_{(\underline{f},\underline{n})} = \underline{f}^t \log \underline{f} + \underline{n}^t \log \underline{n} \tag{56}$$

under the constraints

$$\underline{g} = [H]\underline{f} + \underline{n} \tag{57}$$

$$\underline{1}^t \underline{f} = P \tag{58}$$

This formulation justifies the so-called maximum entropy approach to image restoration. Various methods have been proposed to solve the problem described in (56-58). A review of them can be found in HUANG [3] by FRIEDEN. Most of these methods are iterative and time consuming suggestions are given in (ANDREWS, HUNT) [1] that efficient implementation could be found using hybrid optical/digital systems.

b) <u>Maximum a posteriori (MAP) estimation</u>

This technique introduced by HUNT [18] and developed by TRUSSELL-HUNT [19] is related to the general case where the degradation is modeled as

$$\underline{g} = s\{[H]\underline{f}\} + \underline{n} \tag{56}$$

The MAP estimation consists in maximizing $p(f/g)$. Using the notations introduced in the previous paragraph for the statistics of $\underline{f}$ and $\underline{n}$, and supposing that the signals are gaussian, this procedure amounts to minimizing

$$W(\underline{f}) = (\underline{g} - s\{[H]\underline{f}\})^t \, \phi_n^{-1} \, (\underline{g} - s\{[H]\underline{f}\}) + (\underline{f}-\overline{\underline{f}})^t \, \phi_f^{-1} (\underline{f}-\overline{\underline{f}}) \tag{57}$$

where $\overline{\underline{f}}$ denotes the mean of $\underline{f}$. Using non linear programming techniques requires to compute $\nabla W(\underline{f})$.

$$\nabla W(\underline{f}) = \phi_f^{-1} (\underline{f}-\overline{\underline{f}}) - [H]^t [S] [\phi_n]^{-1} (\underline{g} - s\{[H]\underline{f}\}) \tag{58}$$

where [S] is a diagonal matrix built from the non linearity function s. Setting $\nabla W(\underline{f})$ to zero leads to the non linear equation

$$\hat{\underline{f}} = \bar{\underline{f}} + [\phi_f] [H]^t [S] [\phi_n]^{-1} (\underline{g} - s\{[H]\hat{\underline{f}}\}) \tag{59}$$

where $\hat{\underline{f}}$ appears on both sides of eq(59). Various iterative tecnniques have been developed to find $\hat{\underline{f}}$ and can be found in TRUSSELL-HUNT [19].

CONCLUSION

The main restoration techniques have been presented. Although a lot of efforts have been devoted to this problem, a number of questions stay opened. Restoration for non linear degradation is still too costly to be used on a wide range of applications. Adaptive techniques have blossomed in the last two years and give promising results. Some implementation problems arise at that level too but can be handled efficiently if the processing is linear.

Digital image restoration will probably continue for a while to be strongly influenced by the evolution of hardware components. Further breakthroughs will be found at the convergence of researches for new algorithms as well as new architectures.

REFERENCES

[1] H.C. ANDREWS, B.R. HUNT, Digital Image Restoration",
Prentice-hall Englewood Cliffs, 1977.

[2] W.K. PRATT "Digital Image Processing",
Wiley, New-York, 1979

[3] T.S. HUANG, ed., "Picture Processing and Digital Filtering",
Springer, Berlin, 1979.

[4] R.C. GONZALES, P. WINTZ, "Digital Image Processing",
Addison-Wesley, Reading, 1977.

[5] A. ROSENFELD, A.C. KAK, "Digital Picture Processing"
Academic Press, New-York, 1976.

[6] T.M. CANNON "Digital Image Deblurring by non linear Homomorphic Filtering"
Ph.D. Thesis, University of Utah, 1974.

[7] A.V. OPPENHEIM, ed., "Applications of Digital Signal Processing",
Prentice Hall, Englewood Cliffs, 1979.

[8] G.L. ANDERSON, A.N. NETRAVALI "Image Restoration Based on Subjective criterion",
IEEE Trans. Syst. Man & Cybernetics. Vol 6. p 845-853 (1976).

[9] G. BACKUS, F. GILBERT, "Uniqueness in the Inversion of Inacurrate gross Earth Data"
Phil. Trans. Roy. Soc. London 266 pp. 123-192 (1970)

[10] J.F. ABRAMATIC, L.M. SILVERMAN, "Non linear Restoration of Noisy Images"
IEEE Trans. on PAMI, Vol 4, 2, pp. 141-149 (1982)

[11]   S.A. RAJALA, R.J.P. de FIGUEIREIDO, "Adaptive Non linear
       Image Restoration by a Modified Kalman Filtering Approach"
       Proc. IEEE ICASSP 80, Denver, pp 414-417 (1980).

[12]   F.CLARA, L.M. SILVERMAN, J.F. ABRAMATIC, "Non linear Image
       Restoration : A visual quality Constrained Approach",
       Proc. IEEE ICASSP 82, Paris, pp 2098-2101.

[13]   F. CLARA "Filtrage Adaptatif d'Images couleur avec critère
       Psychovisuel",
       Thèse ing-Doc, Univ. Paris VI - (1980).

[14]   D. BARBA, "Traitement Numérique d'images avec critère psycho-
       visuel de qualité"
       Thèse de Doct. d'Etat, Univ. Paris VI (1981).

[15]   A.A. SAWCHUK, "Space-Variant Image Restoration by Coordinate
       Transformation",
       JOSA, vol 64, N° 2 pp. 138-144 (1974).

[16]   T.G. STOCKHAM, "Image Processing in the Context of a Visual
       Model",
       IEEE Proc. Vol 60, pp 828-841 (1972).

[17]   B.R. FRIEDEN " Restoring with Maximum Likelihood and Maximum
       Entropy",
       JOSA, Vol 62, N° 4 pp. 511-518 (1972).

[18]   B.R. HUNT "Application of constrained least-squares estimation
       to image restoration by digital computer"
       IEEE Trans. Comp. Vol. C 22-9 pp. 805-812 (1973)

[19]   H.J. TRUSSELL, B.R HUNT "Improved methods of maximum a posteriori
       image restoration",
       IEEE Trans.Comp. Vol C 28-1 pp· 57-62 (1979).

IMAGE CODING

Murat KUNT
Laboratoire de
Traitement de Signaux
de l'EPF-Lausanne
16, chemin de Bellerive
CH-1007 Lausanne

1. Introduction

As discussed in the first chapter, the most common way to digitize a picture is to scan it on a cartesian raster for sampling in space and quantize it in brightness. This digital representation is often referred to as the canonical form of a digital image. Generally speaking, it requires a very large number of bits. For example, with a raster of 512 x 512 and 8 bits per picture element (pel), one needs about 2.10^6 bits, to represent a single digital picture, a rather large number ! If digital TV images are considered with the same resolution and 25 frames/s, the rate is then 50.10^6 bits/s. Depending on the particular application, canonical form rates are typically between 10^5 and 10^8 bits per picture or between 10^6 bits/s and 10^{10} bits/s. The large memory and/or channel bandwidth requirements for digital image storage and transmission make it mandatory to use data compression techniques. The statistical aspect of data compression is governed by **Information Theory** and is called *Source coding*. Its aim is to remove the redundancy in the data by assigning short code words to likely messages and longer code words to less likely ones. The only requirement is to be able to reconstruct a faithful duplicate of the original picture. If the decoded picture is somewhat degraded, a fidelity measure should be used. Since almost all decoded pictures are to be viewed by humans, image coding techniques should attempt to take advantage of the properties of the human visual system for higher efficiency.

Image coding methods can be classified in two differents ways. A first classification is related to the distorsions and has two classes : information-lossless methods and information-lossy methods. The techniques of the first category are able to reproduce the original image exactly after decoding without any loss of information, whereas the techniques of the second category introduce some distorsions in the decoded image.

The second classification is related to the way of exploiting the redundancy in the image. There are three categories. In the first one are those methods which are implemented in the image or spatial domain acting directly on the image data. In this case redundancy is related to the **predictability** of data. In the second category, an energy preserving transformation is applied to the entire image which maps the image into the transform domain array such that

maximum information is packed into a minimum number of coefficients.
The third class is made of methods using a combination of the techniques of
the first two categories.

This chapter presents major techniques in these three categories such as
PCM, DPCM, Adaptive DPCM, predictive, runlength coding, bit-plane coding,
synthetic highs coding and transform coding. A brief introduction to fast
linear transforms is also given. For more detailed descriptions, the reader
may refer to [1]-[9]. It will be assumed in the following that images are
sampled and quantized properly and that coarse sampling and/or quantization
cannot be used for redundancy reduction.

2. Spatial coding

2.1 *Pulse code Modulation* (PCM)

Image coding by PCM is nothing more than the digital representation of an
image sampled in space and quantized in brightness. The basic PCM coder
consists of a sampler followed by a quantizer and a code word assigner (fig.1).

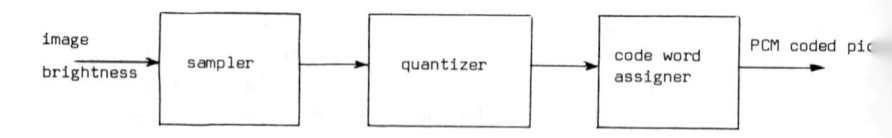

image brightness → sampler → quantizer → code word assigner → PCM coded pic

Fig. 1. Block diagram for PCM coding

Usually, the brightness is quantized to N levels, where N is a power of
two ($N = 2^B$). The value of B is usually 7 or 8. The PCM decoder converts
the binary code words into discrete samples which are then low-pass filtered.
This system is the simplest but does not make any data compression. It is
used mainly as a digitization scheme before the application of more complex
coding techniques.

If B is too small, the decoded picture suffers from quantization noise which
appears as false contours. The visibility of this noise can be reduced by
adding a high-frequency noise to the original signal before quantizing.
A more refined way was proposed by Roberts [10] and is refered to as dithering.
A uniform pseudo-random noise is added to the signal before quantization
whose dynamic range covers exactly one quantization step. The same noise is
substracted at the decoder. This technique may give acceptable quality
pictures with 3 bits per pel, when compared to the original one digitized
with B = 7 or 8.

Image coding

2.2 Delta Modulation and Differential Pulse Code Modulation (DPCM)

In image samples, very often the data sequence has statistical dependency from one sample to the next. A redundancy reduction can be obtained if a prediction of the sample to be coded is made from previously coded information that has been transmitted or stored. Assuming that I_k is the sample to be coded, an estimate $\hat{I}_k$ of I_k is predicted from the previously coded samples. The prediction error is then :

$$e_k = I_k - \hat{I}_k \qquad (1)$$

This error signal will be quantized, coded and transmitted (or stored) instead of I_k. If the quantized version of e_k is e_k^q, the reconstructed value of I_k is given by

$$I_k^r = \hat{I}_k + e_k^q \qquad (2)$$

The error in reproducing I_k^r is equal to the quantization error of e_k. In order to minimize the variance of the prediction error, $\hat{I}_k$ should be the best mean-square estimate of I_k. Furthermore, to avoid accumulation of errors in the reconstructed value I_k^r, the estimate $\hat{I}_k$ is based on the past reproduced values rather than the past input values. This is obtained by feeding back the quantization noise of the coder through the predictor. Figure 2 shows the block diagram of a predictive coding system.

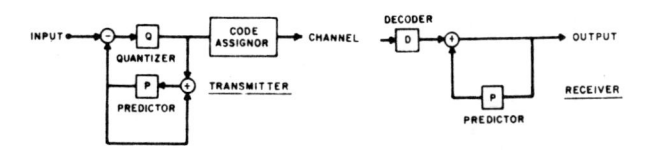

Fig. 2 : Block diagram of a predictive coding and decoding system.

If only the previous pel is used for prediction, a distinction is made between delta modulation and DPCM depending upon the number of levels of the quantizer. The quantizer for delta modulation has only two-levels (1 bit) whereas for DPCM the number of levels is greater than two.

The delta modulation does not require the sampling of the input signal. The predictor integrates the output of the quantizer which is a series of binary pulses. The decoder is also an integrator. There are three disadvantages for delta modulation : a) slope overload, b) granularity noise, c) high sensitivity to channel errors. Whenever there is a large and sudden variation in the input signal, the quantizer cannot respond instantaneously and causes a slope overload. At the other extreme, when the input signal is almost constant, quantizer's output oscillates around a constant value. Slope overload can be reduced by increasing the sampling rate and the granularity noise by low-pass filtering. The average compression achievable with delta modulation is around two. The simplest DPCM coder uses the coded value of the horizontally previous pel as the prediction. The prediction error e_k is usually quantized to eight levels and coded with a 3 or 4 bit code word. Compared to an 8 bits PCM, the compression is around a factor of 2. If the quantizer uses a non linear spacing of levels as opposed to linear quantization, the subjective quality of the reconstructed image is improved substantially. In an 8 level DPCM, the probability of occurence of the quantized levels is not uniform. Small values are more likely than the large ones. Accordingly a variable length code such as Huffman code can achieve greater compression (on the average 2.5 bits/pel). The price paid for this improvement is the necessity of a buffer storage because the code is a variable-length code.

2.3 Predictive coding

As mentioned in the previous paragraph, the general predictive coding scheme makes use of several pels in its predictor (see fig. 2). Predictors can be labeled as linear or non linear depending upon whether the prediction is a linear or non linear combination of the previously coded values. Another classification is based on the location of previous pels used in the predictor one-dimensional predictors use the pels of the same picture line as the pel beeing predict, two-dimensional predicted use other lines as well, finally interframe predictors use pels also from the previously coded frame.

A third classification is based on the adaptivity. Adaptive predictors change their characteristics according to the input data, whereas fixed predictors do not.

Linear predictors use a prediction of the following form

$$\hat{I}_k = \sum_{i=1}^{p} a_i I_{k-i} \qquad (3)$$

where I_{k-i} are previous samples on the same picture line.

Note that here we use the previous input samples. The best set of coefficients $\{a_i\}$ minimizing the mean-square prediction error are the solutions of the equations

$$c(o,1) = \sum_{i=1}^{p} a_i \, c(i,1) \qquad (4)$$

where

$$c(i,j) = \sum_k I_{k-i} I_{k-j} \qquad (5)$$

Image coding

The minimum value of the mean-squarred prediction error is then

$$E_{min} = c(0,0) - \sum_{j=1}^{M} a_j \, c(0,j) \qquad (6)$$

Although not optimum, for reasons discussed above, eq. (3) should be replaced by

$$\hat{I}_k = \sum_{i=1}^{p} a_i \, I^r_{k-i} \qquad (7)$$

to be used as predictor in Fig. 2. This allows the quantization errors to be fed back to the predictor.

Linear prediction can easily be extended to two dimensions. The brightness $I_{k,l}$ at location k,l in image data can be predicted according to the following equation

$$I_{k,l} = \sum_{i=1}^{p} \sum_{j=1}^{q} a_{ij} \, I_{k-i,l-j} \qquad (8)$$

which is "causal" (involving pels in the previous lines and columns). For practical implementation reasons, often p and q are limited to 1. Although the improvement in the mean square prediction error is small compared to the one-dimensional case, the rendition of vertical edges due to two-dimensional prediction is subjectively noticeable. It should also be noted that values of p and q larger than 1 do not give any appreciable improvement. Compared to an original 8 bits/pel PCM image, the compression obtained by two-dimensional prediction is around 2 to 3 bits/pel.

A sequence of images, as it occurs in broadcast television or teleconferencing, can also be predictively coded. In such cases, the pel value at a fixed location k,l varies from frame to frame only in the areas of relative motion. If $I_{k,l,m}$ is the pel value at location k,l in frame m, the interframe difference is given by

$$e_{k,l,m} = I_{k,l,m} - I^r_{k,l,m-1} \qquad (9)$$

One possible technique is to monitor $e_{k,l,m}$ from frame to frame and decide to code it if and only if it exceeds a predetermined threshold. Otherwise, the same pel value from the previous frame is repeated.

The error signal will exceed the threshold only in areas of motion. In the mth frame, the pel value is given by

$$I^r_{k,l,m} = \begin{cases} I^r_{k,l,m-1} + e^r_{k,l,m} & \text{if } |e_{k,l,m}| > \alpha \\ I^r_{k,l,m-1} & \text{otherwise} \end{cases} \qquad (10)$$

This technique is called conditional replenishment and was proposed by
Mounts [10]. For transmission or storage, code words are assigned to
quantized error signal and to its locations. Since the average rate depends
on the extent and duration of the motion, a buffer memory is mendatory to
obtain a steady bit rate. This technique can be improved by noting that the
temporal resolution in stationary areas and spatial resolution in moving areas
can be reduced without too much degradation of the scene. Further improvements
can be included by making the predictor adaptive. One way of doing it is to
compute directional correlations and select a predictor along the direction
of largest correlation.

More efficient predictors are those which take into account the motion.
They are designed in such a way that first an estimate of the translation is
made, then differences are computed with respect to pels in the previous
frame that are accordingly translated. This scheme is called motion compensated
prediction [11], [12]. Performances of this technique are superior to the
frame difference predictor by about 1 or 2 bits/pel.

The second important block of a predictive coder is the quantizer. Compression
is achieved by not quantizing the prediction error as finely as the original
brightness itself. Three types of degradations may result from a poorly
designed quantizer : if small amplitudes are coarsely quantized, the corresponding picture areas look as if a random noise is added (granularity noise).
In contrast, if the largest step of the quantizer is small, for every abrupt
change in the error signal a slope overload occurs. For gradual changes in
the error signal, if the quantizer output oscillates around it, the corresponding area of the image look like a busy edge. Quantizer design can be based
on statistical properties and/or on psychovisual measures. It can be fixed
or adaptive.

2.4 *Interpolative coding*

The basic idea behind interpolative coding is to represent the picture with a
subset of pels and to recover the remaining ones by interpolation. It can
ba applied in both the encoder and the decoder. In the encoder, brightness
values of an original image are approximated by continuous functions within
a permissible error band. The approximation can be made along a scan line or
over areas of the image. Fig. 3 shows one-dimensional zero-order and first-
order interpolations. Higher order polynomial functions (for example cubic
splines) can also be used but the computational complexity is so high that
it does not justify the results. First-order interpolation can give a
compression as low as 1 bit/pel with an admissible error. In the decoder,
interpolation can be done for example between even numbered pels of even
numbered lines in order to recover odd numbered ones yielding a compression
factor of 4.

Image coding

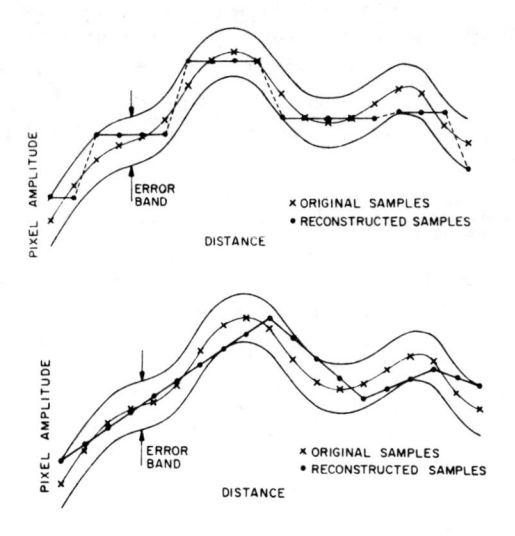

Fig. 3 : Zero-order and first-order interpolation(After [13])

More refined under-sampling and interpolation schemes can be devised yielding higher compressions. The under-sampling should be preceeded by a low-pass filtering in order to avoid aliasing. With some additional complexity, both types of interpolation can be made adaptive. Adaptation is usually made according to the level of picture "activity".

2.5 Synthetic highs coding [14]-[17]

The Mach phenomenon [18], which is a property of the human visual system, results in the subjective enhancement of sharp brightness changes in a picture. This property was first exploited by Schreiber [14] who implemented an edge detection system in one-dimension along scanning lines. Schreiber's system is also known as a "synthetic highs system". This work was extended to two-dimensions and improved by Walpert [15], Graham [16] and Kunt [17].

The sequence of messages for two-dimensional contour coding is selected in the following manner. The original digitized picture is split into two parts : the low pass licture giving the general area brightness without sharp contours and the high-pass picture containing sharp edge information. According to the two-dimensional sampling theorem, the lowpass picture can be represented with very few samples. These samples are the messages characterizing the canonical form of the lowpass picture. The edge detection is performed by using a gradient operator. The application of the two-dimensional gradient (vector) operator on the original picture yields two pictures. These are either the x and y components (rectangular coordinates) or the magnitude and argument (polar coordinates) of the gradient. A nonlinear operation, thresholding, is performed in the gradient magnitude to determine whether an edge point in "important". Finally, the location, magnitude and argument of each selected edge-point gradient is coded for efficient storage or transmission. These variables are the messages characterizing the highpass picture.

A two-dimensional reconstruction filter, whose properties are determined uniquely by the lowpass filter used for lowpass pictures is used to synthesize the two-dimensional high-frequency part from the edge information. This "synthetic highs" signal is then added to the lowpass picture to give the final output. A block diagram of this system is shown in Fig. 4.

Generally, a two-dimensional Gaussian function is chosen as a lowpass filter because of its separability and circular symmetry :

$$l(x,y) = \frac{1}{2\pi\sigma^2} \exp\left\{-\frac{x^2 + y^2}{2\sigma^2}\right\} \quad (11)$$

For this impulse response, 90 % of the energy is at spatial frequencies less than

$$f_c = \frac{1.516}{2\pi\sigma} \quad (12)$$

To limit the processing time and memory requirements on the computer, $\sigma = 4$ pels is chosen. According to the two-dimensional sampling theorem, a sampling rate of $2 f_c = 1/8$ should be sufficient. This leads to a 32x32 array for the lowpass picture. Furthermore, each of the lowpass picture samples can be quantized with less than 8 bits. We used 16 levels (4 bits) for these samples. Hence, the lowpass picture is coded with $4 \times 32 \times 32 = 4096$ bits. Fig. 5 shows an original digitized picture and the corresponding lowpass picture after two-dimensional linear interpolation.

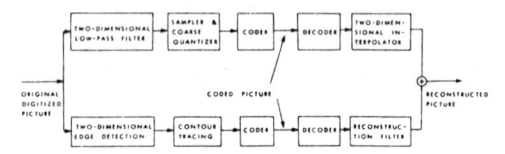

Fig. 4 : Block diagram of a two-dimensional synthetic highs system.

A histogram for the x component of the gradient and the corresponding picture is shown in Fig. 6 where grey is zero, white is positive, and black is negative. Note that a bias of 128 is introduced on this picture for display purposes. This curve indicates two-sided decaying exponential distribution.

Image coding

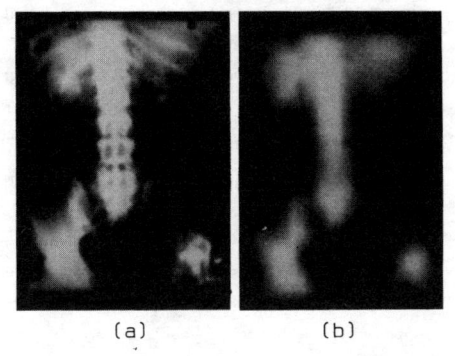

Fig. 5 : An original digitized picture (a) and its lowpass portion after interpolation (b)

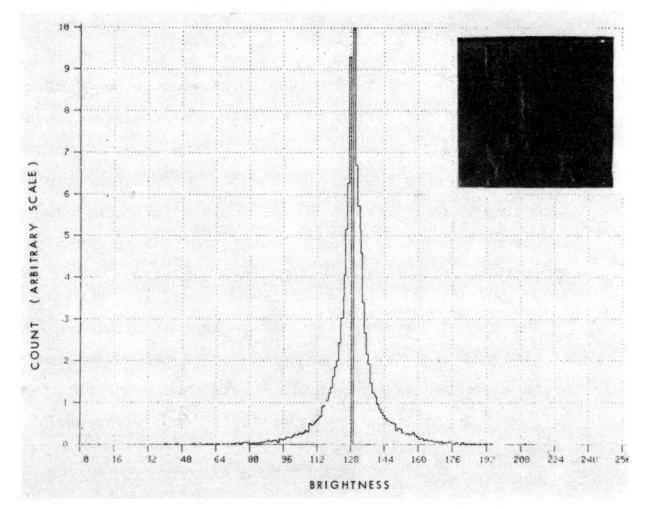

Fig. 6 : The x component of a gradient picture (inset) and its brightness histogram.

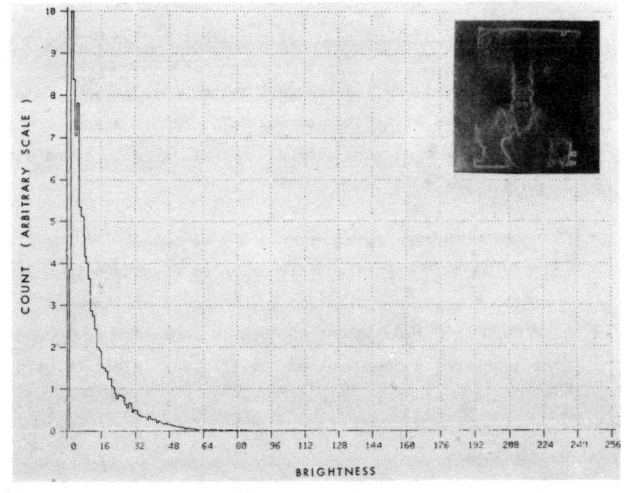

Fig. 7 : Magnitude of a gradient picture (inset) and its brightness histogram

The same behaviour is observed on the y component of the same picture and on several other gradient pictures. Since the magnitude is always positive or zero, according to these results, the histogram of the gradient magnitude is a one-sided decaying exponential, as shown in Fig. 7. The value of the gradient magnitude is large at edge points. A contour in the gradient magnitude is detected by a contour start threshold (CST) whose value can be estimated more or less arbitrarily from the corresponding histogram. Once a contour start point is found, the contour is traced by searching among the neighboring points for the one that has the maximum magnitude exceeding some edge-point threshold (EPT). The same procedure is repeated on the selected point until the contour stops. Since the contour may go both ways from the starting point, the second possible part of the same contour is also searched and traced. All other points, not detected by CST or EPT are erased during the tracing, so that contours are one point thick.

At each contour point three variables need to be coded : location of the contour point, direction and magnitude of the gradient at that point. Because contours tend to be smooth curves, coding the contour direction changes (VDC) from one point to the next on the contour will be more efficient than coding each location separately. As there are only 8 possible directions to move to the next point, a variable-length Huffman code can be generated to code VDC with an average length less than 3 bits.

On the original analog picture the gradient direction on the contour points is perpendicular to the contour. This is not quite true in the digitized picture because of the scanning raster and quantization errors. Another Huffman code is therefore generated to code efficiently the gradient direction changes.

The magnitude of the gradient is quantized to 4 exponentially spaced levels. Furthermore, on the contour, from one point to the next, the value of the gradient magnitude changes slowly. An efficient code can be derived from the joint probability of gradient magnitude on two adjacent contour points.

Coded pictures are thus represented by four sets of data : lowpass picture samples coded with conventional PCM, contour direction changes, gradient direction changes, and gradient magnitude, each coded with an optimum Huffman code derived from their distribution.

For the decoding procedure, x and y components of the gradient are computed first. Since the gradient direction is quantized to 8 levels and the gradient magnitude to 4 levels, x and y components may have only 16 different values. The quantized x-component picture is shown in Fig. 8 where the histogram of the quantized x component of the gradient is also plotted. Note that a bias of 128 is present in this picture for display purposes.

Image coding

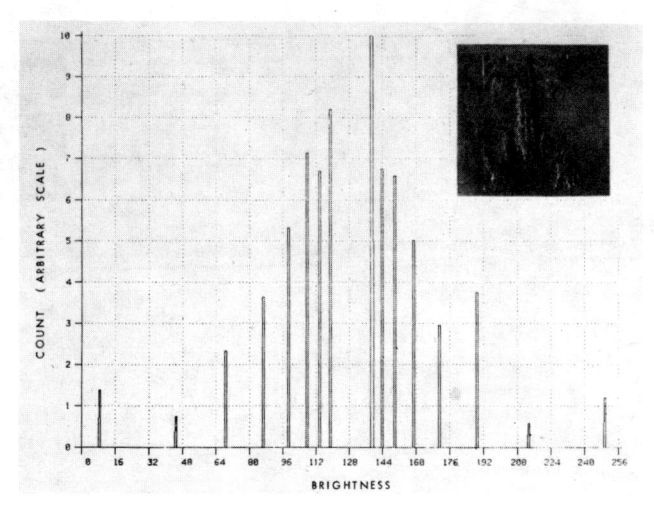

Fig. 8 : Quantized x component of a gradient picture (inset) and its brightness histogram.

The second step is a linear filtering for "synthetic highs" reconstruction. The discrete impulse response $h_x(i,j)$ of the reconstruction filter for the x component of the gradient is given [16] by

$$h_x(i,j) - h_x(i-1,j) = h_x(j,i) - h_x(j-1,i) = \delta_{ij} - l(i,j) \qquad (13)$$

where δ_{ij} is the Kronecker delta and $l(i,j)$ is a sample of the lowpass filter (1). The reconstruction filter for the y component is simply

$$h_y(j,i) = h_x(i,j) \qquad (14)$$

The reconstructed x component of the gradient and the corresponding histogram are shown in Fig. 9. Comparison of Figs. 6 and 9 shows that from a statistical point of view at least the high-frequency part of the picture is very well represented. Of course, this is a necessary but not sufficient condition for a fair reconstruction. Such a statistical behavior cannot take into account textures and structures of a picture. The original and the reconstructed pictures are shown in Fig. 10.

The two-dimensional synthetic highs system exploits elegantly the properties of the human visual system and permits a considerable amount of redundancy reduction. The values of the two thresholds CST and EPT are of primary importance for the compression ratios and for the quality of the reconstructed pictures. For ordinary pictures, CST and EPT can be kept high and thus a high compression ratio can be obtained without too much objectionable distorsion. For specialized pictures such as X-rays containing areas with neither sharp brightness changes nor long contours, very large amounts of texture and detail are lost if these thresholds are kept too high. With a lack of theoretical method, the compromise compression ratio-image quality can best be solved with an empirical cut-and-try procedure.

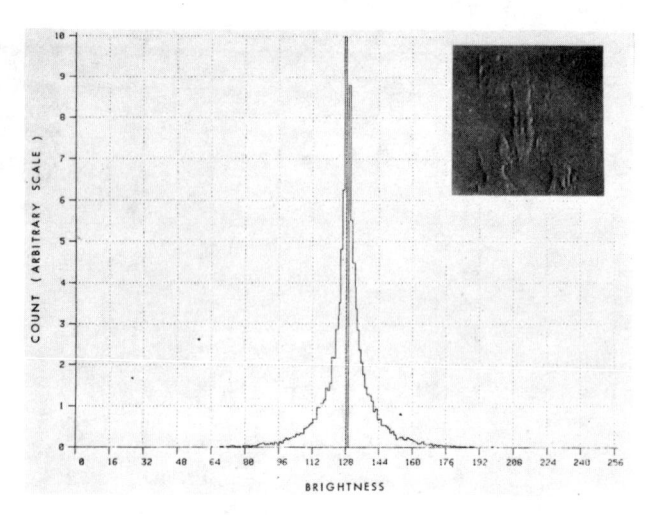

Fig. 9 : Synthetically reconstructed x component of a gradient picture (inset) and its brightness histogram (Compare with Fig. 6).

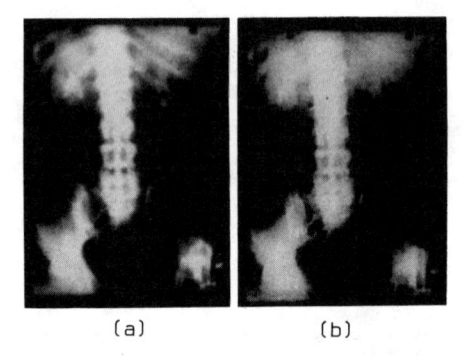

Fig. 10 : An original digitized picture (a) and its reconstructed version obtained with two-dimensional contour coding (b). Compression ratio : C = 8.37.

On the other hand, the tremendous amount of computation required by this technique makes it inappropriate for real-time applications. Even if it is possible to derive the gradient and the lowpass part of the picture optically, the rest of the computation for contour tracing and coding and the memory requirements are not at all negligible.

2.6 *Run length coding*

Run-length coding is one of the most efficient coding techniques for two-level pictures, i.e. facsimile. It can also be used for gray-level picture as it will briefly be explained.

Let's assume in this paragraph that the digital image is quantized with two levels. In other words, each pel is either white or black. The sequence of pels along each line can be viewed as a succession of pel segments of

Image coding

the same value (black or white) with alternating values from one segment to the next. These segments are called runs. Coding their length is sufficient to describe the entire picture. The statistical distribution of run lengths dictates the specific code to be used. Although theoretically it is possible to use an optimum Huffman code, some almost optimum and easily implementable codes can be used advantageously [19], [20]. There are two efficient and commonly used run-length codes, called exponential and logarithmic respectively which will be described here.

The exponential code [3] is optimum for exponential run-length distributions. This code is based on two elementary observations :
1. Run lengths (in bits) are positive integer numbers.
2. Any positive integer can be expressed as modulo another integer. If r denotes the run length and M denotes the modulo base, we can write

$$r = nM + i, \quad n = 0,1,2... \quad (15)$$

and $0 < i < M$. With a given number B of bits, for efficiency it is necessary to be also able to represent a maximum length. Consequently, the modulo base M should be of the form

$$M = 2^B - 1 \quad (16)$$

Equations (15) and (16) indicate that a run of length r is coded with (n+1)B bits. This code is a variable-length code, but each code word is an integer multiple of a B-bits word. This characteristic is very useful for practical implementation. For a given distribution, the optimum value [20] of B is

$$B = \log_2(\overline{r} - 1) + 1 \quad (17)$$

where $\overline{r}$ is the average run length.

In general, the values of B for "zero" run length and for "one" run length are different. They are denoted B_0 and B_1, respectively. Obviously, run-length coding is an information-lossless coding technique.

The compression achievable by the exponential run-length code can be determined as follows. This code uses $(n+1)B_k$ bits to code a run of length r, and of value k(k=0 or 1). Under the assumption of infinite-length picture lines, the average number of code bits per run is given by

$$\sum_{r=1}^{\infty} (n+1)B_k P_k(r) = \sum_{n=0}^{\infty} \sum_{i=1}^{M} (n+1)B_k P_k(nM+i) \quad (18)$$

where $P_k(r)$ is the probability of occurence of a run of length r, and of value k.

The average number of code bits per bit of a given value k, therefore, is

$$K_k = \frac{B_k \sum_{n=0}^{\infty} \sum_{i=1}^{M} (n+1) P_k(nM+i)}{\bar{r}_k} \qquad (19)$$

The global compression on the picture is related to the K_k by

$$C_r = \frac{1}{K_0 P_0 + K_1 P_1} \qquad (20)$$

Where P_0 and P_1 are the probabilities of occurence of zeros and ones in the bit planes. In order to determine the explicit value of C_r, we need to measure the distributions $P_0(r)$ and $P_1(r)$ as well as the probabilities P_0 and P_1.

Fig. 11 shows the histograms of run lengths measured on a weather-map as well as the compression ratio.

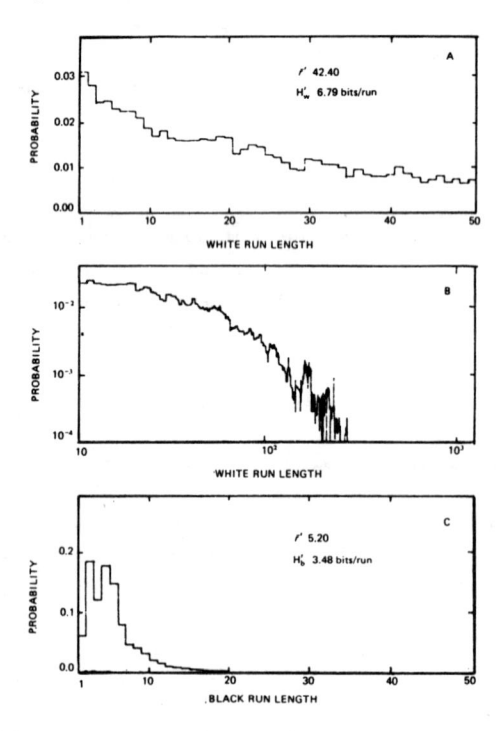

Fig. 11 : Histograms of run lengths measured on a weather map. Compression ratio 3.98 with $B_0 = 7$, $B_1 = 3$.

The logarithmic code [20] uses the same principle as the exponential code, but the value of M in eq. (15) is not fixed. It has the following form

$$M_k = 2^{kB'} \qquad (21)$$

Image coding

where B' is the fixed parameter. A run length is expressed as :

$$r = \sum_{K=0}^{n} 2^{kB'} + i \qquad (22)$$

where n is the maximum value of $_n$k which guarantees a positive value for i. A run whose length is between $\sum_{k=0}^{n} 2^{kB'}$ and $\sum_{k=0}^{n+1} 2^{kB'}$ is coded with a code word having (n+1)(B'+1) bits. Its content is entirely determined by n and i. The comparison of eqs. (15) and (22) indicates that this code is equivalent to an exponential code applied to the logarithm of run lengths. The particular case of B' = 1 is interesting for practical implementation because M_k are then powers of two. The distribution for which the logarithmic code is optimum is given by

$$P_L(r) = (\tfrac{1}{2})^{(n+1)(B'+1)} \qquad (23)$$

There are two ways for applying run length coding to gray level images. The first one consists simply in coding runs of constant gray level by indicating in the code word the value of that level. The second one applies two-level run length coding to the bit planes[†] of a gray level image. Any two level facsimile coding technique can be applied to the bit-planes of a gray level picture. Such a scheme is often referred to as bit-plane coding. Run length coding for facsimile yields an average compression of about 10:1. This figure is reduced to about 3:1 for gray level pictures

2.7 Block coding [21]

Block coding is also basically a two-level facsimile coding technique. It can be used for gray level pictures when applied to bit-planes.

We consider each two-level picture as being composed of a set of juxtaposed rectangular blocks of size n x m. Each pel being either black "1" or white "0", the total number of block configurations, i.e. the number of possible pel arrangements within a block, is obviously 2^{nm}. These form the set of messages characterizing the digitized picture. The transcription of this set of messages into codewords hereafter is called block coding. Fig. 12 shows a basic example of this type of coding. For the highest compression, an optimum Huffman code can be derived for this set of blocks. For blocks larger than 3x3 however, the set of messages becomes very large and the Huffman code is impractical.

[†] If in the canonical representation of a digital picture we isolate the ith bit of each pel, we obtain a two-level pattern called the ith bit-plane. A picture quantized to 2^B levels is completely described by B bit-planes.

The difficulties involved in the optimum coding can be avoided if one can devise suboptimum codes. A statistical analysis of block configurations in several two-level pictures indicates that the all-white block is very likely. Based on this observation and using the results of Dorgelo and Van der Veer [22], a suboptimum code was proposed [19], [23].

The principle of the suboptimum code is the following. The code word for the most likely configuration (in this case all white or all zero block) is simply "0". The code word for other configurations are obtained by the nm bits of the block preceded by the prefix "1". The coding scheme is summarized in Fig. 13. This code is variable-length code with only two possible lengths, namely 1 and nm+1. The source distribution for which this code is optimum is given by

$$P_0(i;n,m) = \begin{cases} P(0;n,m), & \text{for } i = 1 \text{ (all white block)} \\ \dfrac{1-P(0;n,m)}{2^{nm}-1} & \text{for } i = 2,\ldots,2^{nm} \end{cases} \quad (24)$$

where $P(0;n,m)$ is the probability of an all-white block. The average code word length is then

$$\begin{aligned}\overline{l} &= P(0;n,m) + (1+nm)(1-p(0;n,m)) \\ &= nm(1 - P(0;n,m)) + 1 \end{aligned} \quad (25)$$

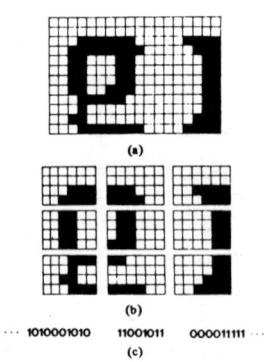

Fig. 12: Examples of block coding. (a) Digital image segment. (b) Messages of 6x4 blocks. (c) Associated codewords.

The compression ratio C_r, i.e. the ratio of the number of bits to represent a given picture before coding to the number of bits after coding can be derived easily if the probability $P(0;n,m)$ is known.

A closed form expression for $P(0;n,m)$ can be derived by using a first-order Markov process model [21] :

$$P(0;n,m) \underset{\sim}{} P(w) \, P(w/w)^{n+m+2} \qquad (26)$$

where $P(w)$ is the probability of occurrence of white pels and $P(w/w)$ is the conditional probability to have a white pel given that the preceeding pel is also white. The compression ratio is then

$$C_r = \frac{n.m}{l} \simeq \frac{nm}{nm(1-P(w) \, P(w/w)^{n+m+2}) + 1} \qquad (27)$$

As described, block coding is information-lossless. It can be made information-lossy by erasing a limited number of black pels in blocks, increasing thus the compression ratio without too much objectionable distorsions. Block coding can also be made adaptive [21], [24]. Adaptation is made by changing the size of blocks according to local "activity" in the image. The average compression ratio obtained with two-dimensional adaptation can be as high as 9 : 1.

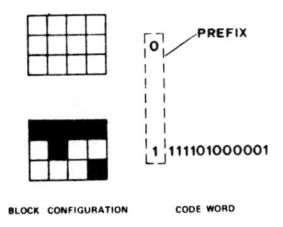

Fig. 13 : Principle of block coding

3. Transform coding [25]

3.1 *Preliminary remarks*

The basic motivation behind transform coding is to transform a set of data into another set of "less correlated" or "more independant" coefficients, before coding. The most commonly used transformations are linear transformations, implemented with fast computational algorithms. Viewed as matrix-vector multiplication, such a transform represents a rotation in a N dimensional space where N is the size of the vector to be transformed.

To illustrate what can be gained by transform coding, consider two consecutive samples $x(k)$ and $x(k+1)$ of a signal. These samples can be, for example, the brightness of two adjacent pels. Assume also that these samples are quantized to 10 levels. A particular configuration $x(k) \cdot x(k+1)$ is one of the $10.10 = 100$ points in a two-dimensional vector-space (see fig. 14). If these two samples are correlated, likely configurations will be around the line segment

$x(k) = x(k+1)$. Such an area is indicated in fig. 14. If a 45° rotation is applied to the coordinate system, in the rotated space the same area is now along $y(k)$ axis. This rotation allows to redistribute the variances in a different manner. The rotational transformation beeing unitary, we have

$$\sigma^2_{x(k)} + \sigma^2_{x(k+1)} = \sigma^2_{y(k)} + \sigma^2_{y(k+1)} \tag{28}$$

with

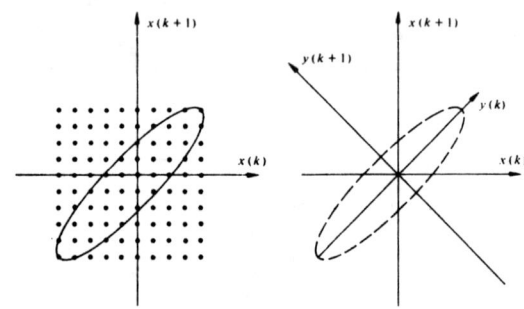

$$\sigma^2_{x(k)} \approx \sigma^2_{x(k+1)} \quad \text{but} \quad \sigma^2_{y(k)} \gg \sigma^2_{y(k+1)}$$

Fig. 14 : Geometrical interpretation of linear transforms

Thus, the dynamic range along $y(k+1)$ will be smaller than that along $x(k+1)$ or $y(k)$. Notice also that the inverse rotation brings everything back to the first coordinate system.

3.2 *Karhunen-Loeve transform*

A linear transform of size N applied to N samples of a signal vector can be expressed as :

$$X(n) = \sum_{k=0}^{N-1} a_{nk} x(k) \quad \text{with } n = 0,\ldots, N-1 \tag{29}$$

For the transformed coefficients to be computed independently, the transform is often required to be orthogonal or orthonormal. Since it should also correspond to a change of coordinate system, it will be unitary. These conditions can be expressed as

$$\sum_{n=0}^{N-1} a_{kn} a^*_{ln} = \delta_{kl} \tag{30}$$

$$\sum_{k=0}^{N-1} a_{kn} a^*_{km} = \delta_{nm} \tag{31}$$

where δ_{kl} is the Kronecker symbol having the value 1 for k=l and zero otherwise.

Image coding

The best linear transformation should lead to uncorrelated coefficients, i.e. to coefficients which satisfy :

$$E[X(n) X^*(m)] = \lambda_n \delta_{nm} \qquad (32)$$

Substituting (29) in (32), we have

$$E[X(n) X^*(m)] = E[\sum_{k=0}^{N-1} a_{nk} x(k) \sum_{l=0}^{N-1} a_{ml}^* x(l)]$$

$$= \sum_{k=0}^{N-1} \sum_{l=0}^{N-1} E[x(k) x^*(l)] a_{nk} a_{ml}^* = \lambda_n \delta_{nm} \qquad (33)$$

The expectation $E[x(k) x^*(l)]$ is, by definition the general entry of the correlation matrix $\phi_x(k,l)$ of the signal. Eq. (33) is then

$$\sum_{k=0}^{N-1} \sum_{l=0}^{N-1} \phi_x(k,l) a_{nk} a_{ml}^* = \lambda_n \delta_{nm} \qquad (34)$$

Comparing this to eq. (30), we have

$$\sum_{k=0}^{N-1} \phi_x(k,l) a_{nk} = \lambda_n a_{nl} \qquad (35)$$

In matrix notation, the equivalent form is

$$\phi_x A_n = \lambda_n A_n \qquad (36)$$

which is the classical eigen-vector, eigen-value problem. The solution of eq. (36) gives the transform matrix which guarantees (32). This transformation is called Karhunen-Loeve transformation. In order to derive this transformation, the correlation matrix $\phi_x(k,l)$ of the signal should be estimated first. Then, a transformation of the type (29) should be computed which requires N^2 operations (additions and multiplications) for N output coefficients. Because of this computational load, the Karhunen-Loeve transformation is not very often used in practice. It gives an indication about the upper-bound of what other transformations, computationally more efficient, should attempt to reach for decorrelating data samples. There are several linear transforms which can be computed with $N \log N$ operations as compared to N^2. The most important ones will be reviewed hereafter.

3.3 General formulation of two-dimensional linear transformations

The general expression of a two-dimensional linear transform is the following

$$X(m,n) = \sum_{k=0}^{N-1} \sum_{l=0}^{N-1} a_{k,l,m,n} \, x(k,l) \tag{37}$$

in which a square array is choosen on the data for simplicity. In this equation $a_{k,l,m,n}$ is the general entry of the kernel $A(k,l,m,n)$. A separable kernel is of the form

$$A(k,l,m,n) = A'(k,m) \, A''(l,n) \tag{38}$$

A linear two-dimensional transform with a separable kernel can be computed with one-dimensional transforms. First, a one-dimensional transformation is applied to each line of the data matrix $x(k,l)$ yielding:

$$X_k(m,n) = \sum_{l=0}^{N-1} a''_{l,n} \, x(k,l) \tag{39}$$

Then a second one-dimensional transform is applied to each column of the matrix $X_k(m,n)$:

$$X(m,n) = \sum_{k=0}^{N-1} a'_{k,m} \, X_k(m,n) \tag{40}$$

Some separable kernels are also symmetrical, i.e. $A'(k,m) = A''(k,m)$.

3.4 General structure of one-dimensional fast linear transformations

If the entries of a transform matrix are not structured in a particular way, there is no possibility, in general, to reduce the total number of operation N^2 required for a transformation. A relatively general structure which can be introduced in a transform matrix is the one induced by successive tensorial products of a set of base matrices. Let B_i denote the ith matrix in this set. We have

$$A_n = \bigotimes_{i=0}^{n-1} B_i = B_{n-1} \otimes B_{n-2} \otimes \ldots \otimes B_0 \tag{41}$$

where $\otimes$ represents tensorial product defined by:

$$A = \begin{bmatrix} a_{11} & a_{12} \\ a_{21} & a_{22} \end{bmatrix} \quad B = \begin{bmatrix} b_{11} & b_{12} \\ b_{21} & b_{22} \end{bmatrix}$$

Image coding

$$C = A \otimes B = \begin{bmatrix} a_{11}b_{11} & a_{11}b_{12} & a_{12}b_{11} & a_{12}b_{12} \\ a_{11}b_{21} & a_{11}b_{22} & a_{12}b_{21} & a_{12}b_{22} \\ a_{21}b_{11} & a_{21}b_{12} & a_{22}b_{11} & a_{22}b_{12} \\ a_{21}b_{21} & a_{21}b_{22} & a_{22}b_{21} & a_{22}b_{22} \end{bmatrix} \qquad (42)$$

If the B_i are of size $\rho \times \rho$, then the size of A_n is $\rho^n \times \rho^n$ with only $n\rho^2$ non redundant entries. It can be shown [25] that A_n can be factored into n matrices of size $\rho^n \times \rho^n$, each of them having only ρ non zero entries per line (Good's theorem) :

$$A_n = \prod_{i=0}^{n-1} C_i = C_{n-1} C_{n-2} \cdots C_0 \qquad (43)$$

Eqs (41) and (43) can be illustrated with the following example with $n = 2$ and $\rho = 2$.

$$A_2 = \begin{bmatrix} b_{1,0,0} & b_{1,0,1} \\ b_{1,1,0} & b_{1,1,1} \end{bmatrix} \otimes \begin{bmatrix} b_{0,0,0} & b_{0,0,1} \\ b_{0,1,1} & b_{0,1,1} \end{bmatrix}$$

$$= \begin{bmatrix} b_{1,0,0} & b_{1,0,1} & 0 & 0 \\ 0 & 0 & b_{1,0,0} & b_{1,0,1} \\ b_{1,1,0} & b_{1,1,1} & 0 & 0 \\ 0 & 0 & b_{1,1,0} & b_{1,1,1} \end{bmatrix} \begin{bmatrix} b_{0,0,0} & b_{0,0,1} & 0 & 0 \\ 0 & 0 & b_{0,0,0} & b_{0,0,1} \\ b_{0,1,0} & b_{0,1,1} & 0 & 0 \\ 0 & 0 & b_{0,1,0} & b_{0,1,1} \end{bmatrix} \qquad (44)$$

A transformation using A_2 as transform matrix can be written as

$$\begin{bmatrix} X(0) \\ X(1) \\ X(2) \\ X(3) \end{bmatrix} = \begin{bmatrix} b_{1,0,0} & b_{1,0,1} & 0 & 0 \\ 0 & 0 & b_{1,0,0} & b_{1,0,1} \\ b_{1,1,0} & b_{1,1,1} & 0 & 0 \\ 0 & 0 & b_{1,1,0} & b_{1,1,1} \end{bmatrix} \begin{bmatrix} b_{0,0,0} & b_{0,0,1} & 0 & 0 \\ 0 & 0 & b_{0,0,0} & b_{0,0,1} \\ b_{0,1,0} & b_{0,1,1} & 0 & 0 \\ 0 & 0 & b_{0,1,0} & b_{0,1,1} \end{bmatrix} \begin{bmatrix} x(0) \\ x(1) \\ x(2) \\ x(3) \end{bmatrix} \qquad (45)$$

where the products should carried out from right to left. A flow chart can be derived from eq. (45) which can be used for designing various fast transform algorithms. Fig. 15 shows the chart of eq. (45). The structure of a flow chart can be modified provided that the information flow is not altered. For example, if the 2nd and 3rd lines of the rightmost matrix in (45) are permuted, the 2nd and 3rd columns of the second matrix should also be permuted to keep the result unchanged. The corresponding flow chart is shown in Fig. 16. The basic structure of this flow chart is known as the the butterfly operation which allows "in-place" computation, i.e. the storage for data is used for intermediary and final results.

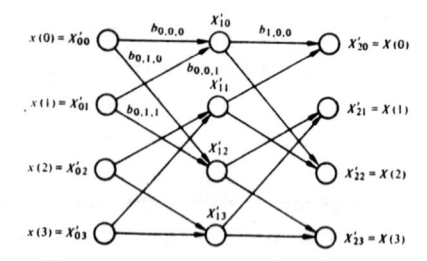

Fig. 15 : Flow chart for a 4x4 fast transform matrix

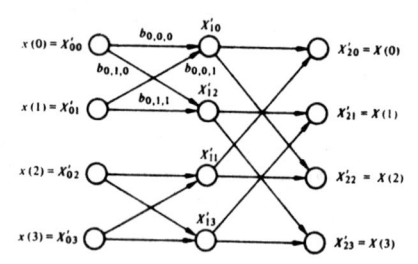

Fig. 16 : Modified flow chart (compare to fig. 15)

Particular fast transforms, such as Fourier, Hadamard, Haar, Cosine, Slant, are obtained by using particular values in base matrices B_i [25].

3.5 Coding strategies

There are several coding strategies in transform coding which will be discussed here. First the dimensionality of the transformation need to be determined. For example, to code a sequence of digital TV images, a three-dimensional 3-D transformation should be used whereas for still pictures, a two-dimensional (2-D) transform is appropriate. It is also possible to use one-dimensional transformations to code, for example, a picture on a line-by-line basis.

The next parameter to be fixed is the size of the transform. Although it is conceivable to transform the entire picture (2-D) or N frames of a NxN digital TV image sequence (3-D), implementation problems make them impractical. Basic limitations are the memory and computation time requirements with grow proportionally to N^2 or N^3. Furthermore, because of the elimination of unimportant transform coefficients before storage or transmission, large transform sizes lead often to more important degradations then small ones. A very commonly used strategy is to subdivide the image into subpictures of size MxM with M < much smaller than N (for example

Image coding

N = 512, M = 32) and to transform each subpicture separately. For 3-D transforms, sub-sequences of size as small as 4x4x4 have been used.

If all the transform coefficients are quantized and coded, the compression ratio might be unity or even less than unity. The important characteristic of these transforms is that all the "important" coefficients are packed into a specific area of the transform domain. One possibility to take advantage of this, is to use a mask covering such an area and to discard the remaining coefficients (they are set to zero). Only those coefficients which are intercepted by the mask are quantized and coded. Considerable compression (up to 10:1) can be obtained depending on the dimensions of the mask. This technique is known as zonal coding. It has the advantage of beeing image independent when the size of the transform and that of the mask are fixed. However, for some images the degradations may be untolerable if the number of important coefficients not intercepted in the mask is relatively large. Another possibility is to put a threshold on the transform coefficients magnitude and set to zero those which are below the threshold. This technique requires also the addresses of the coefficients retained to be coded and transmitted. Run-length coding is very appropriate for this task. This alternative is known as threshold coding.

In both cases, the coefficients which are retained should be quantized. Assuming that these coefficients are gaussian variables, the mean square error is minimized if the average quantization error of each coefficient is the same. This implies that bits are assigned to the coefficients in proportion of the logarithm of their variances. Gaussian assumption is indeed valid, because the coefficients are linear combination of image data (central limit theorem). For better subjective picture quality, the quantizer should be designed to optimize picture quality for a given number of bits. It is suggested that a companding characteristic of the form $\alpha\sqrt{x}$ gives the best subjective results [26].

Transform coding may also be adaptive by matching the parameters of the coder to the statistics of the subpicture being coded. Adaptation can be made at the level of the transform, of bit assignment or of quantizer level assignment. In some cases the coding efficiency can be increased by 30 to 50 percent compared to non adaptive case.

Transform coding is known for its immunity to channel noise. Since an error affect one coefficient, at the decoder, when the inverse transform is computed this error is distributed by the transform over the entire subpicture and hence is less visible.

4. Hybrid coding

Hybrid coding combines predictive coding and transform coding to take advantage from both techniques. The general idea is to transform the picture either line-by-line with 1-D transforms or subpictures with 2-D transforms, then to code each set of coefficients with predictive coders such as DPCM. By doing this one combines the hardware simplicity of DPCM and the robustness of transform coding. Fig. 17 shows the block diagram of a hybrid coder-decoder.

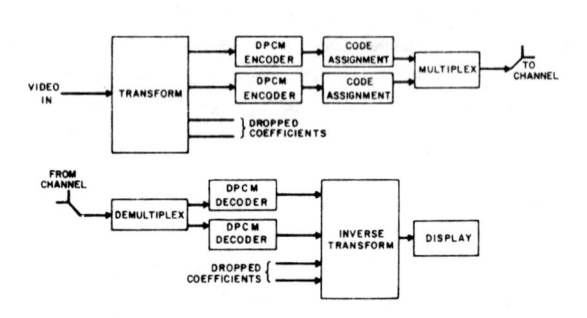

Fig. 17 : Hybrid coder-decoder (after [8])

5. Conclusions

We have discussed in this chapter major and fundamental techniques in image data compression. Methods such as PCM, DPCM, predictive coding, interpolative coding, synthetic highs coding, run-length coding or block coding are implemented in image domain. In contrast, transform coding maps the image data into a set of coefficients in the transform domain where important coefficients are packed in specific areas. Hybrid coding combines both approach for better compression.

Coding in doubtlessly one of the oldest areas of image processing. A huge amount of work and results can be traced over the last 25 years. At the beginning, image coding was governed only by information theory (source coding). Consecutive to simple coder-decoder, more sophisticated adaptive schemes were designed for improving the performances. Since most images or sequence of images are destined to be viewed by humans, it is logical to avoid coding what the eye does not see. A successful attempt in this direction is the synthetic   system which incorporates knowledge from very early stages of the human visual system. Higher compression can be obtained [27] by incorporating in the coder design new knowledge from human visual system as described by Hubel and Wiesel [28].

Because the main applications of image coding are transmission and storage, high compression ratios make it economically possible to store or transmit high resolution-high dynamic range pictures such as X-rays. Efforts in this area are expected in the future.

Image coding

At the other end of the spectra, low resolution-low dynamic range pictures such as digital teleconferencing TV pictures can be transmitted at increasingly, lower rates in real time with increasingly cheaper hardware.

Between these two extreemes, there is a large variety of applications which can benefit from efficient source coding of images.

References

[1] Special issue on Redundancy Reduction, Proc. IEEE, Vol. 55, March 1967.

[2] Special issue on Digital Communication, IEEE Trans. on Communication Tech., Vol. COM-19, part I, Dec. 1971.

[3] T.S. Huang and O.J. Tretiak (Eds), Picture bandwidth compression, New-York, Gordon and Breack, 1972.

[4] Special Issue on Image bandwidth compression, IEEE Trans. Commun., Vol. COM-25, Nov. 1977.

[5] L.D. Davidson and R.M. Gray, (Eds), "Data compression", Stroudsberg PA, Dowden Hutchinson & Ross, 1971.

[6] T.S. Huang, "PCM picture transmission", IEEE Spectrum, Vol. 12, pp 57-60, Dec. 1965.

[7] P. Camana, "Video bandwidth compression : A study in tradeoffs", IEEE Spectrum, pp 24-29, June 1979.

[8] A.N. Netravali and J.O. Limb, "Picture coding : a review", Proc. IEEE Vol. 68, pp 366-406, March 1980.

[9] A.K. Jain, "Image data compression : a review", Proc. IEEE, Vol. 69, pp 349-389, March 1981.

[10] F.W. Mounts, "A video encoding system with conditional picture-element replenishment", Bell Syst. Techn. Journal, Vol. 48, pp 2545-2554, Sept. 1969.

[11] F. Rocca, "Television bandwidth compression utilizing frame-to-frame correlation and movement correlation", in [3]

[12] B.G. Haskell and J.O. Limb, "Predictive vidoe encoding using measured subject velocity", U.S. Patent 3 632 865, Jan. 1972.

[13] W.K. Pratt, "Digital image processing", J. Wiley, New-York, 1978.

[14] W.F. Schreiber, C.F. Knapp and N.D. Kay, "Synthetic highs, an experimental TV bandwidth reduction system", J. Soc. Motion Pict. and TV Engrs., Vol. 68, pp 525-537, August 1959.

[15] G.A. Walpert, "Image bandwidth compression by detection and coding of contours", Ph.D. thesis, Dept. of Electrical Eng., Massachussetts Inst. of Technology, 1970.

[16] D.N. Graham, "Image transmission by two-dimensional contour coding", Proc. IEEE, Vol. 55, No 3, pp 80-91, March 1967.

[17] M. Kunt, "Source coding of x-ray pictures", IEEE Trans. on Biomed. Eng., Vol. BME-25, pp 121-138, March 1978.

[18] F. Ratliff, Mach Bands, San Francisco : Holden-Day, 1965.

[19] M. Kunt, "Comparaison de techniques d'encodage pour la réduction de redondance d'images facsimile à deux niveaux", Ph.D. Dissertation, Department of Electrical Engineering, Ecole Polytechnique Fédérale de Lausanne, June 1974.

[20] H. Meyr, H.G. Rosdolsky and T.S. Huang, "Some results in run-length coding", Proc. 1973 IEEE International Conference on Communication, Paper 48 D.

[21] M. Kunt and O. Johnsen, "Block-coding of graphics : a tutorial review", Proc. IEEE, Vol. 68, No 7, July 1980, pp 770-786.

[22] A.J.G. Dorgelo and H. Van der Veer, "Variable length coding increasing traffic capacity in PDM transmission systems", in Proc. Int. Zurich Seminar on Integrated System for Speech, Video and Data Compression, paper C 6, March 15-17 1972.

[23] F. de Coulon and M. Kunt, "An alternative to run-length coding for black-and-white facsimile", in Proc. Int. Zurich Seminar on Digital Communications, paper C 4, March 12-15, 1974.

[24] O. Johnsen, "Etude de stratégies adaptatives pour la transmission d'images facsimile à deux niveaux", AGEN Mitteil., No 20, pp 41-53, June 1976.

[25] M. Kunt, "Traitement numérique des signaux", Dunod, Paris, 1981.

[26] H.J. Landan and D. Stepian, "Some pomputer experiments in picture processing for bandwidth reduction", BSTJ, Vol. 50, pp 1525-1540, May-June 1971.

[27] M. Kocher and M. Kunt, "A contour-texture approach to picture coding", Proc. ICASSP-82, Paris, 3-5 May 1982, paper IM2.3.

[28] D.H. Hubel and T.S. Wiesel, "Brain mechanism of vision", Scientific American, Vol. 241, No 3, Sept. 1979, pp 130-146.

EDGE EXTRACTION TECHNIQUES

S. Levialdi

Institute of Information Sciences University of Bari, Italy

INRIA-CREST Course on COMPUTER VISION - 1982

1. Introduction

A number of survey papers have appeared, which consider the problem of extracting the edges in a digital image from a general standpoint and describe the most significant techniques developed so far (1,2,3).

The problem of finding edges is involved because of the fuzziness of the concept "edge"; in fact an edge on a digital image is an intuitive concept more than a formal one. Nevertheless, a wide number of authors have tried to define edge elements (in a deterministic way) or candidate edge elements (wint a more heuristic feeling) so as to contain in such definitions the building blocks for a theory of edge extraction which allows the design of algorithms.

In natural language we normally use many terms for naming "edges" like borders, contours, sides, streaks, etc which indicates the variety of interpretations that edges may have according to the image they refer to. The problem of edge definition is further complicated by its discrete nature, i.e. a path of image elements (pixels) which is often disconnected (due to noise) and contains digitization irregularities, light intensity changes, texture variations, etc. In short, many grey level variations of neighboring pixels have no edge significance and viceversa: some continuous grey level regions may contain edge pixels.

For the above reasons, two simple approaches for the definition of edges may appear rather simplistic: an edge exists on an image if there are two homogeneous regions of different intensity value. The edge elements will be those on the border between regions. The other approach considers the local variation of intensity and assumes that an edge element must reside in the proximity of such a variation. In both cases, what seems correct to intuition, the fact of associating homogeneity and discontinuity as edge cues, remains, in practice, very far from achieving the goal.

As mentioned in (3), a number of features may be used for classifying edge detectors. Speed is relevant for deciding whether the edge is obtained in the required time for the application under consideration. As a consequence of this evaluation, the algorithm may be reformulated if it can be speeded up by means of a different computer architecture (SIMD machine, array, etc). The requested memory is also important especially when working with large images which may cause memory shortage. Isotropy, namely the property by which the algorithm detects the edge regardless of its orientation. Static and dynamic response (with regards to the grey level variation range), noise immunity which is indeed very hard to obtain in general terms. Universality, which means its independence from the specific problem; level and quality which represent the usefulness of the output (for interpretation) and the geometrical similarity (for line drawings) of the result.

The mentioned features stem from different fields: filter theory, computer science, artificial intelligence and computer graphics showing the complex mixture of theory and technology which is required for obtaining a reliable, efficient edge extractor for a wide class of images.

One possible way to classify the edge extractors, from the point of view of their originating approach is the following one: local, regional, global, heuristic, dynamic and using relaxation.

2. Local methods

They are the oldest (1962)(4) and, for some specific cases may be useful.

Generally, they are based on an approximation of an operator which is originally applied on continuous functions and must be adapted to the case of digital images.

Within the realm of continuous two dimensional functions, the gradient $f(x,y)$ is given by $\nabla f(x,y) = (\frac{\partial f}{\partial x} + \frac{\partial f}{\partial y})$, its magnitude will be

$$|\nabla f(x,y)| = \sqrt{(\frac{\partial f}{\partial x})^2 + (\frac{\partial f}{\partial y})^2}$$

and orientation of the gradient vector $\alpha = \tan^{-1}(\frac{\partial f}{\partial x}) / (\frac{\partial f}{\partial y})$

But when operating on the discrete, like on digital images, x, y and $f(x,y)$ are non negative integer numbers so that the partial derivatives must be approximated with finite differences along the two orthogonal directions x and y, so obtaining

$\nabla_x f(x,y) = f(x,y) - f(x-1,y)$

$\nabla_y f(x,y) = f(x,y) - f(x,y-1)$

x-1,y	x,y
x-1,y-1	x,y-1

and for any orientation we will have

$\nabla f(x,y) = f(x,y) \cos\alpha + f(x,y) \sin\alpha$

and finally the digital approximation to the gradient of $f(x,y)$ will be given by

$$|\nabla f(x,y)| = \sqrt{\nabla_x f(x,y)^2 + \nabla_y f(x,y)^2}$$

Since this expression may be cumbersome, it generally happens that the digital gradient is considered to be either the sum of the absolute values of the two directional increments or the maximum between these same two increments:

$$|\nabla_x f(x,y)| + |\nabla_y f(x,y)| \quad \text{or} \quad \max(|\nabla_x f(x,y)|, |\nabla_y f(x,y)|)$$

These approximations are dependent on orientation(5).

In some cases other neighbors are also used (the eight neighbors) and we then have

$$f_{l,m}(x,y) = \max \{f(x,y) - f(l,m)\}$$

where l,m are the coordinates of the eight (or four) neighbors of the pixel located at x,y.

Another practical approximation which is often used, is due to Roberts(6), and may be written as

$$f(x,y) = \max(|f(x,y) - f(x+1,y+1)|, |f(x+1,y) - f(x,y+1)|)$$

x,y+1	x+1,y+1
x,y	x+1,y

Since this approximation computes the finite differences about an ideal element located at $(x+1/2, y+1/2)$, it may be considered an approximation to the continuous gradient at that position.

If a three by three neighborhood is considered there are two well known approximations to the gradient, one due to Prewitt(7) and the other to Sobel(8), they are both given below with the correponding labelled elements on the image.

x,y+2	x+1,y+2	x+2,y+2
x,y+1	x+1,y+1	x+2,y+1
x,y	x+1,y	x+2,y

Prewitt

$$\left(\left| \sum_{y}^{y+2} f(x,y) - \sum_{y}^{y+2} f(x+2,y) \right| - \left| \sum_{x}^{x+2} f(x,y) - \sum_{x}^{x+2} f(x,y+2) \right| \right)$$

The <u>Sobel</u> operator introduces weights in the summation of the values of the elements as shown in the expression below

$$\{ |f(x,y)+2f(x,y+1)+f(x,y+2)-(f(x+2,y)+2f(x+2,y+1) +$$
$$+ f(x+2,y+2))| - |f(x,y)+2f(x+1,y)+f(x+2,y)-(f(x,y+2)+$$
$$+ 2f(x+1,y+2)+f(x+2,y+2))| \}$$

which have been written in carthesian notation for uniformity reasons but will be re-written as masks, i.e., as configurations of values which will be multiplied by the corresponding values of the pixels and then added or subtracted as the sign of the coefficient in the mask suggests (see also)(9).

Roberts

$$H_1 = \begin{vmatrix} 0 & -1 \\ 1 & 0 \end{vmatrix} \qquad H_2 = \begin{vmatrix} -1 & 0 \\ 0 & 1 \end{vmatrix}$$

Prewitt

$$H_1 = \begin{vmatrix} 1 & 0 & -1 \\ 1 & 0 & -1 \\ 1 & 0 & -1 \end{vmatrix} \qquad H_2 = \begin{vmatrix} -1 & -1 & -1 \\ 0 & 0 & 0 \\ 1 & 1 & 1 \end{vmatrix}$$

Sobel

$$H_1 = \begin{vmatrix} 1 & 0 & -1 \\ 2 & 0 & -2 \\ 1 & 0 & -1 \end{vmatrix} \qquad H_2 = \begin{vmatrix} -1 & -2 & -1 \\ 0 & 0 & 0 \\ 1 & 2 & 1 \end{vmatrix}$$

This new form is easier to follow since we may observe all the weights (simultaneously) on the corresponding positions of the neighbourhood. Since these last local approximations to gradients have more than two levels (if compared to the initial digital gradient) they are sometimes called 3-level gradient (for values -1,0,1) and 5-level gradient (for values -2,-1,0,1,2).

Moreover, there is also another well known gradient operator, due to Kirsch(10), which uses stronger weights:

Kirsch

$$H_1 = \begin{vmatrix} 3 & 3 & -5 \\ 3 & 0 & -5 \\ 3 & 3 & -5 \end{vmatrix} \quad H_2 = \begin{vmatrix} 3 & -5 & -5 \\ 3 & 0 & -5 \\ 3 & 3 & 3 \end{vmatrix}, \text{ etc. and the}$$

other rotated versions of the masks which will be a total of 8 as with the 3 and 5-level masks. Finally, the compass gradient should also be mentioned, it is made of 8 masks, two of which are given below.

Compass gradient

$$H_1 = \begin{vmatrix} 1 & 1 & -1 \\ 1 & -2 & -1 \\ 1 & 1 & -1 \end{vmatrix} \text{ and } H_2 = \begin{vmatrix} 1 & -1 & -1 \\ 1 & -2 & -1 \\ 1 & 1 & 1 \end{vmatrix}$$

In all cases, the output of the digital gradient, $0_{h,y}$ one for each pixel of the image, will be given by the maximum value obtained after applying all the rotated versions of a mask:

i.e. $\quad 0_{x,y} = \max_i \{|G_i(x,y)|\}$

(for i = 1,8)

and the orientation (α) of the gradient will correspond to the $0_{x,y}$ which was found to be the maximum.

An exhasperated version of this approach is the one using the digital version of the Laplacian, namely of

$$\nabla^2 f(x,y) = \frac{\partial^2 f}{\partial x^2} + \frac{\partial^2 f}{\partial y^2}$$

which may be approximated on the discrete, for x,y,f(x,y) integer values, as

$$\nabla^2 f(x,y) = f(x+1,y) + f(x-1,y + f(x,y-1) +$$
$$+ f(x,y+1) - 4f(x,y).$$

Edge extraction techniques

The graphic interpretation of the central pixel and its four adjacent neighbors helps in identifying the considered elements and, as before, the masks with the corresponding weights give the most direct way for what the digital Laplacian really does:

$$DL_4(x,y) = \begin{vmatrix} 0 & 1 & 0 \\ 1 & -4 & 1 \\ 0 & 1 & 0 \end{vmatrix}$$

If the eight neighbors are considered we will have

$$DL_8 = \begin{vmatrix} 1 & 1 & 1 \\ 1 & -8 & 1 \\ 1 & 1 & 1 \end{vmatrix}$$

and if only positive values are requested, the absolute value of the summation may be taken.

The problem arising with this operator is that points will be deleted (for the 4-neighbor case) 4 times as strongly as an edge, a line end will be 3 times as strong and a line, 2 times as strong since when the weighing of the neighbors is performed, its number will influence the obtained value.

All the methods which are gradient or Laplacian based are intrinsically noise sensitive, since a small number of pixels having a different grey value from the background will be extracted independently from their organization, i.e. whether they lie on a line (edge) or whether they are a small dot-like cluster. These observations are certainly not new but we may say that these consequences are inevitable because the edge's main cue is considered to be the grey level difference that may be measured between adjacent pixels on the image.

It is difficult to compare the performance of all these local operators, whether they are gradients, laplacians or templates, in a way which is general enough and therefore does not depend on the particular images chosen for this test.

Different authors have approached this problem (the evaluation problem) and, for instance in(11) four local operators have been considered: Prewitt, Kirsch, 3-level and 5-level simple masks and they have been applied to three different images: a toy tank, a girl's face and a satellite image so as to have a variety of edges to extract. As mentioned before, the mask which will provide the largest output is retained and thresholded: a binary edge map will be finally obtained. One general edge detection scheme (suggested by Robinson(12)) may be seen in the block diagram below.

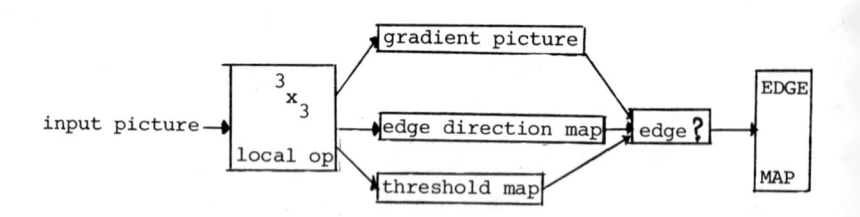

```
                            gradient picture
                         ↗                    ↘
                    3x3                          EDGE
input picture →  local op → edge direction map → edge ? →
                         ↘                    ↗          MAP
                            threshold map
```

In this approach, the evidence of the gradient, its magnitude, its direction and a local threshold are combined to build up an edge knowl edge base after which a binary decision will be taken so as to produce an edge map.

If we now consider the 5-level simple masks their advantages may be summarized in the next five points (11)
1) they approximate the partial derivatives along both axis (x,y);
2) the 0-weights in the center help to suppress line jitter;
3) the use of only four masks (the ones having the 0s along the four principal directions) computes the analog of the gradient of the image, the sign of the output will establish either one direction (for instance positive sign, East direction) or the opposite one (negative sign, West

direction);

4) a defocussing action may be simulated on the same 3 by 3 neighborhood by using $M_o = 1/16 \begin{vmatrix} 1 & 2 & 1 \\ 2 & 4 & 2 \\ 1 & 2 & 1 \end{vmatrix}$ and convolving the image with it so as to compute a local threshold value (a weighted average);

5) the 5-level masks have a greater sensitivity to the diagonal directions than to the horizontal and vertical ones, so compensating the human visual response which has a smaller acuity for the diagonal directions.

One important aspect of building the edge knowledge base is the use of local connectivity information which might be simply stated as (12):

<u>if</u> the direction of the central point is k (for k = 0, 1, ... 7 as in the Freeman chain coding scheme)

and

<u>if</u> the directions of the preceeding and succeeding edge vectors are k, k±1 (modulo 8)

<u>then</u> the edges are connected

In this way, we will require that the edge vectors exceed a threshold <u>and</u> are locally connected if they must be considered as edges. In practice by adding the local connectivity condition about 3/4 of the pixels disappear and noise is greatly reduced; the details are more prominent and spurious edges are cancelled.

The same happens for the Kirsch operator and a fixed threshold, when the connectivity tests are applied the results improve in a significant way.

Another parameter which may help in evaluating the edge's strength is the edge activity index (EAI) which is defined as the ratio of the maximum gradient magnitude with respect to the average magnitude of the gradients along the 8 compass directions.

The evaluation of the local techniques for edge extraction is not only connected to the post processing operations which are applied for improving the results but should also consider some of the effects that directly descend from the structure of the masks: weights and configuration. For instance, some work has been done(5) on the influence of the orientation of an edge on its detection and a comparison of the different masks on the basis of this feature has been carried out, both when considering the magnitude and the square root of the maximum output for each pixel with a given mask. If we take a look at the reported curves in(5) we may see that the Sobel and Roberts operators are worse than the template matching operators (Kirsch, Compass-Gradient, 3and 5-level simple masks): the amplitude of the gradient increases (from a normalized value of 1 to 1.5) with the variation of the angle a in radians (from 0 to 1) for the Sobel operator and decreases (from 1 to .3) as the angle increases in the same range as before. Another interesting result relates the real and the detected orientation for each mask; from the figure we may see that the best accuracy is obtained with the Sobel operator whilst the template matching gives a step edge for the transfer functions between the orientations (real and detected). We may then consider that the sensitivity may be described both as the amplitude response in terms of the orientation of the edge and as the ratio between real and detected orientations.

Another important feature of an edge detector is its independence from the edge position (translation independence). All the considered(11) operators (Kirsch, Sobel, Compass Gradient and Prewitt) exhibit a fall, as the edge is displaced, with a steep descent around the normalized value .7 of the width of the pixel until the displacement reaches 1.5 where the output of the operator is practically zero.

Other tests were performed on vertical and diagonal edges on which white noise was added with signal to noise ratios 1 and 10. All the detectors have been tried (Sobel, Prewitt, Kirsch, 3-level and 5-level, Compass Gradient) and it was shown, on a statistical basis, that the best operators are Sobel and Prewitt.

3. Regional methods

Another and, different, approach has been suggested by Hueckel[13] who considers what a theoretical edge should look like (within a circular diameter) and then finds, for each region, how distant from this configuration we are, by using a best fit criterion such as the least squares one.

As briefly mentioned above, a circular neighborhood is considered, having an area D in F(x,y), two grey levels are assumed to exist in this area: b, and b+d which is higher and therefore represents a darker region. The analysis of this region involves looking for a two dimensional step function S such that

$$S(x,y,c,s,r,b,d) = \begin{cases} b & \text{if } cx + sy > r \\ b + d & \text{if } cx + sy < r \\ 0 & \text{otherwise} \end{cases}$$

the functional equation to be solved is

$$\int_D (F(x,y) - S(x,y,c,s,r,b,d))^2 \, dx \, dy = \min$$

The Fourier coefficients for F and S are

$$f_i = \int_D F(x,y) \, H_i(x,y) \, dx\,dy \quad S_i(c,s,r,d,b) =$$

$$= \int_D S(x,y,c,s,r,b,d) \, H_i(x,y) \, dx\,dy$$

$$\lambda(c,s,r,b,d) = (f_i - S_i(csrdb))^2 = \min.$$

Hueckel's operator solves this functional equation and finds a set of functions (H_i functions) which approximate the series expansion only considering the first 8 base functions (H_0 ... H_7) which may be graphically represented by the figures shown below in which + and - are the signs of the functions and the lines denote the zero crossing in the D area. Note the resemblance with the receptor fields in the cat's cortex(14).

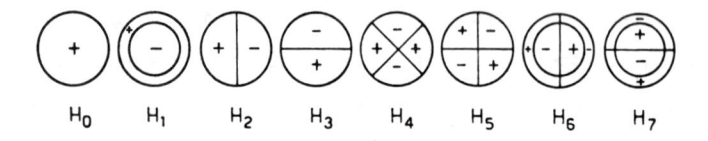

H_0 H_1 H_2 H_3 H_4 H_5 H_6 H_7

The expansion of the grey values is done on an orthogonal basis and since only the low frequencies are considered, high frequency noise is reduced. The solution is interesting because it gives a method which is orientation invariant.

4. Global methods

A completely different attitude has been taken by people who were experts in signal analysis and digital filtering and therefore viewed the problem of edge extraction as one of filtering the image so that only the edge remains and all the rest is elliminated.

For instance (Modestino, Fries(15)) suggest an image model and then perform a linear shift invariant spatial filtering operation on the image in such a way that the mean square estimation error is minimized

$$I_e = E\{e^2(\underline{x})\} = E\{\ell(\underline{x}) - \hat{\ell}(\underline{x})\}^2$$

The parameters that enter in the system transfer function $H_o(r)$ are the following:

ρ correlation coefficient for the spatial evaluation of the random amplitude process- the degree of correlation across an edge-

Edge extraction techniques

ξ the signal-to-noise ratio of the edge structure with respect to the background noise

λ the number of events per sampled distance (of edges per unit distance)

r the neighborhood distance

If $\lambda_1 = \lambda_2 = \lambda$

then $H_o(r) = \dfrac{r^2 e^{-r^2/2}}{\dfrac{r^2 + 2(1-\rho)^2 \lambda^2 (r^2 + (1-\rho)^2 \lambda^2)^{1/2}}{8(1-\rho)\lambda \xi} + 1}$; $r \geq 0$

$\xi \simeq \dfrac{\sigma^2}{\sigma_n^2}$ S/N, ratio the model is completely defined in terms of ρ, λ and ξ.

Edge detectors designed on this basis prove quite efficient for a wide range of images. The digital approximation to the analog filter described above is $H_o(z_1 x_2)$ whose frequency response approximates $H_o(r)$.

A window ($e^{-r^2/2}$) is chosen, empirically, for obtaining a frequency weighting of a Laplacian kind so as to avoid ringing effects on the edges. A two dimensional infinite impulse response filter (IIR) is then considered, of the kind

$$H_o(z_1 z_2) = \dfrac{\sum_{i=0}^{Mb} \sum_{J=0}^{Nb} b_{ij} z_1^{-i} z_2^{-j}}{1 + \sum_{\substack{i=0 \\ i=j}}^{Ma} \sum_{\substack{j=0 \\ i \neq j}}^{Na} a_{ij} z_1^{-i} z_2^{-j}} \quad ; \quad z_i = e^{j\omega i} \quad (i = 1,2)$$

this expression may be rewritten in terms of coefficients $a_{10} a_{11} b_{11}$ which may be found for each set of values of ρ, λ and ξ.

For every point in a 3 by 3 neighborhood, pairs of grey values in opposition are compared and if the grey level difference exceeds a threshold than an edge is declared; if there are many pairs exceeding this difference the largest one is chosen. The final expression is

$$H_o(z_1 z_2) = A \left(\frac{1 - 1/2(b_{11} + 1)(z_1^{-1} + z_2^{-1}) + b_{11} z_1^{-1} z_2^{-1}}{1 + a_{10}(z_1^{-1} + z_2^{-1}) + a_{11} z_1^{-1} z_2^{-1}} \right) ;$$

where A is the gain.

The authors have applied this filter to an image having different f and λ (events/sample dist.) and $\int$ from ∞ to 10 and 3 db. A Laplacian was next applied on a 4 by 4 image and then thresholded; if the noise was high (low signal to noise ratio) the Laplacian did not improve significantly the results. Nevertheless, the filtered image does show the edges neatly as may be seen from the figures in(15).

The examples contain an X ray photo of the chest with a filter having λ = 0.0125 and f = .5 and $\int$ = 3db, very good results were obtained for a wide class of images.

Finally the authors conclude by saying that this filter is an optimum compromise between simplicity, generality and noise immunity. As we have seen, this approach does not include any knowledge in the process of the edge extraction, it is completely independent from the kind of image that must be processed.

5. Heuristic methods

Eleven years ago, (Kelly, (16) suggested an approach, called "planning" which borrowed some ideas form the planning concept introduced by Minsky(1963). Although Kirsch in 1957 had already suggested to defocuss the original image for later processing, this concept is used in Kelly's approach as the first step for finding edge elements; the next stage uses the coordinates of the found pixels and PLANs future action on this basis.

Edge extraction techniques

An algorithm, given below, is designed for the extraction of horizontal edges. The output from the first step will be used for a search in a tree on reasonable assumptions (no vertical edges may be present in the middle of a horizontal edge, etc.). A head, the object used by Kelly, suggests that the top and the sides should be firstly located and then an attempt to connect them should be made, etc.

Algorithm for horizontal edge extraction:

```
                    if      b > h
a   b   c                   then
d   e   f                       if [ min(a,b,c)-max(g,h,i) ] > 1
g   h   i                           then edge
                                else    no  edge
                    else if b < h   then
                                if [ max(a,b,c)-min(g,h,i) ] > 1
                                    then edge
                                else    no  edge
                    else no edge
```

Returning to the search, since the shoulders should be under the head an attempt is made to locate outward lines, finally the indentations that represent the ears are searched.

The input to the plan follower is the original picture and the list structure containing the small head outline. The output will be a new list structure with the coordinates of the accurate outline of the head. A narrow band is searched between the successive points in the plan and then connected by the program. The band is 16 pixels wide (the original image was reduced by 8), several edges may be contained in this band, whenever parallel edges are found, the external one (with respect to the center of the head) is chosen; a similar edge detector will be used on the large image.

The quality of this program is difficult to measure but it is certainly fast and easy to debug. It is far better to work with about 600 pixels than with 40.000 (from a 256x256 input image). The time, on a PDP10 machine, for locating the head contour, is of about 6 seconds whilst without planning, the program requires about 234 seconds. At this point one may suppose that if more knowledge is accumulated in the list, better results will follow, but a heavier burden is layed on the planner, more information to process and therefore slower processing will result. According to Kelly his program is a good compromise between high speed and amount of knowledge in the program. A good feature to add to the program would be a <u>language</u> for expressing this knowledge so that a head might be described on a relatively high level, directly to the planner.

An evolution of this approach is the one due to (Martelli(17)), who uses another heuristic scheme for finding edges in noisy images. He firstly defines a microedge (a pair of pixels having different grey values):

a = (i,j)   and   b = (i,j+1)

	j	j+1
i	a ▼	b
i+1		

edge ab

An edge is a sequence of edge elements which starts from the top row (Martelli's hypothesis), contains no loops, ends in the bottom row and no element will have an "upwards" direction. In this way an edge is a path in the graph that must be found optimally between a first node and a goal node. If MAX is the highest grey level of the image, the cost of (expanding) a node $n = a_r a_s$ will be given by $C(n) = (MAX - g(a_r) + g(a_s))$ where g is the grey level at a; C(n) is low if a has a high grey level and a_s a low grey level; the cost of getting to state n from some initial root node is the sum of all the costs encountered along the path from the sarting node to n; g(n) is low if the path to n is of high contrast.

Davis (in his review paper, (1)) has noted that the cost in terms of grey level differences only, is unfortunate since it is difficult to evaluate the goodness of an edge on the basis of the contrast level only, moreover the space representation (nodes associated to micro edges) is not a very good one for a proper definition of the heuristic function. The criticsism to the method, which has come from different sources, is concentrated on the choice of the two components of the function (the e-valuation function) which are supposed to be additive. It is obviously a difficult problem to choose a good heuristic function and, moreover, to adequately weigh the global with the local information for choosing the next node in the path. If the contrast is low, a local gradient would still find the edge whilst Martelli's method would not.

6. Dynamic methods

Within a very comprehensive work in image analysis of bone marrow cells (Lemkin, (18)) there is an interesting approach to the detection of a boundary which seems useful to report here. The basic idea is to use a sequence of thresholds (in proximity of an optimal one extracted from considerations on the grey level histogram of the cell image) and to consider the "stable regions" of darker pixels which have small variations whilst the image is thresholded at different values. The main idea is to extract a boundary trace (hence Boundary Trace Transform) which will then support the evidence of an external boundary and of objects (generally darker than the boundary) within such a boundary. The trace image T_{ti} is defined as follows

$T_{ti}(x,y)$ = b   if (x,y) is a boundary pixel defined for the image G_i segmented at threshold t

= h   if (x,y) is a hole pixel defined for G_i such that $G_i(x,y) < t$ inside a boundary

= 0   otherwise

A trace image contains information about the boundary and about the holes which may be inside the boundary. The boundary trace transform builds up an image BTT(T) such that after starting from a 0 value for all its pixels (initialization) every element which has a b value in its trace image (for a certain threshold value) will be incremented by a constant value $q = \frac{\text{max value}}{\text{threshold range}}$. After all the thresholds have been tested, an image will result which has some pixels with very high values (the so called stable pixels) which correspond to the boundary. In practical cases, the extraction of boundaries is severely complicated both by the existence of artifacts and of pixels having a high grey level and not belonging to the contour.

7. Relaxation methods

A more recent approach (1976) to edge extraction and to image analysis in general is the one employing relaxation techniques. Its main advantages are that within each iteration of the process the future value of a pixel does not depend on any previous decision taken in the same iteration and, at the same time, the results obtained at the previous iterations are collected and gradually improve the reliability of the values obtained for each pixel, for instance in edge extraction procedures. The main idea (Rosenfeld(14)) is to use an initial (perhaps fuzzy) knowledge about the fact that a pixel may belong to an edge, the same is done for the probabilities at neighboring points, and if there is supporting evidence (which means high probabilities that the neighbors of this pixel have compatible assignments to the fact that the pixel belongs to the edge) then the probability of that pixel of belonging to the edge is increased. Conversely, if there is contradictory evidence (that the assignments of neighboring pixels are incompatible with the fact that the pixel belongs to the edge) then the probability that the pixel is an edge pixel is decreased. This process is done simultaneously for every pixel, and every probability corresponding to each pixel is varied, the process may

Edge extraction techniques

be iterated a number of times and during the iteration the probabilities converge to a stable value. If the probability for a pixel is high, then this pixel will belong to the edge class, otherwise it will not, and this is the output of the relaxation process.

In particular(19), a set of objects $A_1...A_n$ is considered which must be classified probabilistically into $C_1 ... C_m$ classes, which may be m for n objects, in general m≠n and the objects may be curves, lines, regions, etc.

Each A_i has a well defined set of neighbors (A_j) and then to each A_i a probability vector $(P_{i1}...P_{im})$ is associated which gives an estimate of the probability that A_i belongs to class $C_1 ... C_m$. Initially, the estimate of p_{ik} depends on a conventional analysis (gradient, for instance) for estimating the probability that a pixel belongs to an edge. Then an assumption is made that for each pair of neighboring objects (A_i, A_j) and each pair of classes (C_h, C_k) a measure of compatibility between A_i belonging to class C_h and A_j belonging to class C_k may be defined.

As an example, $\boxed{\begin{array}{c} A_i \\ A_j \end{array}}$ are vertically adjacent pixels and C_h, C_k can be the classes of vertical edges having their dark sides on their right and on their left, respectively, (see figure below).

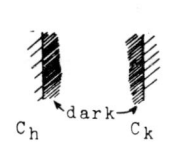

We will have that $A_i \in C_h$ and $A_j \in C_h$ are highly compatible, the same will be true for

C_h dark C_k

$A_i \in C_k$ and $A_j \in C_k$ but for

$A_i \in C_h$ and $A_j \in C_k$ they will not

be highly compatible but, just the opposite, highly incompatible.

A compatibility measurement r, has been introduced, ranging from -1 to 1 and having the value 0 for an irrilevant (don't care) situation. The compatibility variable is $r(A_i, C_h, A_j, C_k)$ which has four arguments: the pair of pixels and the pair of classes. A numerical example from (19), illustrated below, is given to clarify matters; there are three objects which are neighbors of one object and their probability vectors are written for the three classes as well as their compatibilities.

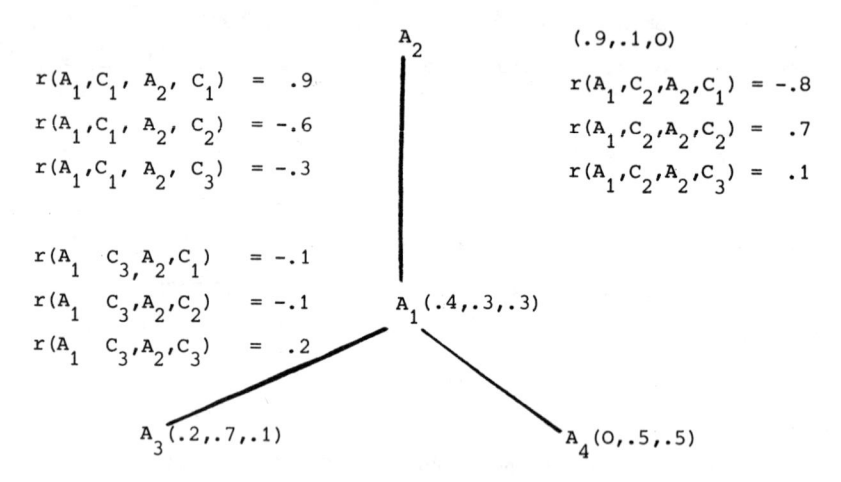

$r(A_1, C_1, A_2, C_1) = .9$

$r(A_1, C_1, A_2, C_2) = -.6$

$r(A_1, C_1, A_2, C_3) = -.3$

$r(A_1, C_3, A_2, C_1) = -.1$

$r(A_1, C_3, A_2, C_2) = -.1$

$r(A_1, C_3, A_2, C_3) = .2$

A_2 (.9, .1, 0)

$r(A_1, C_2, A_2, C_1) = -.8$

$r(A_1, C_2, A_2, C_2) = .7$

$r(A_1, C_2, A_2, C_3) = .1$

$A_1 (.4, .3, .3)$

$A_3 (.2, .7, .1)$

$A_4 (0, .5, .5)$

The iterations of the relaxation process will adjust the probabilities on the basis of the values of the probabilities of the neighboring pixels and their associated compatibility values. The rules may be explicitly stated as follows:

1. if p_{ij} (probability of pixel j to belong to class k) is high and if $r(A_i C_h A_j C_k)$ is close to +1 then p_{ih} should be increased. (Let us remember that p_{ih} is the probability that pixel i belongs to class h),

2. if p_{jk} is low and $r(A_i C_h A_j C_k)$ is close to -1 then P_{ik} should be decreased,

3. if p_{jk} is low and $r(A_i C_h A_j C_k)$ is close to 0, then p_{ih} should not change significantly.

A compact way of expressing these three rules is by means of the product:

$P_{jk} r(A_i C_h A_j C_k)$ which will indicate, for all A_j, C_k whether the value of p_{ih} should be incremented or decremented. The contribution of the neighbors (A_j) may be expressed by

$$\sum_j P_{jk} \cdot r(A_i C_h A_j C_k)$$

which will range, in value, between -1 and +1 since $\sum_j P_{jk} = 1$ and $-1 \leq r \leq +1$

In order to have a positive value in all cases for the contribution expression, weights may be introduced which, when added, will sum 1 and then the net contribution at each iteration may be rewritten as:

$$P_{ih}(1 + \sum_j s_j \sum_k P_{jk} \cdot r(A_i C_h A_j C_k))$$

since the interior part ranges between -1 and +1 the full expression will range between 0 and +2. In this way the value of p_{ih} will always be multiplied by a non-negative number, after all the p_{ih}'s have been updated, the expression may be renormalized by dividing each p_{ih} by

$$\sum_{h=1}^{m} P_{ih}$$

so that their total sum will add to 1.

Let us now describe the process along the time scale, where t is the time variable and $p_{ih}^{(t)}$ is the value of p_{ih} at the tth iteration; the value of the net increment will be given by:

$$q_{ih}^{t} = \sum_j s_j \sum_k p_{jk}^{(t)} \cdot r(A_i, C_h, A_j, C_k)$$

and the updated value of p_{jk} at time instant t+1 will become:

$$p_{jk}^{(t+1)} = p_{ih}^{(t)}(1 + q_{ih}^{(t)} / \sum_h (1 + q_{ih}^{(t)}))$$

The independence of the $(A_j C_k)$'s from the p_{ih}'s has been assumed and although this works well in some cases, a cooperative process in which neighboring probabilities affect the central p, may be introduced and lead to more refined models of the general relaxation procedure which is the basis of this approach.

The classes that interest us in this work are those concerning pixels that belong to the edge or do not (edge class, no-edge class). Therefore, the probability adjustment scheme is such that the edge probability for a pixel is incremented in accordance with the neighboring edge probabilities, if the orientations and the neighboring directions are such that the edges reinforce each other, the orientation might be adjusted in the direction of the strongest edge probability.

$A_1 \ldots A_n$ are image pixels and $\theta_1 \ldots \theta_{m-1}$ are the classes which correspond to edges having specific orientations, we will add two more classes: edge and no edge. The starting point is the assignment of a probability p_i at A_i according to the magnitude of the gradient at A_i (for instance) and estimate θ_i perpendicular to the gradient direction. The p_i's are adjusted in accordance with the probabilities p_j and orientations θ_j of edges at neighboring pixels A_j that extend or contradict the evidence of an edge along θ_i at A. If θ_{ij} is the direction from A_i, to A_j use an increment proportional to $|\cos(\theta_i - \theta_{ij})|$ and to $\cos(\theta_i - \theta_j)$;
the dark side will be assumed always on the clockwise direction so as to distinguish edges with the same direction and opposite senses

$$\cos(\theta_i - \theta_j) = \cos \pi = -1$$

and they will decrement one another's probabilities.

Finally, the edge reinforcement technique used to reduce or supress noise in the image is performed by assigning an initial edge probability to a pixel given by

Edge extraction techniques

$$p_e^{(0)} = \frac{\text{grad magn at pixel } p}{\max_{\text{pixQ}} \text{grad magn at } Q}$$

the maximum value is taken over the whole image.

For any point (x,y) and neighboring pixel (uv) let α be the edge slope at x,y and β the edge slope at (u,v) and $D = \max(|x-u|, |y-v|)$, in(19) the compatibilities were defined as follows:

for the pairs of class assignments of pixels (x,y) and (u,y)

$$r(\text{edge}, \text{edge}) = \cos(\alpha - \gamma)\cos(\beta - \gamma)/2^D$$

$$r(\text{edge}, \text{no edge}) = \min(0, -\cos(2\alpha - 2\gamma)/2^D$$

$$r(\text{no edge}, \text{edge}) = (1-\cos(2\beta - 2\gamma))/2^{D+1}$$

$$r(\text{no edge}, \text{no edge}) = 1/2^D$$

these compatibilities have the following properties: parallel or perpendicular edges have no effect on one another, collinear edges reinforce on one another (inversely proportional to the distance), non edges collinear with edges weaken them, non edges alongside edges have no effect on them, edges alongside nonedges strengthen them; edges collinear with non edges have no effect and nonedges reinforce one another.

The convergence of the relaxation process has been proved (although the proof holds for sufficient conditions rather than for necessary conditions), see Zucker(20) and the results of a large number of experiments using relaxation may be found in Davis-Rosenfeld(21), or more recently in Eklundh, Yamamoto, Rosenfeld(22).

In conclusion, the relaxation method is based on an initial estimate of the class probabilities for each object and then, by iteration, in updating such probabilities on the basis of the probabilities of the neighboring objects and their mutual compatibilities.

A report(23) discusses the difficulties that arise when using relaxation techniques, they stem from the problem of adequate knowledge representation which can be expressed as "... how do we represent what we know about our problem in terms of an interactive local process?" (Davis

Rosenfeld,(23)). The compatibility coefficients, the neighborhood model and the formalization of the interactivity between the properties of the pixels, are the three main points that should be adequately taken care of in any relaxation process. Finally, the control of the process, how many times should it be iterated and how can the partial results be evaluated still remains an open question. Both convergence and unambiguity have been analyzed and recently, the consistency of both the initial labelling and the compatibilities (assumed model) has been fully investigated (Peleg (24)).

8. Conclusion

A small number of techniques have been illustrated here in order to convey the difficulties of the problem which starts at the very beginning: what is an edge? More sophisticated approaches combine knowledge on the shape of the objects with the output of edge extractors and test for consistency. Other possibilities include using the information obtained from region growing algorithms which are designed for image segmentation and therefore also aim at detecting region borders, again this might be combined with local edge extractors to increase the confidence level of the found edge elements.

Most of the material of this lecture comes from (2) which was designed for the purpose of discussing a number of edge detection techniques.

New approaches to edge detection consider differences in texture, results from research on low level vision, available information on shape and other knowledge sources.

The possibility of performing in parallel, on the whole image at a time, a set of local operations designed for local variation detection, or for region extraction or global filtering (this is much harder to achieve on most arrays of processors which have restricted neighborhoods) or a graph search (also not trivial) or dynamic techniques and relaxation (ideal for the SIMD class of machines) is appealing and results in speed-up factors which range from 10^3 to 10^5.

It is for this reason that algorithms for edge detection are also studied from the point of view of their computational complexity as well as from their direct efficiency in solving the task.

REFERENCES

1) L.S.Davis, "A Survey of Edge Detection Techniques" CGIP, vol. 4,3,1976, pp.248-270.

2) S.Levialdi, "Finding the edge", NATO ASI on Digital Image Processing, 1980, edits. J.C.Simon, R.M.Haralick, 105-148.

3) N.Zavidovique, "Les Methodes Elementaires de detection de contour dans les images digitalisées", Univ. de Technologie de Compiegne, 1980.

4) W.S.Holmes, "Design of a photointerpretation automation", Proc. Fall Joint Comp. Conf., pp. 27-35, 1962.

5) E.S.Deutsch, J.R.Fram, "A quantitative study of the orientation bias of some edge detector schemes", IEEE Transactions on Comp., vol. C-27, 3, pp. 205-213, 1978.

6) L.G.Roberts, "Machine perception of three dimensional solids", Optical and Electrooptical Information processing, edits. J.Tippett, D. Berkowitz, L.Clapp, C.Koester, A.Vnaderburgh, MIT Press, pp.159-197, 1965.

7) J.M.S.Prewitt, "Object enhancement and extraction" in Picture Processing and Psychopictorics, edits. A.Rosenfeld, B.Lipkin, Academic Press, pp. 75-149, 1970.

8) J.M.Tennenbaum, A.C.Kay, T.Binford, G.Falk, J.Feldman, G.Grape, R.Paul, K.Pingle, I.Sobel, "The Stanford hand-eye project", Proc. IJCAI, edits. D.A.Walker, L.M.Norton, pp. 521-526, 1969.

9) F.C.A. Groen, Course on Pattern Recognition and Image Processing, Delft University of Technology, Department of Applied Physics, pp.130-144, 1878.

10) R.A.Kirsch, L.Cahn, C.Ray, G.H.Urban, "Experiments in processing pictorial information with a digital computer", Proc. Eastern Joint Comp. Conf., pp. 221-229, 1957.

11) I.E.Abdou, W.K.Pratt, "Quantitative design and evaluation of enhancement/thresholding edge detectors", Proc. IEEE, vol.67, 5, pp.753-763, 1979.

12) G.S.Robinson, "Edge detection by compass gradient masks", CGIP, vol. 6,5, pp. 429-501, 1977.

13) M.H.Hueckel, "An operator which locates edges in digital pictures", Journal ACM, 18, pp.113-125, 1971.

14) D.H.Hubel, T.N.Wiesel, "Receptive fields of single neurons in the cat's striate cortex", J.Physiol., vol. 148, pp.574-591, 1959.

15) J.W.Modestino, R.W.Fries, "Edge detection in noisy images using recursive digital filtering", CGIP, vol. 6, 5, pp.409-433, 1977.

16) M.D.Kelly, "Edge detection in pictures by computer using planning", in *Machine Intelligence*, edits. B.Meltzer, D.Michie, Ediburgh UP, pp.397-409, 1971.

17) A. Martelli, "Edge detection using heursitic search methods", Proc. Ist IJCPR, Washington, pp.375-388, 1973.

18) P.Lemkin, "The boundary trace transform", CGIP, vol. 9,2, pp.150-165, 1979.

19) A. Rosenfeld, "Iterative methods in image analysis", Pattern Recognition, vol. 10, pp.181-187.

20) S.W.Zucker, "Scene labeling by relaxation operations", IEEE Trans. on Syst. Man and Cybernetics, 8, pp.41-48, 1976.

21) L.S.Davis, A.Rosenfeld, "Application of relaxation labeling, 2: spring-loaded template matching", Proc. 3rd IJCPR, Coronado, 76CH1140-3C, pp.591-597, 1976.

22) J.O.Eklund, H.Yamamoto, A.Rosenfeld, "Relaxation methods in multispectral pixel classification", IEEE Trans. PAMI, (in press).

23) L.S.Davis, A.Rosenfeld, "Cooperating processes for low-level vision: a survey". Technical Rep. TR-123, University of Texas, Dep. of Computer Sciences, 1980.

24) S.Peleg, "Monitoring relaxation algorithms using labeling", Technical Rep. TR-842, University of Maryland, Computer Science Centre, College Park, 1979.

IMAGE TEXTURE SURVEY

Robert M. HARALICK
Virginia Polytechnic Institute and State University
Blacksburg, Virginia 24061

ABSTRACT

In this survey we review the image processing literature on the various apporaches and models investigators have used for texture. These include statistical approaches of autocorrelation function, optical transforms, digital transforms, textural edgeness, structural element, gray tone co-occurrence, run lengths, and auto-regressive models. We discuss and generalize some structural approaches to texture based on more complex primitives than gray tone. We conclude with some structural-statistical generalizations which apply the statistical techniques to the structural primitives.

1.0 INTRODUCTION

Texture is an important characteristic for the analysis of many types of images. It can be seen in all images from multi-spectral scanner images obtained from aircraft or satellite platforms (which the remote sensing community analyzes) to microscopic images of cell cultures or tissue samples (which the bio-medical community analyzes). Despite its important and ubiquity in image data, a formal approach or precise definition of texture does not exist. The texture discrimination techniques are, for the most part, ad-hoc. In this paper we survey, unify, and generalize some of the extraction techniques and models which investigators have been using to measure textural properties.

The image texture we consider is non-figurative and cellular. We think of this kind of texture as an organized area phenomena. When it is decomposable, it has two basic dimensions on which it may be described. The first dimension is for describing the primitives out of which the image texture is composed, and the second dimension is for the description of the spatial dependence or interaction between the primitives of an image texture. The first dimension is concerned with tonal primitives or local properties, and the second dimension is concerned with the spatial organization of the tonal primitives.

Tonal primitives are regions with tonal properties. The tonal primitive can be described in terms such as the average tone, or maximum and minimum tone of its region. The region is a maximally connected set of pixels having a given tonal property. The tonal region can be evaluated in terms of its area and shape. The tonal primitive includes both its gray tone and tonal region properties.

An image texture is described by the number and types of its primitives and the spatial organization or layout of its primitives. The spatial organization may be random, may have a pairwise dependence of one primitive on a neighboring primitive, or may have a dependence of n primitives at a time. The dependence may be structural, probabilistic, or functional (like a linear dependence).

To characterize texture, we must characterize the tonal primitive properties as well as characterize the spatial inter-relationships between them. This implies that texture-tone is really a two-layered structure, the first layer having to do with specifying the local properties which manifest themselves in tonal primitives and the second layer having to do with specifying the organization among the tonal primitives. We, therefore, would expect that methods designed to characterize texture would have parts devoted to analyzing each of these aspects of texture. In the review of the work done to date, we will discover that each of the existing methods tends to emphasize one or the other aspect and tends not to treat each aspect equally.

2.0 REVIEW OF THE LITERATURE ON TEXTURE MODELS

There have been eight statistical approaches to the measurement and characterization of image texture : autocorrelation functions, optical transforms, digital transforms, textural edgeness, structural elements, spatial gray tone co-occurrence probabilities, gray tone run lengths, and auto-regressive models. An early review of some of these approaches is given by Hawkins (1970). The first three of these approaches are related in that they all measure spatial frequency directly or indirectly. Spatial frequency is related to texture because fine textures are rich in high spatial frequencies while coarse textures are rich in low spatial frequencies.

An alternative to viewing texture as spatial frequency distribution is to view texture as amount of edge per unit area. Coarse textures have a small number of edges per unit area. Fine textures have a high number of edges per unit area.

The structural element approach of Serra (1974) and Matheron (1967) uses a matching procedure to detect the spatial regularity of shapes called structural elements in a binary image. When the structural elements themselves are single resolution cells, the information provided by this approach is the autocorrelation function of the binary image. By using larger and more complex shapes, a more generalized autocorrelation can be computed.

The gray tone spatial dependence approach characterizes texture by the co-occurrence of its gray tones. Coarse textures are those for which the distribution changes only slightly with distance and fine textures are those for which the distribution changes rapidly with distance.

The gray level run length approach characterizes coarse textures as having many pixels in a constant gray tone run and fine textures as having few pixels in a constant gray tone run.

The auto-regressive model is a way to use linear estimates of a pixel's gray tone given the gray tones in a neighborhood containing it in order to characterize texture. For coarse textures, the coefficients will all be similar. For fine textures, the coefficients will have wide variation.

The power of the spatial frequency approach to texture is the familiarity we have with these concepts. However, one of the inherent problems is in regard to gray tone calibration of the image. The procedures are not invariant under even a linear translation of gray tone. To compensate for this, probability quantizing can be employed. But the price paid for the invariance of the quantized images under monotonic gray tone transformations is the resulting loss of gray tone precision in the quantized image. Weszka, Dyer and Rosenfeld (1976) compare the effectiveness of some of these techniques for terrain classification. They conclude that spatial frequency approaches perform significantly poorer than the other approaches.

The power of the structural element approach is that it emphasizes the shape aspects of the tonal primitives. Its weakness is that it can only do so for binary images.

The power of the co-occurrence approach is that it characterizes the spatial inter-relationships of the gray tones in a textural pattern and can do so in a way that is invariant under monotonic gray tone transformations. Its weakness is that it does not capture the shape aspects of the tonal primitives. Hence, it is not likely to work well for textures composed of large-area primitives.

The power of the auto-regressive linear estimator approach is that it is easy to use the estimator in a mode which synthesizes textures from any initially given linear estimator. In this sense, the auto-regressive approach is sufficient to capture everything about a texture. Its weakness is that the textures it can characterize are likely to consist mostly of micro-textures.

2.1 THE AUTOCORRELATION FUNCTION AND TEXTURE

From one point of view, texture relates to the spatial size of the tonal primitives on an image. Tonal primitives of larger size are indicative of coarser textures ; tonal primitives of smaller size are indicative of finer textures. The autocorrelation function is a feature which tells about the size of the tonal primitives.

We describe the autocorrelation function with the help of a thought experiment. Consider two image transparencies which are exact copies of one another. Overlay one transparency on top of the other and with a uniform source of light, measure the average light transmitted through the double transparency. Now translate one transparency relative to the other and measure only the average light transmitted through the portion of the image where one transparency overlaps the other. A graph of these measurements as a function of the (x, y) translated positions and normalized with respect to the (0, 0) translation depicts the two-dimensional autocorrelation function of the image transparency.

Let I(u, v) denote the transmission of an image transparency at position (u, v). We assume that outside some bounded rectangular region $0 \leq u \leq L_x$ and $0 \leq v \leq L_y$, the image transmission is zero. Let (x, y) denote the x-translation and y-translation, respectively. The autocorrelation function for the image transparency d is formally defined by :

$$\rho(x,y) = \frac{\frac{1}{(L_x - |x|)(L_y - |y|)} \int_{-\infty}^{\infty}\int I(u,v) I(u + x, v + y) du\, dv}{\frac{1}{L_x L_y} \int_{-\infty}^{\infty}\int I^2(u,v) du\, dv},$$

$$|x| < L_x \text{ and } |y| < L_y$$

If the tonal primitives on the image are relatively large, then the autocorrelation will drop off slowly with distance. If the tonal primitives are small, then the autocorrelation will drop off quickly with distance. To the extent that the tonal primitives are spatially periodic, the autocorrelation function will drop off and rise again in a periodic manner. The relationship between the autocorrelation function and the power spectral density function is well known : they are Fourier Transforms of one another (Yaglom, 1962).

The tonal primitive in the autocorrelation model is the gray tone. The spatial organization is characterized by the correlation coefficient which is a measure of the linear dependence one pixel has on another.

An experiment was carried out by Kaiser (1955) to see of the autocorrelation function had any relationship to the texture which photointerpreters see in images. He used a series of seven aerial photographs of an Artic region and determined the autocorrelation function of the images with a spatial correlator which worked in a manner similar to the one envisioned in our thought experiment. Kaiser assumed the autocorrelation function was circularly symmetric and computed in only as a function of radial distance. Then for each image, he found the distance d such that the autocorrelation function ρ at d took the value $\frac{1}{e}$: $\rho(d) = \frac{1}{e}$.

Kaiser then asked 20 subjects to rank the seven images on a scale from fine detail to coarse detail. He correlated the rankings with the distances corresponding to the $(\frac{1}{e})$th value of the autocorrelation function. He found a correlation coefficient of .99. This established that at least for his data set, the autocorrelation function and the subjects were measuring the same kind of textural features.

Kaiser noticed, however, that even though there was a high degree of correlation between $\rho^{-1}(\frac{1}{e})$ and subject rankings, some subjects put first what $\rho^{-1}(\frac{1}{e})$ put fifth. Upon further investigation, he discovered that a relatively flat background (indicative of high frequency or fine texture) can be interpreted as a fine textured or coarse textured area. This phenomena is not unusual and actually points out a fundamental characteristic of texture : it cannot be analyzed without a reference frame of tonal primiti-

being stated or implied. For any smooth gray tone surface, there exists a scale such that when the surface is examined, it has no texture. Then as resolution increases, it takes on a fine texture and then a coarse texture. In Kaiser's situation, the resolution of his spatial correlator was not good enough to pick up the fine texture which some of his subjects did in an area which had a weak but fine texture.

2.2 ORTHOGONAL TRANSFORMATIONS

Spatial frequency characteristics of two-dimensional images can be expressed by the autocorrelation function or by the power spectra of those images. Both may be calculated digitally and/or implemented in a real-time optical system.

Lendaris and Stanley (1969, 1970) used optical techniques to perform texture analysis on a data base of low altitude photographs. They illuminated small circular sections of those images and used the Fraunhoffer diffraction pattern to generate features for identifying photographic regions. The major discriminations of concern to these investigators were those of man-made versus natural scenes. The man-made category was further subdivided into roads, intersections of roads, buildings and orchards.

Feature vectors extracted from these diffraction patterns consisted of forty components. Twenty of the components were mean energy levels in concentric annular rings of the diffraction pattern and the other twenty components were mean energy levels in 9°-wedges of the diffraction pattern. Greater than 90 % classification accuracy was reported using this technique.

Cutrona, Leith, Palermo and Porcello (1969) present a review of optical processing methods for computing the Fourier transform. Goodman (1968), Preston (1972) and Shulman (1970) also present in their books comprehensive reviews of Fourier optics. Swanlund (1971) discusses the hardware specifications for a system using optical techniques to perform texture analysis.

Gramenopolous (1973) used a digital Fourier transform technique to analyze aerial images. He examined subimages of 32×32 pixels and determined that for an (ERTS) image over Phoenix, spatial frequencies between 3.5 and 5.9 cycles/km contained most of the information required to discriminate among terrain types. An overall classification accuracy of 87 % was achieved using image categories of clouds, water, desert, farms, mountain, urban, river bed and cloud shadows. Horning and Smith (1973) used a similar approach to interpret aerial multispectral scanner imagery.

Bajcsy (1972, 1973) and Bajcsy and Lieberman (1974, 1976) computed the two-dimensional power spectra of a matrix of square image windows. They expressed the power spectrum in a polar coordinate system of radius (r) versus angle (a). As expected, they determined that directional textures tend to have peaks in the power spectrum along a line orthogonal to the principle direction of the texture. Blob-like textures tend to have peaks in the power spectrum at radii (r) comparable to the sizes of the blobs. This work also shows that texture gradients can be measured by determining the trends of relative maxima of radii (r) and angles (a) as a function of the position of the image window whose power spectrum is being analyzed. For example, as the power peaks along the radial direction tend to shift towards larger values of r, the image surface becomes more finely textured. In general, features based on Fourier power spectra have been shown to perform more poorly than features based on second order gray level co-occurenc statistics (Haralick, Shanmugam and Dinstein, 1973) or those based on first order statistics of gray level differences (Weska, Dyer and Rosenfeld, 1976). Presence of aperture effects has been hypothesized to account for part of the unfavorable performance by Fourier features compared to space-domain gray level statistics (Dyer and Rosenfeld, 1976), although experimental results indicate that this effect, if present, is minimal.

Transforms other than the Fourier Transform can be used for texture analysis. Kirvida (1976) compared the fast Fourier, Hadamard and Slant Transforms for textural features on aerial images of Minnesota. Five classes (i.e., hardwood trees, conifers, open space, city and water) were studied using 8×8 subimages. A 74 % correct classification rate was obtained using only spectral information. This rate increased to 98.5 % when textural information was also included in the analysis. These researchers reported no significant difference in the classification accuracy as a function of which transform was employed.

Pratt (1978) and Pratt, Faugeras and Gagalowicz (1978) suggest measuring texture by the coefficients of the linear filter required to decorrelate an image and by the first four moments of the gray level distribution of the decorrelated image. They have shown promising preliminary results.

The linear dependence which one image pixel has on another is well known and can be measured by the autocorrelation function. This linear dependence is exploited by the autoregression texture characterization and synthesis model developped by McCormick and Jayaramamurthy (1974) to synthesize textures. McCormick and Jayaramamurthy used the Box and Jenkins (1970) time series seasonal analysis method to estimate the parameters of a given texture. These estimated parameters and a given set of starting values were then used to illustrate that the synthesized texture was close in appearance to the given texture. Deguchi and Morishita (1978), Tou, Kao and Chang (1976) and Tou and Chang (1976) used similar techniques.

The autoregressive model for texture synthesis begins with a randomly generated noise image. Then, given any sequence of K synthesized gray level values in its immediately past neighborhood, the next gray level value can be synthesized as a linear combination of those values plus a linear combination of the previous L random noise values. The coefficients of these linear combinations are the parameters of the model. Texture analysis work based on this model requires the identification of these coefficient values from a given texture image.

2.3 GRAY TONE CO-OCCURRENCE

Textural features can also be calculated from a gray level spatial co-occurrence matrix. The co-occurrence $P(i, j)$ of gray tone i and j for an image I is defined as the number of pairs of neighboring resolution cells (pixels) having gray levels i and j, respectively. The co-occurrence matrix can be normalized by dividing each entry by the sum of all of the entries in the matrix. Conditional probability matrices can also be used for textural feature extraction with the advantage that these matrices are not affected by changes in the gray level histogram of an image, only by changes in the topological relationships of gray levels within the image.

Apparently Julesz (1962) was the first to use co-occurrence statistics in visual human texture discrimination experiments. Darling and Joseph (1968) used statistics obtained from nearest-neighbor gray-level transition probability matrices to measure texture using spatial intensity dependence in satellite images taken of clouds. Bartels and Wied (1975), Bartels, Bahr and Wied (1969) and Wied, Bahr and Bartels (1970) used one-dimensional co-occurrence statistics for the analysis of cervical cells. Rosenfeld and Troy (1970), Haralick (1971) and Haralick, Shanmugan and Dinstein (1973) suggested the use of spatial co-occurrence for arbitrary distances and directions. Galloway (1975) used gray level run length statistics to measure texture. These statistics are computable from co-occurrence assuming that the image is generated by a Markov process. Chen and Pavlidis (1978) used the co-occurrence matrix in conjunction with a split and merge algorithm to segment an image at textural boundaries. Tou and Chang (1977) used statistics from the co-occurrence matrix, followed by a principal components eigenvector dimensionality reduction scheme (Karhunen-Loeve expansion) to reduce the dimensionality of the classification problems.

Statistics which Haralick, Shanmugan and Dinstein (1973) computed from such co-occurrence matrices have been used to analyze textures in satellite images (Haralick and Shanmugan, 1974). An 89 % classification accuracy was obtained. Additional applications of this technique include the analysis of microscopic images (Haralick and Shanmugan, 1973), pulmonary radiographs (Chien and Fu, 1974) and cervical cell, leukocyte and lymph node tissue section images (Pressman, 1976a, 1976b).

Haralick (1975) illustrates a way to use co-occurrence matrices to generate an image in which the value at each resolution cell is a measure of the texture in the resolution cell's neighborhood. All of these studies produced reasonable results on different textures. Conners and Harlow (1976 concluded that this spatial gray level dependence technique is more powerful than spatial frequency (power spectra), gray level difference (gradient and gray level run length methods (Galloway, 1975) of texture quantitation.

2.4 MATHEMATICAL MORPHOLOGY

A structural element and filtering approach to texture analysis of binary images was proposed by Matheron (1967) and Serra and Verchery (1973). This approach requires the definition of a structural element (i.e., a set of pixels constituting a specific shape such as a line or square) and the generation of binary images which result from the translation of the structural element through the image and the erosion of the image by the structural element. The textural features can be obtained from the new binary images by counting the number of pixels having the value 1. This mathematical morphology approach of Serra and Matheron is the basis of the Leitz Texture Analyser (TAS) (Muller and Hunn, 1974 ; Muller, 1974 ; Serra, 1974). A broad spectrum of applications has been found for this quantitative analysis of microstructures method in materials science and biology.

Watson (1975) summarizes this approach to texture analysis and we now give a precise description. Let H, a subset of resolution cells, be the structural element. We define the translate of H by row column coordinates (r, c) as H(r, c) where

$$H(r,c) = \{(i,j) | \text{for some } (r',c') \in H, i = r+r', j = c+c'\}$$

Then the erosion of F by the structural element H, written F⊖H, is defined as

$$F \ominus H = \{(m,n) | H(m,n) \subseteq F\}$$

The eroded image J obtained by eroding F with structural element H is a binary image where pixels take the value 1 for all resolution cells in F⊖H. Textural properties can be obtained from the erosion process by appropriately parameterizing the structural element (H) and determining the number of elements of the erosion as a function of the parameter's value. Theoretical properties of the erosion operator as well as other operators are presented by Matheron (1975), Serra (1978) and Lantuejoul (1978). The importance of this approach to texture analysis is that properties obtained by the application of operators in mathematical morphology can be related to physical properties of the materials imaged.

2.5 GRADIENT ANALYSIS

Rosenfeld and Troy (1970) and Rosenfeld and Thurston (1971) regard textures in terms of the amount of "edge" per unit image area. An edge can be detected by a variety of local mathematical operators which essentially measure some property related to the gradient of the image intensity function. Rosenfeld and Thurston used the Roberts gradient and then computed, as a measure of texture for any image window, the average value of the Roberts gradient taken over all of the pixels in the window. Sutton and Hall (1972) extend this concept by measuring the gradient as a function of the distance between pixels. An 80 % classification accuracy was achieved by applying this textural measure in a pulmonary disease identification experiment.

Related approaches include Triendl (1972) who, smoothes the image using 3 × 3 neighborhoods, then applies a 3 × 3 digital Laplacian operator and finally smoothes the image with an 11 × 11 window. The resulting texture parameters obtained from the frequency filtered image can be used as a discriminatory textural feature. Hsu (1977) determines edgeness by computing variance-like measures for the intensities in a neighborhood of pixels. He suggests the deviation of the intensities in a pixel's neighborhood from both the intensity of the central pixel and from the average intensity of the neighborhood. The histogram of a gradient image was used to generate textural parameters by Landeweerd and Gelsema (1978) to measure texture properties in the nuclei of leukocytes. Rosenfeld (1975) generates an image whose intensity is proportional to the edge per unit area of the original image. This transformed image is then further processed by gradient transformations prior to textural feature extraction.

For example, mosaic texture models tessellate a picture into regions and assign a gray level to the region according to a specified probability density function (Schacter, Rosenfeld and Davis, 1978). Among the kinds of mosaid models are the Occupancy Model (Miles, 1970), Johnson-Mehl Model (Gilbert, 1962), Poisson Line Model (Miles, 1969) and Bombing Model (Switzer, 1967). The mosaic texture models seem readily adaptable to numerical analysis and their properties seem amenable to mathematical analysis.

3.0 STRUCTURAL

Pure structural models of texture presume that textures consists of primitives which appear in quasi-periodic spatial arrangements. Descriptions of these primitives and their placement rules can be used to describe textures (Rosenfeld and Lipkin, 1970). The identification and location of a particular primitive in an image may be probabilistically related to the identification and distribution of primitives in its neighborhood.

Carlucci (1972) suggests a texture model using primitives of line segments, open polygons and closed polygons in which the placement rules are given syntactically in a graph-like language. Zucker (1976a, 1976b) conceives of a real texture to be the distorsion of an ideal texture. Zucker's model, however, is more of a competance based model than a performance model. Lu and Fu (1978) and Tsai and Fu (1978) use a syntactic approach to texture.

In the remainder of this section, we discuss some structural-statistical approaches to texture models. The approach is structural in the sense that primitives are explicity defined. The approach is statistical in that the spatial interaction, or lack of it, between primitives is measured by probabilities.

We classify textures as being weak textures, or strong textures. Weak textures are those which have weak spatial-interaction between primitives. To distinguish between them it may be sufficient to only determine the frequency with which the variety of primitive kinds occur in some local neighborhood. Hence, weak texture measures account for many of the statistical textural features. Strong textures are those which have non-random spatial interactions. To distinguish between them it may be sufficient to only determine, for each pair of primitives, the frequency with which the primitives co-occur in a specified spatial relationship. Thus, our discussion will center on the variety of ways in which primitives can be defined and the ways in which spatial relationships between primitives can be defined.

3.1 PRIMITIVES

A primitive is a connected set of resolution cells characterized by a list of attributes. The simplest primitive is the pixel with its gray tone attribute. Sometimes it is useful to work with primitives which are maximally connected sets of resolution cells having a particular property. An example of such a primitive is a maximally connected set of pixels all having the same gray tone or all having the same edge direction.

Gray tones and local properties are not the only attributes which primitives may have. Other attributes include measures of shape of connected region and homogeneity of its local property. For example, a connected set of resolution cells can be associated with its length or elongation of its shape or the variance of its local property.

3.2 SPATIAL RELATIONSHIPS

Once the primitives have been constructed, we have available a list of primitives, their center coordinates, and their attributes. We might also have available some topological information about the primitives such as which are adjacent to which. From this data, we can select a simple spatial relationship such a adjacency of primitives or nearness of primitives and count how many primitives of each kind occur in the specified spatial relationship.

More complex spatial relationships include closest distance or closest distance within an angular window. In this case, for each kind of primitive situated in the texture, we could lay expanding circles around it and locate the shortest distance between it and every other kind of primitive. In this case our co-occurrence frequency is three-dimensional, two dimensions for primitive kind and one dimension for shortest distance. This can be dimensionally reduced to two dimensions by considering only the shortest distance between each pair of like primitives.

3.3 WEAK TEXTURE MEASURES

Tsuji and Tomita (1973) and Tomita, Yachida and Tsuji (1973) describe a structural approach to weak texture measures. First a scene is segmented into atomic regions based on some tonal property such as constan gray tone. These regions are the primitives. Associated with each primitive is a list of properties such as size and shape. Then they make a histogram of size property or shape property over all primitives in the scene. If the scene can be decomposed into two or more regions of homogeneous texture, the histogram will be multi-modal. If this is the case, each primitive in the scene can be tagged with the mode in the histogram to which it belongs. A region growing/cleaning process on the tagged primitives yields the homogeneous textural region segmentation.

If the initial histogram modes overlap too much, a complete segmentation may not result. In this case, the entire process can be repeated with each of the then so far found homogeneous texture region segments. If each of the homogeneous texture regions consists of mixtures of more than one type of primitive, then the procedure may not work at all. In this case, the technique of co-occurrence of primitive properties would have to be used.

Zucker, Rosenfeld and Davis (1975) used a form of this technique by filtering a scene with a spot detector. Non-maxima pixels on the filtered scene wer thrown out. If a scene has many different homogeneous texture regions, the histogram of the relative max spot detector filtered scene will be multi-modal. Tagging the maxima with the modes they belong to and region growing/cleaning thus produced the segmented scene.

The idea of the constant gray level regions of Tsuji and Tomita or the spots of Zucker, Rosenfeld and Davis can be generalized to regions which are peaks, pits, ridges, ravines, hillsides, passes, breaks, flats and slopes (Toriwaki and Fukumura, 1978 ; Peucker and Douglas, 1975). In fact, the possibilities are numerous enough that investigators doing experiments will have a long working period before understanding will exhaust the possibilities. The next three subsections review in greater detail some specific approaches and suggest some generalizations.

3.3.1 EDGE PER UNIT AREA

Rosenfeld and Troy (1970) and Rosenfeld and Thurston (1971) suggested the amount of edge per unit area for a texture measure. The primitive here is the pixel and its property is the magnitude of its gradient. The gradient can be calculated by any one of the gradient neighborhood operators. For some specified window centered on a given pixel, the distribution of gradient magnitudes can then be determined. The mean of this distribution is the amount of edge per unit area associated with the given pixel. The image in which each pixel's value is edge per unit area is actually a defocussed gradient image. Triendl (1972) used a defocussed Laplacian image. Sutton and Hall (1972) used such a measure for the automatic classification of pulmonary disease in chest X-rays.

Ohlander (1975) used such a measure to aid him in segmenting textured scenes. Rosenfeld (1975) gives an example where the computation of gradient direction on a defocussed gradient image is an appropriate feature for the direction of texture gradient. Hsu (1977) used a variety of gradient-like measures.

3.3.2 RUN LENGTHS

The gray level run lengths primitive in its one-dimensional form is a maximal collinear connected set of pixels all having the same gray level. Properties of the primitive can be length of run, gray level, and angular orientation of run. Statistics of these properties were used by Galloway (1975) to distinguish between textures.

In the two-dimensional form, the gray level run length primitive is a maximal connected set of pixels all having the same gray level. These maximal homogeneous sets have properties such as number of pixels, maximum or minimum diameter, gray level, angular orientation of maximum or minimum diameter. Maleson at al. (1977) has done some work related to maximal homogeneous sets and weak textures.

3.3.3 RELATIVE EXTREMA DENSITY

Rosenfeld and Troy (1970) suggest the number of extrema per unit area for a texture measure. They define extrema in a purely local manner allowing plateaus to be considered extrema. Ledley (1972) also suggests computing the number of extrema per unit area as a texture measure.

Mitchell, Myers and Boyne (1977) suggest the extrema idea of Rosenfeld and Troy except they proposed to use true extrema and to operate on a smoothed image to eliminate extrema due to noise. See also Carlton and Mitchell (1977) and Ehrich and Foith (1976, 1978).

One problem with simply counting all extrema in the same extrema plateau as extrema is that extrema per unit area is not sensitive to the difference between a region having few large plateaus of extrema and many single pixel extrema. The solution to this problem is to only count an extrema plateau once. This can be achieved by locating some central pixel in the extrema plateau and marking it as the extrema associated with the plateau. Another way of achieving this is to associate a value 1/N for every extrema in a N-pixel extrema plateau.

In the one-dimensional case, there are two properties that can be associated with every extrema : its height and its width. The height of a maxima can be defined as the difference between the value of the maxima and the highest adjacent minima. The height (depth) of a minima can be defined as the difference between the value of the minima and the lowest adjacent maxima. The width of a maxima is the distance between its two adjacent minima. The width of a minima is the distance between its two adjacent maxima.

Two-dimensional extrema are more complicated than one-dimensional extrema. One way of finding extrema in the full two-dimensional sense is by the iterated use of some recursive neighborhood operators propagating extrema values in an appropriate way. Maximally connected areas of relative extrema may be areas of single pixels or may be plateaux of many pixels. We can mark each pixel in a relative extrema region of size N with the value h indicating that it is part of a relative extrema having height h or mark it with the value h/N indicating its contribution to the relative extrema area. Alternatively, we can mark the most centrally located pixel in the relative extrema region with the value h. Pixels not marked can be given the value 0. Then for any specified window centered on a given pixel, we can add up the values of all pixels in the window. This sum divided by the window size is the average height of extrema in the area. Alternatively we could set h to 1 and the sum would be the number of relative extrema per unit area to be associated with the given pixel.

Going beyond the simple counting of relative extrema, we can associate properties to each relative extrema. For example, given a relative maxima, we can determine the set of all pixels reachable only by the given relative maxima and not by any other relative maxima by monotonically decreasing paths. This set of reachable pixels is a connected region and forms a mountain. Its border pixels may be relative minima or saddle pixels.

The relative height of the mountain is the difference between its relative maxima and the highest of its exterior border pixels. Its size is the number of pixels which constitute it. Its shape can be characterized by features such as elongation, circularity and symmetric axis. Elongation can be defined as the ratio of the larger to small eigenvalue of the 2 × 2 second moment matrix obtained from the $\binom{x}{y}$ coordinates of the border pixels (Bachi, 1973 ; Frolov, 1975). Circularity can be defined as the ratio of the standard deviation to the mean of the radii from the region's center to its border (Haralick, 1975). The symmetric axis feature can be determined by thinning the region down to its skeleton and counting the number of pixels in the skeleton. For regions which are elongated, it may be important to measure the direction of the elongation or the direction of the symmetri axis.

3.4 STRONG TEXTURE MEASURES AND GENERALIZED CO-OCCURRENCE

Strong texture measures take into account the co-occurrence between texture primitives. On the basis of Julesz (1975) it is probably the case that the most important interaction between texture primitives occurs as a two-way interaction. Textures with identical second and lower order interactions but with different higher order interactions tend to be visually similar.

The simplest texture primitive is the pixel with its gray tone property. Gray tone co-occurrence between neighboring pixels was suggested as a measure of texture by a number of researchers as discussed in Section 2.6. All the studies mentioned there achieved a reasonable classification accuracy of different textures using co-occurrences of the gray tone primitive.

The next more complicated primitive is a connected set of pixels homogeneous in tone (Tsuji and Tomita, 1973). Such a primitive can be characterized by size, elongation, orientation and average gray tone. Useful texture measures include co-occurrence of primitives based on relationships of distance or adjacency. Maleson et al. (1977) suggests using region growing techniques and ellipsoidal approximations to define the homogeneous regions and degree of co-linearity as one basis of co-occurrence. For example, for all primitives of elongation greater than a specified threshold, we can use the angular orientation of each primitive with respect to its closest neighboring primitive as a strong measure of texture.

Relative extrema primitives were proposed by Rosenfeld and Troy (1970), Mitchell, Myers and Boyne (1977), Ehrich and Foith (1976), Mitchell and Carlton (1977), and Ehrich and Foith (1978). Co-occurrence between relative extrema was suggested by Davis et al. (1978). Because of their invariance under any monotonic gray scale transformation, relative extrema primitives are likely to be very important.

It is possible to segment an image on the basis of relative extrema (for example, relative maxima) in the following way : label all pixels in each maximally connected relative maxima plateau with a unique label. Then label each pixel with the label of the relative maxima that can reach it by a monotonically decreasing path. If more than one relative maxima can reach it by a monotonically decreasing path, then label the pixel with a special label "c" for common. We call the regions so formed the descending components of the image.

Co-occurrence between properties of the descending components can be based on the spatial relationship of adjacency. For example, if the property is size, the co-occurrence matrix could tell us how often a descending component of size s_1 occurs adjacent to or nearby to a descending component of size s_2 or of label "c".

To define the concept of generalized co-occurrence, it is necessary to first decompose an image into its primitives. Let Q be the set of all primitives on the image. Then we need to measure primitive properties such as mean gray tone, variance of gray tones, region, size, shape, etc. Let T be the set of primitive properties and f be a function assigning to each primitive in Q a property of T. Finally, we need to specify a spatial relation between primitives such as distance or adjacency. Let $S \subseteq Q \times Q$ be the binary relation pairing all primitives which satisfy the spatial relation. The generalized co-occurrence matrix P is defined by :

$$P(t_1, t_2) = \frac{\#\{(q_1, q_2) \in S | f(q_1) = t_1 \text{ and } f(q_2) = t_2\}}{\#S}$$

$P(t_1, t_2)$ is just the relative frequency with which two primitives occur with specified spatial relationship in the image, one primitive having property t_1 and the other primitive having property t_2.

Zucker (1974) suggests that some textures may be characterized by the frequency distribution of the number of primitives any primitives has related to it. This probability p(k) id defined by :

$$p(k) = \frac{\#\{(q \in Q | \#S(q) = k\}}{\#Q}$$

Although this distribution is simpler than co-occurrence, no investigator appears to have used it in texture discrimination experiments.

4.0 CONCLUSION

We have surveyed the image processing literature on the various approaches and models investigators have used for textures. For microtextures, the statistical approach seems to work well. The statistical approaches have included autocorrelation functions, optical transforms, digital transforms, textural edgeness, structural element, gray tone co-occurrence, and autoregressive models. Pure structural approaches based on more complex primitives than gray tone seems not to be widely used. For macrotextures, investigators seem to be moving in the direction of using histograms of primitive properties and co-occurrence of primitive properties in a structural-statistical generalization of the pure structural and statistical approaches.

Acknowledgement : This is a condensed version of a paper "Statistical and Structural Approaches to Texture" which appeared in the May 1978 issue of IEEE Proceedings.

REFERENCES

1. Bachi, Roberto, "Geostatistical Analysis of Territories", Proceedings of the 39th Session-Bulletin of the International Statistical Institute, Vienna, 1973.

2. Bajcsy, R., Computer Identification of Textured Visual Scenes, Stanford Univ., Palo Alto, Calif., 1972.

3. Bajcsy, R., "Computer Description of Textured Surfaces", Third Int. Joint Conf. on Artificial Intelligence, Stanford, Calif., Aug. 20-23, 1973, pp. 572-578.

4. Bajcsy, R. and L. Lieberman, "Computer Description of Real Outdoor Scenes", Proceedings of Second International Joint Conference on Pattern Recognition, Copenhagen, Denmark, August 1974, pp. 174-179.

5. Bajscy, R. and L. Lieberman, "Texture Gradient as a Depth Cue", Computer Graphics and Image Processing, Vol. 5, N° 1, 1976, pp. 52-67.

6. Bartels, P., G. Bahr, and G. Weid, "Cell Recognition from Line Scan Transition Probability Profiles", Acta Cytol., Vol. 13, 1969, pp. 210-217.

7. Bartels, P.H., and Wied, G.L., "Extraction and Evaluation of Information From Digitized Cell Image", (In) Mammalian Cells : Probes and Problems, U.S. NTIS Technical Information Center, Springfield, Va., 1975, pp. 15-28.

8. Box, J.E., and G.M. Jenkins, Time Series Analysis, Holden-Day, San Francisco, California, 1970.

9. Carlton, S.G. and O. Mitchell, 'Image Segmentation Using Texture and Grey Level", Pattern Recognition and Image Processing Conference, Troy, New York, June 1977, pp. 387-391.

10. Carlucci, L., "A Formal System for Texture Languages", Pattern Recognition, Vol. 4, 1972, pp. 53-72.

11. Chen, P. and T. Pavlidis, "Segmentation by Texture Using a Co-occurrence Matrix and a Split-and-Merge Algorithm", Technical Report 237, Princeton University, Princeton, New Jersey, January 1978.

12. Chien, Y.P. and K.S. Fu, "Recognition of X-Ray Picture Patterns", IEEE Transactions on Systems, Man, and Cybernetics, Vol. SMC-4, N° 2, March 1974, pp. 145-156.

13. Conners, Richard W. and Charles A. Harlow, "Some Theoretical Considerations Concerning Texture Analysis of Radiographic Images", Proceedings of the 1976 IEEE Conference on Decision and Control, 1976.

14. Cutrona, L.J., E.N. Leith, C.J. Palermo, and L.J. Porcello, "Optical Data Processing and Filtering Systems", IRE Transactions on Information Theory, Vol. 15, N° 6, June 1969, pp. 386-400.

15. Darling, E. M. and R.D. Joseph, "Pattern Recognition from Satellite Altitudes", IEEE Transactions on Systems, Man, and Cybernetics, Vol. SMC-4, March 1968, pp. 38-47.

16. Davis, L., S. Johns, and J.K. Aggarwal, "Texture Analysis Using Generalized Co-Occurrence Matrices", Pattern Recognition and Image Processing Conference, Chicago, Illinois, May 31-June 2, 1978.

17. Deguchi, K., and Morishita, I., 1978, "Texture Characterization and Texture-Based Image Partitioning Using Two-Dimensional Linear Estimation Techniques", IEEE Trans. on Computers, Vol. 27, N° 8, pp. 739-745.

18. Dyer, C. and A. Rosenfeld, "Fourier Texture Features : Suppression of Aperature Effects", IEEE Transactions on Systems, Man, and Cybernetics, Vol. SMC-6, N° 10, October 1976, pp. 703-705.

19. Ehrich, Roger and J.P. Foith, "Representation of Random Waveforms by Relational Trees", IEEE Transactions on Computers, Vol. C-25, N° 7, July 1976, pp. 725-736.

20. Ehrich, Roger and J.P. Foith, "Topology and Semantics of Intensity Arrays", Computer Vision, Hanson and Riseman (Editors), Academic Press, New York, 1978.

21. Frolov, Y.S., "Measuring the Shape of Geographical Phenomena : A History of the Issue", Soviet Geography : Review and Translation, Vol. XVI, N° 10, December 1975, pp. 676-687.

22. Galloway, M.M., "Texture Analysis Using Gray Level Run Lengths", Computer Graphics and Image Processing, Vol. 4, 1975, pp. 172-179.

23. Gilbert, E., "Random Subdivisions of Space into Crystals", Annals Math. Stat., Vol. 33, pp. 958-972.

24. Goodman, J.W., Introduction to Fourier Optics, McGraw-Hill Book Company, New York, 1968.

25. Gramenopoulos, Nicholas, "Terrain Type Recognition Using ERTS-1 MSS Images", Symposium on Significant Results Obtained from the Earth Resources Technology Satellite, NASA SP-327, March 1973, pp. 1229-1241.

26. Haralick, R.M., "A Texture-Context Feature Extraction Algorithm for Remotely Sensed Imagery" Proceedings of the 1971 IEEE Decision and Control Conference, Gainesville, Florida, December 15-17, 1971, pp. 650-657.

27. Haralick, R.M., "A Textural Transform for Images", Proceedings of the IEEE Conference on Computer Graphics, Pattern Recognition, and Data Structure, Beverly Hills, California, May 14-16, 1975.

28. Haralick, R.M. and K. Shanmugam, "Computer Classification of Reservoir Sandstones", IEEE Transactions on Geoscience Electronics, Vol. GE-11, N° 4, October 1973, pp. 171-177.

29. Haralick, R.M. and K. Shanmugam, "Combined Spectral and Spatial Processing of ERTS Imagery Data", Journal of Remote Sensing of the Environment, Vol. 3, 1974, pp. 3-13.

30. Haralick, R.M., K. Shanmugam, and I. Dinstein, "Textural Features for Image Classification", IEEE Transactions on Systems, Man, and Cybernetics, Vol. SMC-3, N° 6, November 1973, pp. 610-621.

31. Hawkins, Joseph K., "Textural Properties for Pattern Recognition", Picture Processing and Psychopictorics, Bernic Sacks Lipkin and Azriel Rosenfeld (Editors), Academic Press, New York, 1970.

32. Horning, R.J. and J.A. Smith, "Application of Fourier Analysis to Multispectral/Spatial Recognition", Management and Utilization of Remote Sensing Data ASP Symposium, Sioux Falls, South Dakota, October 1973.

33. Hsu, S., "A Texture-Tone Analysis for Automated Landuse Mapping with Panchromatic Images", Proceedings of the American Society for Photogrammetry, March 1977, pp. 203-215.

34. Julesz, Bela, "Visual Pattern Discrimination", IRE Transactions on Information Theory, Vol. 8, N° 2, February 1962, pp. 84-92.

35. Julesz, Bela, "Experiments in the Visual Perception of Texture", April 1975, 10 pages.

36. Kaiser, H., "A Quantification of Textures on Aerial Photographs", Boston University Research Laboratories, Technical Note 121, 1955, AD 69484.

37. Kirvida, L., "Texture Measurements for the Automatic Classification of Imagery", IEEE Transactions on Electromagnetic Compatibility, Vol. 18, N° 1, February 1976, pp. 38-42.

38. Landeweerd, G.H., and Gelsema, E.S., "The Use of Nuclear Texture Parameters in the Automatic Analysis of Leukocytes", Pattern Recognition, Vol. 10, 1978, pp. 57-61.

39. Lantuejoul, C., "Grain Dependence Test in a Polycristalline Ceramic", Quantitative Analysis of Microstructures in Materials Science, Biology, and Medecine, J.L. Chernant (Editor), Dr. Riederer-Verlag, GmbH, Stuttgart, 1978, pp. 40-50.

40. Ledley, R.S., "Texture Problems in Biomedical Recognition", (In) Proc. of the 1972 IEEE Conf. on Decision and Control and the 11th Symposium on Adaptive Processes, December 13-15, 1972, New Orleans, La.

41. Lendaris, G. and G. Stanley, "Diffraction Pattern Sampling for Automatic Pattern Recognition", SPIE Pattern Recognition Studies Seminar Proceedings, June 9-10, 1969, pp. 127-154.

42. Lendaris, G.G. and G.L. Stanley, "Diffraction Pattern Samplings for Automatic Pattern Recognition", Proceedings of the IEEE, Vol. 58, N° 2, February 1970, pp. 198-216.

43. Lu, S.Y. and K.S. Fu, "A Syntactic Approach to Texture Analysis", Computer Graphics and Image Processing, Vol. 7, 1978, pp. 303-330.

44. Maleson, Joseph, C. Brown, and J. Feldman, "Understanding Natural Texture", Computer Science Department, University of Rochester, New York, September 1977.

45. Matheron, G., Eléments Pour Une Théories des Milieux Poreux, Masson, Paris, 1967.

46. Matheron, G., Random Sets and Integral Geometry, Wiley and Sons, Inc., New York, 1975.

47. McCormick, B.H. and S.N. Jayaramamurthy, "Time Series Model for Texture Synthesis", International Journal of Computer and Information Sciences", Vol. 3, N° 4, December 1974, pp. 329-343.

48. Miles, R., "Random Polygons Determined by Random Lines in the Plane", Proc. National Academy of Sciences, USA, Vol. 52, 1969 pp. 901-907, 1157-1160.

49. Miles, R., "On the Homogeneous Planar Poisson Point-Process", Math. Biosciences, Vol. 6, 1970, pp. 85-127.

50. Mitchell, Owen, Charles Myers, and William Boyne, "A Max-Min Measure for Image Texture Analysis", IEEE Transactions on Computers, Vol. C-25, N° 4, April 1977, pp. 408-414.

51. Mitchell, O.R. and S.G. Carlton, "Image Segmentation Using a Local Extrema Texture Measure", Special Issue of Pattern Recognition, June 1977.

52. Muller, W., "The Leitz-Texture-Analyzing System", Laboratory for Applied Microscopy, Science and Technical Information, Suppl. I., N° 4, Wetzler, West Germany, April 1974, pp. 101-136.

53. Muller, W., and Hunn, W., "Texture Analyzer System", Industrial Research, 1974, pp. 49-54.

54. Ohlander, R., "Analysis of Natural Scenes", Ph.D. Dissertation, Carnegie Melon University, Pittsburgh, Pennsylvania, 1975.

55. Peucker, T. and D. Douglas, "Detection of Surface-Specific Points by Local Parallel Processing of Discrete Terrain Elevation Data", Computer Graphics and Image Processing, Vol. 4, N° 4, December 1975, pp. 375-387.

56. Pratt, W.K., "Image Features Extraction", (In) Digital Image Processing, 1978, pp. 471-513.

57. Pratt, W.K., Faugeras, O.D., and Gagalowicz, A., "Visual Discrimination of Stochastic Texture Fields", IEEE Trans. on Systems, Man, and Cybernetics, Vol. SMC-8, N° 11, November 1978, pp. 796-804.

58. Pressman, N.J., "Markovian Analysis of Cervical Cell Images", J. Histochem. Cytochem, Vol. 24, N° 1, 1976a, pp. 138-144.

59. Pressman, N.J., "Optical Texture Analysis for Automated Cytology and Histology : A Markovian Approach", Dissertation, Lawrence Livermore Laboratory Report UCRL-52155, Livermore, Calif., 1976b.

60. Preston, K., Coherent Optical Computers, McGray-Hill Book Company, New York, 1972.

61. Rosenfeld, Azriel, "A Note on Automatic Detection of Texture Gradients", IEEE Transactions on Computers, Vol. C-23, N° 10, October 1975, pp. 988-991.

62. Rosenfeld, A. and B.S. Lipkin, "Texture Synthesis", Picture Processing and Psychopictorics, Lipkin and Rosenfeld (Editors), Academic Press, New York, 1970, pp. 309-345.

63. Rosenfeld, A. and M. Thurston, "Edge and Curve Detection for Visual Scene Analysis", IEEE Transactions on Computers, Vol. C-20, N° 5, May 1971, pp. 562-569.

64. Rosenfeld, Azriel and Eleanor Troy, "Visual Texture Analysis", Technical Report 70-116, University of Maryland, College Park, Maryland, June 1970. Also in Conference Record for Symposium on Feature Extraction and Selection in Pattern Recognition (IEEE Publication 70C-51C), Argonne, Illinois, October 1970, pp. 115-124.

65. Schacter, B.J., Rosenfeld, A. and Davis, L.S., "Random Mosaic Models for Textures", IEEE Trans. on Systems, Man, and Cybernetics, SMC-8, N° 9, 1978, pp. 694-702.

66. Serra, J., "Theoretical Bases of the Leitz Texture Analyses System", Leitz Scientific and Technical Information, Supplement 1, 4, April 1974, Wetzlar, Germany, pp. 125-136.

67. Serra, J., "One, Two, Three, ..., Infinity", Quantitative Analysis of Microstructures in Materials Science, Biology, and Medecine, J.L. Chernant (Editor), Dr. Riederer-Verlag GmbH, Stuttgart. 1978, pp. 9-24.

68. Serra, J. and G. Verchery, "Mathematical Morphology Applied to Fibre Composite Materials", Film Science and Technology, Vol. 6, 1973, pp. 141-158.

69. Shulman, A.R., "Optical Data Processing", John Wiley & Sons, Inc., New York, 1970.

70. Sutton, R. and I. Hall, "Texture Measure for Automatic Classification of Pulmonary Disease", IEEE Transactions on Computers, Vol. C-21, N° 1, 1972, pp. 667-676.

71. Swanlund, G.D., "Design Requirements for Texture Measurements", (In) Proc. Two Dimensional Digital Signal Processing Conf., October 1971.

72. Switzer, P., "Reconstructing Patterns for Sample Data", Annals Math. Stat., Vol. 38, 1967, pp. 138-154.

73. Tomita, F., M. Yachida, and S. Tsuji, "Detection of Homogeneous Regions by Structural Analysis", Proceedings of the Third International Joint Conference on Artificial Intelligence, 1973, pp. 564-571.

74. Toriwaki, J. and T. Fukumura, "Extraction of Structural Information from Grey Pictures", Computer Graphics and Image Processing, Vol. 7., N° 1, 1978, pp. 30-51.

75. Tou, J.T. and Y.S. Chang, "An approach to Texture Pattern Analysis and Recognition", Proceedings of the 1976 IEEE Conference on Decision and Control, 1976.

76. Tou, J.T. and Y.S. Chang, "Picture Understanding by Machine Via Textural Feature Extraction", Proceedings of the 1977 IEEE Conference on Pattern Recognition and Image Processing, Troy, New York, June 1977.

77. Tou, J.T., D.B. Kao, and Y.S. Chang, "Pictorial Texture Analysis and Synthesis", Third International Joint Conference on Pattern Recognition, Coronado, California, August 1976.

78. Triendl, E.E., "Automatic Terrain Mapping by Texture Recognition", Proceedings of the Eighth International Symposium on Remote Sensing of Environment, Environmental Research Institute of Michigan, Ann Arbor, Michigan, October 1972.

79. Tsai, W.H., and Fu, K.S., "Image Segmentation and Recognition by Texture Discrimination : A Syntactic Approach", (In) Fourth Int. Joint Conf. on Pattern Recognition, 1978.

80. Tsuji, S; and F. Tomita, "A Structural Analyzer for a Class of Textures", Computer Graphics and Image Processing, Vol. 2, 1973, pp. 216-231.

81. Watson, G.S., Geological Society of America Memoir 142, 1975, pp. 367-391.

82. Weska, J., C. Dyer, and A. Rosenfeld, "A Comparative Study of Texture Measures for Terrain Classification", IEEE Transactions on Systems, Man, and Cybernetics, Vol. SMC-6, N° 4, April 1976, pp. 269-285.

83. Wied, G., G. Bahr, and P. Bartels, "Automatic Analysis of Cell Images", Automated Cell Identification and Cell Sorting, Wied and Bahr (Editors), Academic Press, New York, 1970, pp. 195-360.

84. Yaglom, A.M., Theory of Stationary Random Functions, Prentice-Hall, Inc., New Jersey, 1962.

85. Zucker, S.W., "Toward a Model of Texture", Computer Graphics and Image Processing, Vol. 5, 1976a, pp. 190-202.

86. Zucker, S.W., "On the Structure of Texture", Perception, Vol. 5, 1976b, pp. 419-436.

87. Zucker, S., On the Foundations of Texture : A Transformational Approach", Technical Report TR-331, University of Maryland, College Park, Maryland, September 1974.

MOTION: ANALYSIS OF TIME-VARYING IMAGERY

A. Rosenfeld
Computer Vision Laboratory, Computer Science Center, University of Maryland, College Park, MD 20742

1 *INTRODUCTION*

This chapter discusses the processing and analysis of time sequences of images. In a sequence of images of a scene ("frames") taken at closely spaced times, there may be changes due to the motions of objects in the scene, or due to motion of the sensor relative to the scene. Even in the latter case, different parts of the image will be affected differently, depending on the distances of the corresponding parts of the scene from the sensor. Scene changes resulting from sensor motion are known as "optical flow".

Given a sequence of images, there are various ways of associating motion vectors with the pixels or the regions in the images. Such methods will be reviewed in Sections 2-3. Conversely, if we know the motion vector at each pixel, and the parts of the scene are not moving relative to one another, we can determine the motion of the sensor relative to the scene, and the relative ranges of the scene pixels relative to the sensor, as discussed in Section 4.

The literature on analysis of time-varying imagery is growing very rapidly. For reviews of large segments of this literature see Martin & Aggarwal (1978), Nagel (1978), and Nagel (1981a). Some recent collections of papers are Badler & Aggarwal (1979), Aggarwal & Badler (1980), Snyder (1981), and Huang (1981). Two papers in Huang (1981) deal with the enhancement (specifically, noise cleaning) and compression of time-varying imagery; we will not discuss these topics further here.

2 *MOTION ESTIMATION: REGION-BASED METHODS*

We first consider methods of estimating the motions of regions in the images. Such methods may be used in cases where the images can be relatively reliably segmented into regions, e.g., into objects on a background.

If the motion is large relative to the object size, so that the occurrences of a given object in two successive frames are far apart, the key problem is to determine which objects in the two frames correspond. This may not be trivial, since a segmented object may change in size or shape from frame to frame because its range and orientation have changed, or because of the effects of noise on the segmentation process. Objects may even appear or disappear between one frame and the next, due to their occlusion or disocclusion (by the edges of the frame or by other objects) or due to noise. The task of pairing off corresponding objects is a matching task which can be carried out using mathematical programming techniques to minimize a suitable cost function based on object dissimilarity and motion irregularity, or using relaxation matching methods (as discussed in a subsequent chapter).

Once the objects have been paired off, a shape matching process can be used to estimate the motion of each object. This involves finding the motion (translation and rotation) that, when applied to the version of the object in the first frame, best brings it into coincidence with the version in the second frame. If 3D rotation is not allowed, this can be determined by a search through the possible combinations of planar translation, planar rotation, and scale change. If such rotation is allowed, on the other hand, the task is more difficult, because the shape of the object as it appears in the image may change in an unpredictable way (if the 3D shape of the object is not known), and it is not clear how to allow for the effects of this.

We next consider region-based approaches in which the motion is small relative to the region (or object) size, so that there is substantial overlap between the successive images of the same object. Since the objects are assumed to be easily distinguishable from the background, by comparing the two images we can extract "difference regions" composed of pixels that belong to the object in the first image and to the background in the second image (Type I) or vice versa (Type II). An examination of these regions and their relationships to the object provides useful information about the motion. For example, if there is a Type II region adjacent to the object along one side of it, say on its right, and a Type I region adjacent to the background along the opposite (i.e., left) side of the object, this is an indication that the object is moving from left to right; the Type II region consists of background pixels that were covered up by the object when it moved from its position in the first image to its position in the second, and the Type I region similarly consists of background pixels that were uncovered. If the object is completely surrounded by a Type I region, it has grown smaller (and so has uncovered background pixels on all sides) from the first image to the second and so must be receding; and similarly if it is completely surrounded by a Type II region, it must be approaching. Of course, the shapes of the difference regions ideally should also be consistent with these interpretations. For further discussion of this method of motion analysis see Jain (1981). Note that if the object is rotating as well as translating, it becomes more difficult to derive simple conclusions from the Type I and Type II regions.

3 MOTION ESTIMATION: PIXEL-BASED METHODS

We now discuss methods of motion estimation that do not pre suppose that the image is easily segmentable into regions. Such methods attempt to determine motion vectors for pixels, not for entire regions.

In analogy with Section 2, let us first suppose that the motion is large relative to the pixel size, i.e., if P and P' are corresponding pixels in two successive frames, we expect them to be located several pixels apart. In this situation, we can attempt to locate P' by matching a neighborhood of P with the second image in the vicinity of P; we assume that P' is at the position where the match is strongest. (In principle, P' can be located to subpixel precision by fitting a surface to the match values and finding the maximum of the surface.) Depending on the types of motion that are possible, the matching process may involve searching over a range of positions, orientations, and scales; and if substantial 3D rotation is allowed, matching may not work, since the pattern of gray levels in the vicinity of P may change unpredictably from frame to frame. Note that matching will not give reliable results (i.e., there will not be a clearly defined match peak) if the image is smooth in the vicinity of P; it is best applied at points P where the image is strongly patterned. If the pattern at P is not isotropic, e.g., if P lies on an edge, the results will be reliable only as regards the component of motion in the direction across the edge, since the match peak will not be sharp in the direction along the edge. For further discussion of matching methods in time-varying imagery analysis, see Thompson & Barnard (1981) and Nagel (1981b).

The neighborhood matching approach just described assumes that P and its neighbors are all moving in the same way. The larger the neighborhood used, the less likely this will be; in particular, if the neighborhood contains an edge, it may be the edge of a moving object, so that the motion on the two sides of the edge is not the same. On the other hand, if the neighborhood used is too small, or if it does not contain edges, it becomes difficult to obtain reliable matches. We shall now consider a different approach to motion estimation which makes use of very small neighborhoods (immediate neighbors only), and which is appropriate when the motion from one frame to the next is on the order of the pixel size.

From calculus we recall that

$$\frac{df}{dt} = \frac{\partial f}{\partial x}\frac{dx}{dt} + \frac{\partial f}{\partial y}\frac{dy}{dt} \tag{1}$$

Let f be the image gray level, which is a function of position (x,y) and times (t). Assume that the derivatives can be approximated by differences, so that $\frac{df}{dt}$ is the change in gray level at a given pixel from one frame to the next (i.e., in unit time), and $\frac{\partial f}{\partial x}$ and $\frac{\partial f}{\partial y}$ are the differences in gray level between the given pixel and its horizontal and vertical neighbors (assumed to be at unit distance away). Then (1) defines a constraint on the x and y components of the velocity (at the given pixel), $\frac{dx}{dt}$ and $\frac{dy}{dt}$, in terms of the observed quantities $\frac{df}{dt}$, $\frac{\partial f}{\partial x}$, and $\frac{\partial f}{\partial y}$.

Note that if $\frac{\partial f}{\partial x}$ and $\frac{\partial f}{\partial y}$ are (close to) zero, (1) gives us no (accurate) information about $\frac{dx}{dt}$ and $\frac{dy}{dt}$; this corresponds to the intuitive observation that motion cannot be accurately determined at a pixel where the gray level is constant or slowly varying. If just one of them, say $\frac{\partial f}{\partial x}$, is (close to) zero, (1) becomes $\frac{df}{dt} \doteq \frac{\partial f}{\partial y}\frac{dy}{dt}$, so that $\frac{dy}{dt}$ is determined ($= \frac{df}{dt}/\frac{\partial f}{\partial y}$), while $\frac{dx}{dt}$ is unknown. This corresponds to the intuitive remark that at an edge, say in the horizontal direction ($\frac{\partial f}{\partial x} \doteq 0$, $\frac{\partial f}{\partial y} \neq 0$), we can detect motion in the direction across the edge, but not in the direction along the edge. In general, let g and ℓ denote the gradient and level directions at the given pixel, i.e., $g = \tan^{-1}(\frac{\partial f}{\partial y}/\frac{\partial f}{\partial x})$ is the direction in which the gray level is changing most rapidly, and ℓ is the perpendicular direction, in which there is no change. Then if we use g and ℓ instead of x and y as the spatial coordinates in (1), we have $\frac{df}{dt} \doteq \frac{\partial f}{\partial g}/\frac{dg}{dt}$, so that we can determine the component of motion in the gradient direction ($= \frac{df}{dt}/\frac{\partial f}{\partial g}$), but not in the level direction - i.e., the component across the edge, but not that along the edge.

Since (1) gives only one constraint on the velocity vectors $(\frac{dx}{dt}, \frac{dy}{dt})$, if we want to do motion estimation based on (1) we must impose an additional constraint. One approach is to assume that the velocity varies smoothly from pixel to pixel, so that its rate of change (as measured by the magnitude of the gradient or Laplacian) is small. We can then use iterative methods to find a velocity field that satisfies the constraint of (1) and has minimum rate of change. Examples of velocity field estimation by this method are shown in Figure 1 (Horn & Schunck, 1981). Note that the assumption of smooth variation is reasonable within a given object, where rotation and scale change should only give rise to small differences in velocity from one pixel to the next; but it is false at object boundaries.

Figure 1. Examples of iterative velocity field estimation. (a) Translation of (0.5,1) pixels per unit time: results after 1,4,16, and 32 iterations. (b) Rotations of a cylinder and a sphere (5° per unit time): results after 32 iterations. From Horn & Schunck, 1981.

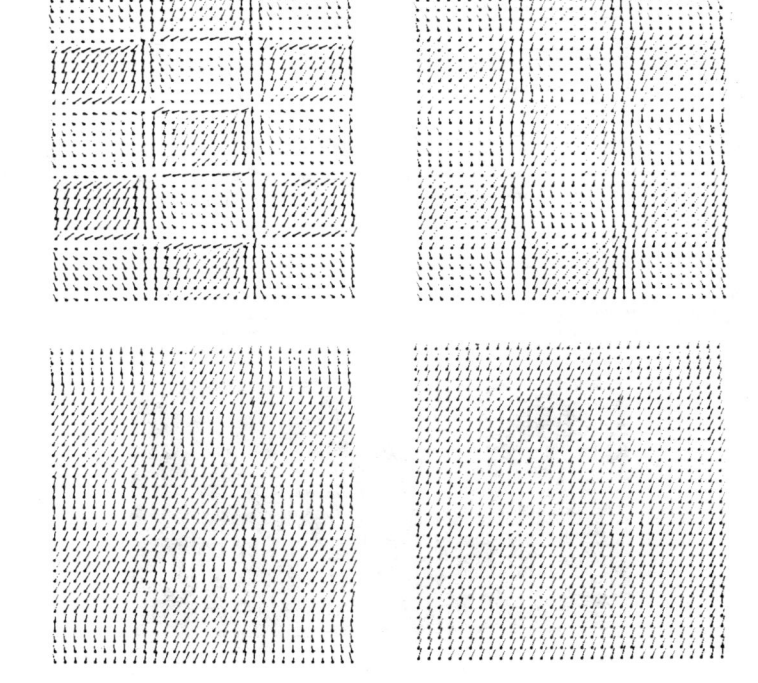

(a)

Another way to use (1) to estimate velocities is to observe that at a corner (a sharply turning edge), (1) provides information about the components of velocity in two directions (across the two edges that meet at the corner), so that the velocity at a corner is completely determined. We can then "propagate" the velocity estimate along the edges (again using the assumption that the velocity changes slowly from point to point) so as to obtain a consistent velocity field for the entire object border that agrees with the velocities at the corners. An example of a velocity field computed in this way is shown in Figure 2 (Wu et al., 1982). If desired, one could then propagate these velocities into the object interior using

Figure 1, (cont'd.)

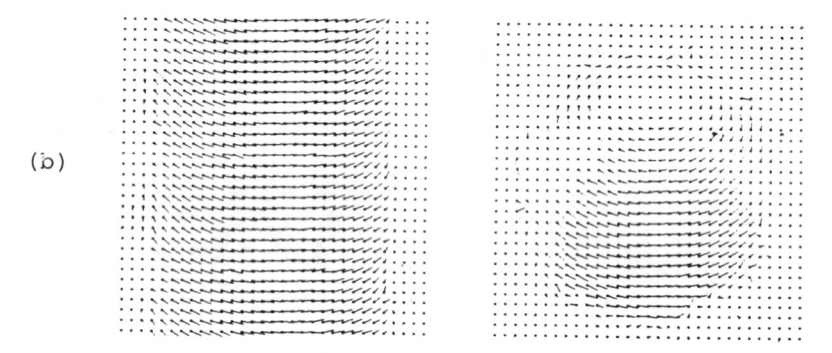

Figure 2. Velocity estimation along contours. (a) Image of a moving car (two frames), with corner points marked on first frame. (b) Velocity field along contours. From Wu, Sun, & Davis, 1982.

Motion: Analysis of time-varying imagery

the method of Horn & Schunck (1981). Note that this method requires explicit detection of edges (and corners), and so might be called an "edge-based" method.

If we know the velocity vectors at a small set of pixels that all belong to the same rigid object, the 3D motion of the object is completely determined. It has been shown (Ullman, 1981; see also Webb & Aggarwal, 1981) that it suffices to know the velocities for 5 pixels at a given time, or for 4 pixels at two successive times, in order to determine the object motion. If the motion is of some special type, of course, fewer pixels suffice; for example, if it is a 2D translation, the velocity vector is the same at every pixel of the object, and we need to know the velocity only for one pixel.

When velocities are estimated at individual pixels either by neighborhood matching or by using space and time derivatives, the estimates will generally be noisy, since they are based on small amounts of information (i.e., small neighborhoods). In the case where the motion of some particular object is a two-dimensional translation, the pixels belonging to that object will all have the same velocity; thus cluster detection (in velocity vector space) can be used to detect the existence of such an object, and its velocity can then be estimated more accurately by averaging the velocity estimates of its pixels. Similarly, if we have estimated the velocity components perpendicular to edges, we can perform

Figure 2 (cont'd.)

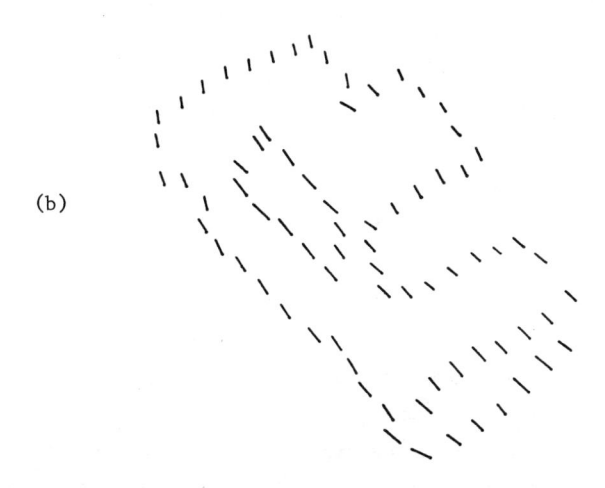

(b)

clustering in (speed, slope) space to detect sets of edge elements that have consistent velocities, and so (probably) belonging to the border of a translating object. On the use of clustering and Hough transform approaches for velocity estimation see Thompson & Barnard (1981).

Velocity estimates at the pixels of an image can be used, in conjunction with the gray level (or spectral, or textural) properties of the pixels, as an aid in segmenting the image. This in turn permits the use of region-level velocity estimation, so that cooperation among processes of segmentation and motion estimation (both pixel-based and region-based) becomes possible in principle.

4 SCENE STRUCTURE FROM SENSOR MOTION

When the sensor is moving relative to a three-dimensional scene, the pixels in the scene have different velocity vectors, depending on the distances of the corresponding scene points from the sensor and their relationships to its instantaneous axis of rotation. In this section, we discuss how, if these velocity vectors are known, the motion of the sensor and the distances of the scene points from the sensor can be determined. We assume in this section that the scene itself is stationary. For further details on this topic see Prazdny (1980, 1981).

Let us first consider the translational component of the sensor motion. This can be decomposed into a translation along the optical axis (approaching or receding from the scene) and a translation perpendicular to the optical axis. The latter translation simply causes the pixels to shift, all in the same direction, by amounts that depend on the distances of the corresponding scene points from the sensor. The translation along the axis, on the other hand, causes the image to expand or shrink, depending on whether the sensor is approaching or receding. This results in pixel velocity vectors that all pass through a common point, the "focus of expansion" (or contraction), where again the lengths of the vectors depend on distance from the sensor. Thus if we know the translational component of sensor motion, it is easy to determine the relative distance from the sensor to every point of the scene. The distances are only relative, because the effects of absolute distance and of the speed of the sensor are indistinguishable.

The rotational component of the sensor motion has more complicated effects. It too can be decomposed into two parts: rotation about the optical axis, and rotation of the optical axis about the cur-

rent true axis of rotation of the sensor. The first part causes the
images of the scene points to move along circles centered on the optical
axis. The second part is equivalent, at any given moment, to the optical axis scanning across the scene, so that the perspective of the image
changes. This has the effect of changing the scale of the image, where
the rate of change of scale is constant along each line perpendicular
to the (current) line of scan; the lines toward which the optical axis
is moving are demagnified, while those away from which it is moving are
magnified. Under this type of motion, the image points move along hyperbolae.

The effects of the translational and rotational components
of sensor motion can be regarded as independent. Based on this, a
method of decomposing the sensor motion into these components is described
in Prazdny (1981). We recall that if the motion were a pure translation,
the velocity vectors would all pass through a common point (or would all
be parallel, when the translation is perpendicular to the optical axis).

Let v_0 be any one of the velocity vectors, and let P_i be the points at
which the other vectors v_i intersect v_0; then the scatter of the points
P_i (e.g., as measured by their standard deviation) is a measure of the
extent to which the motion is not a translation. Given any set of
rotational parameters, we can adjust the velocity vectors to compensate for this (assumed) rotation. We can thus search through the set
of possible rotational parameters and find a set such that, when we compensate for that particular rotation, the scatter of the points P_i is
a minimum. These parameters must then be the actual rotational parameters, and the compensated velocity vectors represent the translational component of the motion. From this translational component,
the relative distances from the sensor to the points of the scene can
be determined, as already pointed out. An example of this approach to
determining the translational component of sensor motion is shown in
Figure 3.

5 CONCLUDING REMARKS

Much research is currently in progress on the analysis and
processing of time-varying imagery. This section has introduced a few
of the main themes of this work, but many other approaches are under

investigation. Real-time analysis of image sequences is rapidly becoming practical, and a wide variety of techniques are now available for application to this task.

REFERENCES

Aggarwal, J.K. & Badler, N.I., guest eds. (1980). Special Issue on Motion and Time-Varying Imagery. IEEE Trans. Pattern Analysis Machine Intelligence, $\underline{2}$, 493-588.
Badler, N.I., & Aggarwal, J.K. eds. (1979), Abstracts of the Workshop on Computer Analysis of Time-Varying Imagery (Philadelphia, PA).
Horn, B.K.P., & Schunck, B.G. (1981). Determining optical flow. Artificial Intelligence, $\underline{17}$, 185-203.
Huang, T.S., ed. (1981). Image Sequence Analysis. Berlin: Springer.
Jain, R. (1981). Dynamic scene analysis using pixel-based processes. Computer, $\underline{14}$, no. 8, 12-18.
Martin, W.N., & Aggarwal, J.K. (1978). Dynamic scene analysis. Computer Graphics Image Processing, $\underline{7}$, 356-374.
Nagel, H.H., (1978). Analysis techniques for image sequences. Proc. 4th Intl. Joint Conf. on Pattern Recognition, Kyoto, Japan, 186-211.
Nagel, H.H., (1981a). Image sequence analysis: what can we learn from applications? In Image Sequence Analysis, ed. T.S. Huang, pp. 19-228.
Nagel, H.H., (1981b). Representation of moving rigid objects based on visual observations. Computer, $\underline{14}$, no. 8, 29-39.

Figure 3. Estimating the translational component of sensor motion. (a) Velcity field resulting from curvilinear motion of observer relative to two planes. (b) Estimate of translation component; the +'s are estimated positions of the common point of intersection of the vectors. From Prazdny, 1982.

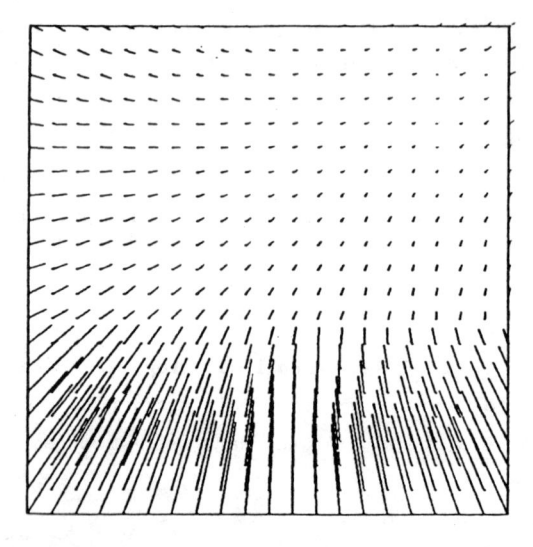

(a)

Prazdny, K. (1980). Egomotion and relative depth map from optical flow. Biological Cybernetics, 36, 87-102.
Prazdny, K. (1981). Determining the instantaneous direction of motion from optical flow generated by a curvilinearly moving observer. Proc. IEEE Conf. on Pattern Recognition and Image Processing, Dallas, Texas, 109-114.
Snyder, W.E., guest ed. (1981). Computer Analysis of Time-Varying Images. Computer, 14, no. 8.
Thompson, W.B., & Barnard, S.T. (1981). Lower-level estimation and interpretation of visual motion. Computer, 14, no. 8, 20-28.
Ullman, S. Analysis of visual motion by biological and computer systems. Computer, 14, no. 8, 57-69.
Webb, J.A., & Aggarwal, J.K. (1981). Visually interpreting the motion of objects in space. Computer, 14, no. 8, 40-46.
Wu, Z., Sun, H., & Davis, L.S. (1982). Determining velocities by propagation. Proc. 6th Intl. Joint Conf. on Pattern Recognition, Munich, Federal Republic of Germany.

Figure 3 (cont'd.)

(b)

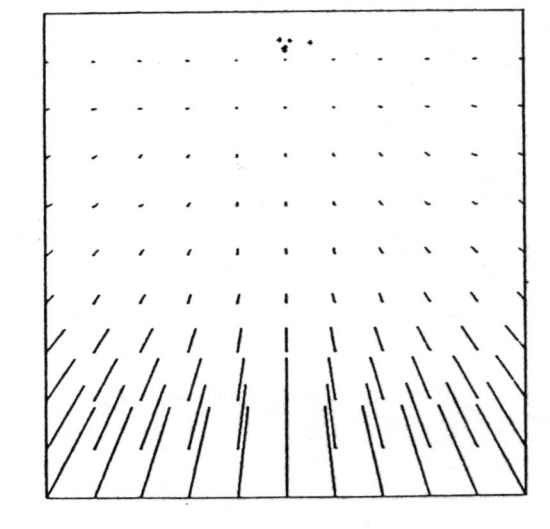

"INTRINSIC IMAGES": DERIVING THREE-DIMENSIONAL INFORMATION
ABOUT A SCENE FROM SINGLE IMAGES

A. Rosenfeld
Computer Vision Laboratory, Computer Science Center, University of Maryland, College Park, MD 20742

1 *INTRODUCTION*

The basic goal of image analysis is to construct a description of a scene on the basis of information extracted from images. Since the image is only a two-dimensional projection of the scene, constructing descriptions based on a single image is often quite nontrivial. The brightness at a point of the image depends on the illumination, reflectivity, surface slope, and distance of the corresponding point in the scene, and it is not possible to unambiguously determine these "intrinsic" scene properties from a given array of image brightnesses. Under suitable simplifying assumptions, however, constraints on these properties can be computed even from a single image.

This chapter defines a general paradigm for image analysis, and discusses the importance of three-dimensional interpretation in the analysis of many types of scenes. It then illustrates how three-dimensional information can sometimes be derived from a single image -- e.g., from the brightness variations in the image, if an appropriate model for the illumination and reflectivity is assumed; or from the two-dimensional shapes of contours in the image, if certain types of symmetry or extremality are assumed. For general reviews of image analysis see Rosenfeld & Kak (1982) and Rosenfeld (1981), and for a treatment emphasizing 3D aspects see Barrow & Tenenbaum (1981a). Many specific methods of deriving 3D information from an image are treated in Brady (1981); references on specific methods will be given later.

2 *IMAGE ANALYSIS TASKS*

The general goal of image analysis is to construct a description of the scene from which the image was obtained. The following are some typical examples of image analysis tasks. If the "scene" is a page containing written or printed characters, the desired description might be the code representations (e.g., ASCII) of the characters, in sequence. If the scene is a photomicrograph of a blood smear, the description might be a list of the numbers of various types of blood cells that are present. If the scene is a chest x-ray, the description might consist of a set of measurements specifying the size and shape of the heart or the "texture" of the lung tissue, or the locations of spots that might be tumors. In the case of an aerial or satellite image of terrain, the description might take the form of a map showing the different terrain types, land use classes, crop types, etc. that are present. Note that in all these examples, the three-dimensional nature of the scene may not play an important role; characters are two-dimensional, photomicrographs are essentially thin cross-sections, x-rays are projections, and terrain seen from a high altitude is essentially flat. Three-dimensional information plays a more crucial role in images obtained from ordinary viewing distances, e.g., showing traffic in a street or industrial parts in a bin (where the desired information might be the types and positions of the vehicles or parts). Here the three-dimensional nature of the scene becomes very significant, since the shape of an object as it appears in the image depends on its 3D orientation relative to the sensor, and objects can also partially occlude one another.

3 2D IMAGE ANALYSIS

As the examples just given illustrate, describing a scene involves identifying objects or regions that appear in it - characters, blood cells, the heart or lungs, tumors, terrain types, vehicles, industrial parts. A simplified paradigm for the image analysis process, where we have for the moment ignored the 3D aspects, is shown in Figure 1.

Figure 1. Two-dimensional image analysis.

Image	→	Features Regions	→	Properties (Shape, Texture) Relations	→	Identifications
	Segmentation		Description		Model Matching	

The first step, segmentation, extracts features (edges, curves, corners, etc.) or regions from the image; the aim is to extract parts that are compatible with the desired description of the scene. Segmentation techniques will be discussed in greater detail in later chapters.

The second step creates a description of the image in terms of properties of and relationships among the parts. This is an "objective" description involving such properties as color, texture, size, shape, etc. (Further details on shape and texture will be given in later chapters.) The results of this step can be represented by a graph or similar relational structure in which the nodes represent the parts, labelled with lists of property values, and the arcs represent relationships among parts.

In the third step, the image description is compared with a set of "models", i.e., generic descriptions that characterize classes of images. Establishing a correspondence between the image structure and a model structure gives us a "semantic" description of the image in terms of the types of objects that occur in the scene. Further material on relational structures and their comparison will be given in later chapters.

It should be pointed out that the models may be probabilistic, e.g., they may specify probability densities of property values for objects belonging to given classes; in this case, choosing the correct model involves Bayesian inference (discussed in an earlier chapter). The descriptions will often be hierarchical, e.g., describing the image in terms of parts, the parts in terms of subparts, etc.; in this case, a model might take the form of a set of "grammatical" rules, and the model matching process becomes a form of "parsing" in accordance with these rules, as will be discussed in a later chapter.

Aside from the fact that it ignores the 3D aspect, the paradigm shown in Figure 1 is also simplified in various other respects. It describes a straightforward staged process (segmentation, description, model matching), with no provision for feedback between the stages - e.g., if the measured properties are inconsistent with the model, the segmentation should be reconsidered. It also assumes that we are analyzing a single, static image; analysis of image sequences ("time-varying imagery") will be treated in a later chapter.

4 3D SCENE ANALYSIS

For scenes in which the 3D aspects are important, the 2D analysis paradigm just described is likely to be inadequate. The regions extracted by segmenting the image may not correspond to objects or parts of objects in the scene. Even if the regions are correct, the properties measured for the regions depend on the 3D orientations of the objects, which makes model matching difficult (models for the objects in all possible orientations would be needed).

A better approach in the 3D case is to determine the 3D structure of the scene before attempting to segment and describe the image. When such 3D information is available, segmentation into objects (not merely into arbitrary regions) becomes much easier, and measurement of object properties can take the orientations of the objects into account.

3D information can be derived in a number of ways. If we have two or more images taken from different positions, stereomapping techniques can be used to determine the 3D shapes of the surfaces that appear in the images; this will be discussed in a later chapter. If we have a range sensor in addition to (or instead of) a brightness sensor, it gives us 3D shape immediately. In either case, segmentation and description techniques can be applied directly to the range data, as will be discussed in a later chapter. In this section, we consider the more difficult problem of deriving 3D information about the scene from a single image on the basis of simple assumptions about the nature of the scene, e.g., on the basis of models for the illumination, the object reflectivity, etc.

Some specific methods of deriving 3D information from single images will be described in the next three sections. Figure 2 shows

Figure 2. Three-dimensional image analysis.

Image	→	3D Representation	→	3D Features and Surfaces	→	3D Properties and Relations	→	Identifications
Local (2D) Feature Extraction				Segmentation		Description		Model Matching

a possible paradigm for 3D image analysis, in which a 3D representation is first derived from the image, and processes of segmentation, description, etc. are then applied to this representation. As in the 2D case, provision should also be made for feedback between the stages.

5 SHAPE FROM SHADING

The reflectivity of a surface, i.e., the fraction of an incident ray of light (of a given wavelength) that emerges from a given point P in a given direction, depends on the angles that the incident ray i and the emergent ray e make with the normal n to the surface at P. In many cases, we can express this dependency in terms of a reflectance function of the form $r(\theta_i, \theta_e)$, where θ_i, θ_e are the angles that i and e make with n, respectively. For perfectly <u>specular</u> reflection, we have r=1 if $\theta_i = \theta_e$ and i and e are both coplanar with n, and r=0 otherwise. At the other extreme, for perfectly <u>diffuse</u> or "Lambertian" reflection, r depends only on θ_i and not on θ_e, and we have $r = \rho \cos\theta_i$, where $0 \leq \rho \leq 1$ is a constant.

Suppose the surface is illuminated by a small ("point") light source in a known position. If the source and observer are not too close to the surface, the directions θ_i and θ_o to the source and observer are essentially the same for two neighboring surface points P and P'. The brightnesses of the surface at P and P' are given by the intensities of the emergent rays from these points. Thus the change in brightness from P to P' gives us a constraint on how the surface normal has rotated as we move from P to P'.

For smooth surfaces, iterative methods can be used to derive surface orientation information from these constraints, with the aid of boundary conditions which specify the orientations of the normals at some points of the surface. For example, at an occluding contour on a smooth surface, the direction θ_o to the observer must be tangential to the surface, so that the surface normal lies in the plane perpendicular to θ_o and is orthogonal to the occluding edge; thus occluding contours are one possible source of the needed boundary conditions.

Figure 3 illustrates how surface orientation (i.e., 3D

Figure 3. Shape from shading. (a) Input image. (b) Array of estimated surface normals: initially, and after 30 iterations. From Ikeuchi & Horn, 1981.

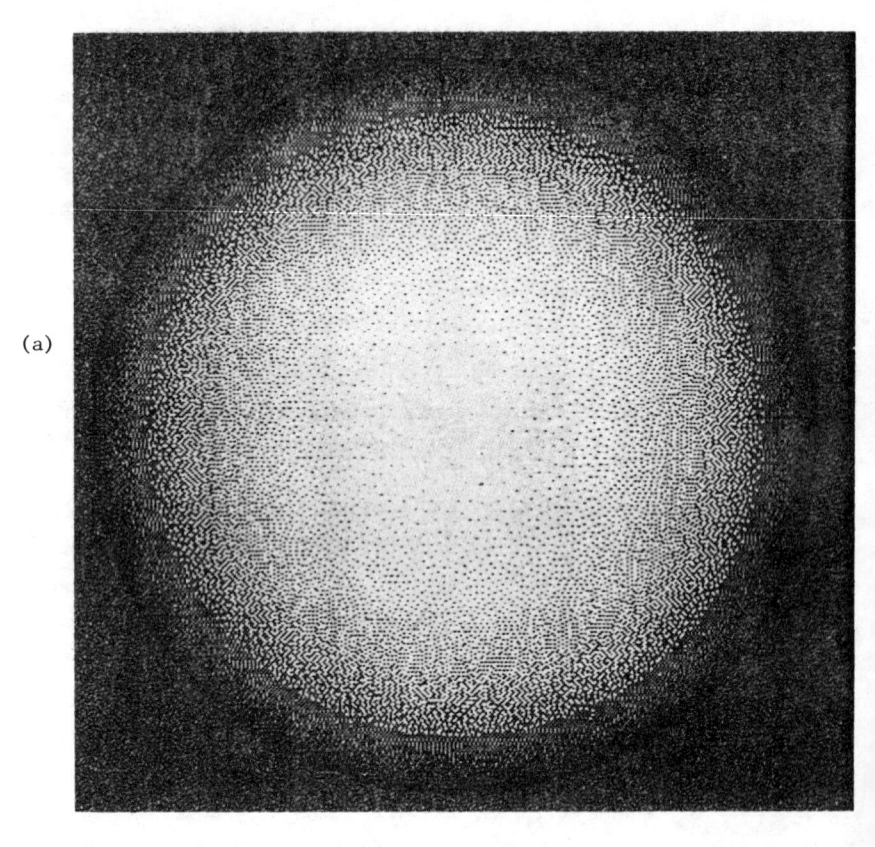

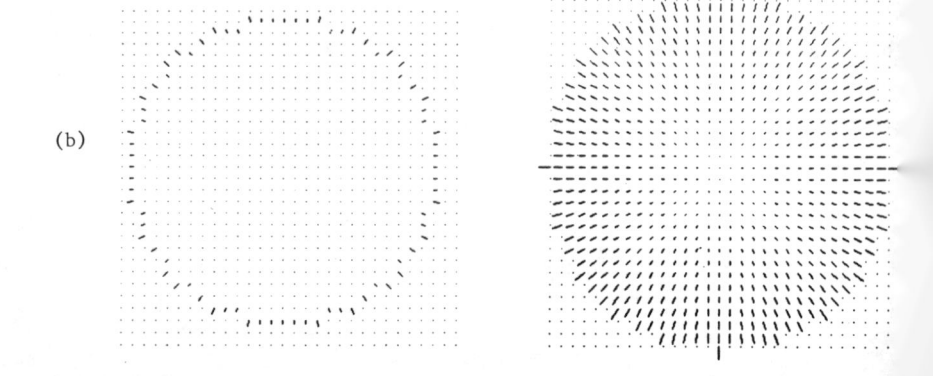

(a)

(b)

shape) can be derived from brightness variations ("shading") using this method. The input image is shown in Figure 3a. An iterative method was used to estimate surface orientation subject to the boundary conditions derived from the occluding edge. Figure 3b shows the sampled array of surface normals (displayed as line segments of unit length projected onto the image plane) obtained in this way at various stages of the iteration process.

Another way of obtaining surface orientation from brightness information is to illuminate the surface successively from two or more directions. The change in brightness at a given point P under this change in illumination gives us information about the orientation of the surface normal n at P relative to these directions, and this allows us to determine n. For further details on this "photometric stereo" method, and on the topic of shape from shading in general, see Ikeuchi & Horn (1981) and Woodham (1981).

6 SHAPE FROM EDGES OR SURFACE CONTOURS

An edge, i.e., an abrupt brightness change, in an image can have several possible causes in the original scene. It might be due to an abrupt change in illumination, i.e., it might be the edge of a shadow. Alternatively, it might be due to a change in reflectivity (i.e., in surface material), to a change in surface orientation (e.g., an edge of a polyhedron), or to a change in range (i.e., an occluding edge at which one surface is partially visible behind another). It would be useful to distinguish these types of edge from one another; for example, as we saw in Section 5, occluding edges of smooth surfaces provide useful boundary conditions for solving the shape from shading problem. Ideally, it should be possible to distinguish among the various types of edges by careful analysis of the brightness variations in their vicinities - e.g., a shadow edge might not be very sharp; a convex orientation edge might have a highlight on it; and so on. For further discussion of this topic see Binford (1981). Edges belonging to the same object may also be recognizable by virtue of having similar patterns of brightness variations (Kanade, 1981).

The 2D shapes of edges also provide important information about 3D surface shape (Barrow & Tenenbaum, 1981b), as illustrated in

Figure 4. Here a line drawing of a set of occluding edges conveys strong clues about the 3D shapes of the surfaces. If we make suitable assumptions about the nature of the surface, its shape can in fact be determined from the 2D shape of the occluding edge. For example, we might assume that the actual occluding edge is the most nearly planar and most uniformly curved space curve that could give rise to the observed edge in the image. We might further assume that the surface has the least possible curvature of all surfaces containing this space curve (i.e., it is a "soap film" surface), or is as uniformly curved as possible.

The 2D shapes of surface contours, i.e., curves that lie on the surface, also provide strong clues to surface shape (Stevens,

Figure 4. Example showing how a drawing of a small set of edges (primarily occluding edges) strongly conveys 3D information about surfaces. From Barrow & Tenenbaum, 1981.

1981), as illustrated in Figure 5. Here the key assumptions seem to involve stipulating that the principal normal to the curve not lie in the plane of the surface, and that the curve not be of least possible curvature.

7 SHAPE FROM 2D SHAPE AND FROM TEXTURE

Various global types of assumptions can also be used to deduce 3D shape from 2D shape (i.e., from shape in the image). For example (Kanade, 1981), suppose that the observed 2D shape has "skewed symmetry", i.e., it has a family of parallel chords whose midpoints are collinear (but not necessarily on a line perpendicular to the chords), as shown in Figure 6. It may be plausible to assume that

Figure 5. Example showing how surface contours convey a strong impression of 3D surface shape. From Stevens, 1981.

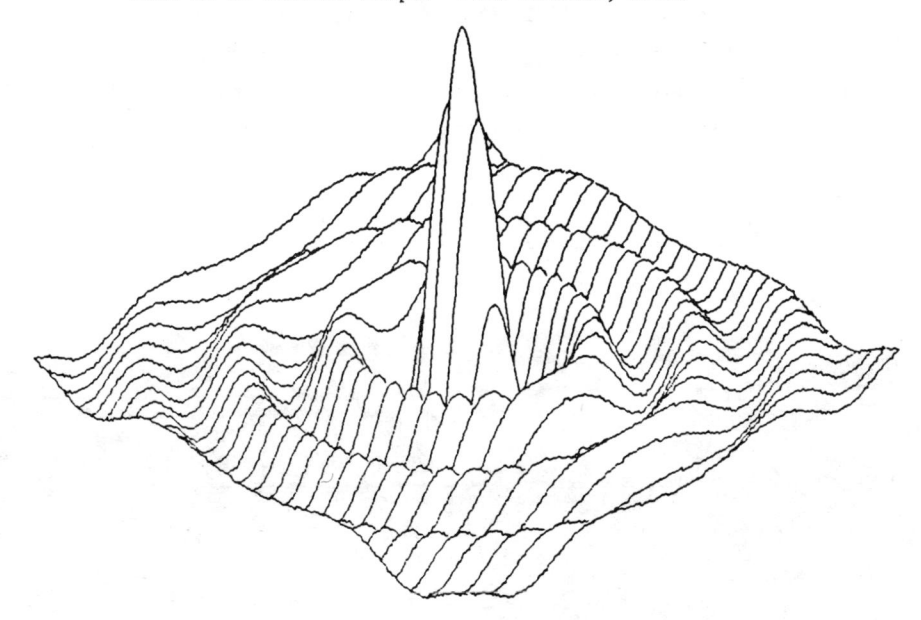

Figure 6. Examples of "skew symmetry". From Kanade, 1981.

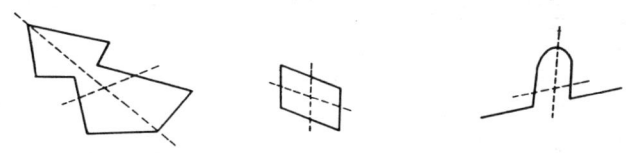

this arises from a perspective view of a symmetric shape. As another
example, it may be plausible that parallel lines in the 2D image arise
from parallel lines in space. In general, if an observed 2D shape could
have arisen from a more symmetrical shape by a 3D rotation (e.g., an
ellipse could be the image of a tilted circle), one might assume that
this is actually the case.

An analogous argument can be used to deduce 3D surface orientation from anisotropies in the texture of a region in the image (Witkin, 1981). If we assume that the surface texture is actually isotropic, the orientations of microedges in the texture should be uniformly distributed (and similarly, the spacings between facing pairs of edges should be independent of orientation). If the observed distribution is nonuniform, in such a way that the nonuniformity could have arisen from 3D rotation of an isotropically textured surface, we can assume that the surface is actually oriented in that way. (We do not really have to assume that the texture is isotropic, but only that its nonisotropy does not mimic the effects of 3D rotation.) This method of deriving surface orientation from textural anisotropy is illustrated in Figure 7.

Figure 7. Surface orientation estimation from texture anisotropy. The ellipse is the projection of a circle lying on a surface having the estimated orientation. From Witkin, 1981.

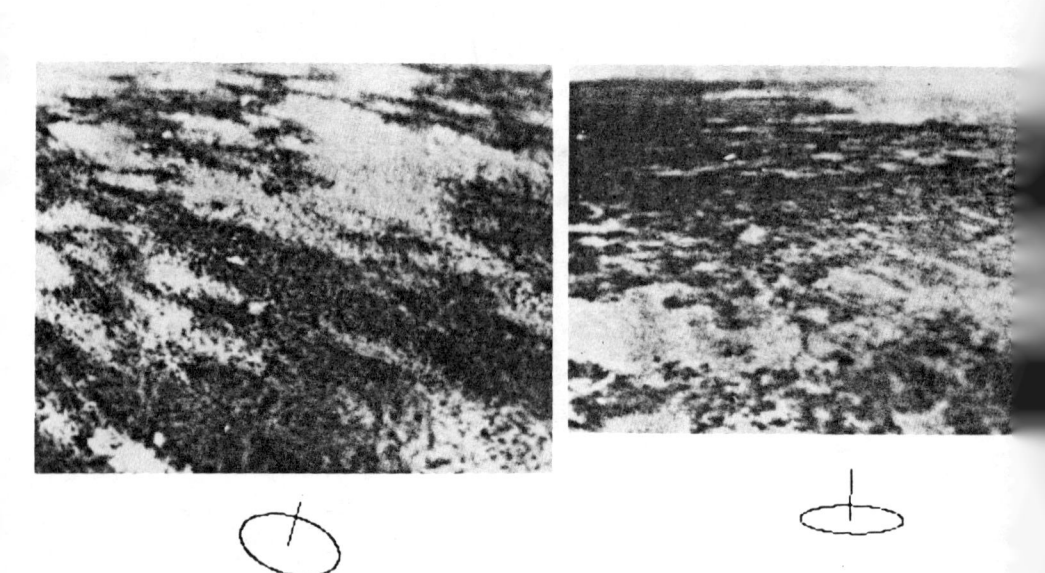

8 CONCLUDING REMARKS

In this chapter we have discussed the general problem of image analysis, with emphasis on the importance of making 3D information explicit when we analyze certain types of scenes. We have described a number of methods of plausibly inferring 3D shape from an image, based on brightness variations ("shading"), on the 2D shapes of edges or contours, or on 2D orientation distributions. Other aspects of image analysis, including segmentation (based on either 2D or 3D information), property measurement, and model matching, will be treated in subsequent chapters.

REFERENCES

Barrow, H.G. & Tenenbaum, J.M. (1981a). Computational vision. Proc. IEEE 69, 572-595.
Barrow, H.G. & Tenenbaum, J.M. (1981b). Interpreting line drawings as three-dimensional surfaces. Artificial Intelligence 17, 75-116.
Binford, T.O. (1981). Inferring surfaces from images. Artificial Intelligence 17, 205-244.
Brady, J.M., ed. (1981). Special Volume on Computer Vision. Artificial Intelligence 17.
Ikeuchi, K. & Horn, B.K.P. (1981). Numerical shape from shading and occluding boundaries. Artificial Intelligence 17, 141-184.
Kanade, T. (1981). Recovery of the three-dimensional shape of an object from a single view. Artificial Intelligence 17, 409-460.
Rosenfeld, A. (1981). Image pattern recognition. Proc. IEEE 69, 596-605.
Rosenfeld, A. & Kak, A.C. (1982). Digital Picture Processing (second edition), volume 2. New York: Academic Press.
Stevens, K.E. (1981). The visual interpretation of surface contours. Artificial Intelligence 17, 47-73.
Witkin, A.P. (1981). Recovering surface shape and orientation from texture. Artificial Intelligence 17, 17-45.
Woodham, R.J. (1981). Analysing images of curved surfaces. Artificial Intelligence 17, 117-140.

DIGITAL GEOMETRY: GEOMETRIC PROPERTIES OF SUBSETS OF
DIGITAL IMAGES

A. Rosenfeld
Computer Vision Laboratory, Computer Science Center, University of Maryland, College Park, MD 20742

1 INTRODUCTION

Geometric properties of subsets of a digital image play important roles in the description of the image. However, the measurement of such properties is complicated by the discrete nature of the image. In this chapter, we discuss how to define such concepts as connectedness, curvature, convexity, and elongatedness for digital image subsets. We also consider the generalization of these concepts to three-dimensional discrete arrays.

A general treatment of "digital geometry" in the two-dimensional case can be found in Rosenfeld & Kak (1982). Some additional references to topics not covered there, particularly as regards the three-dimensional case, will be given in later sections.

In this chapter, S denotes a subset of a digital image. We assume that S is specified by a one-bit "overlay" image which has 1's at the pixels in S and 0's elsewhere.

2 CONNECTEDNESS

We first consider the topological property of connectedness. An understanding of connectedness is important if we want to correctly count objects in an image, since "object" usually means "connected object."

Any pixel P (ignoring border effects) has four horizontal and vertical neighbors and four diagonal neighbors. For brevity, we call the former the 4-neighbors of P, and we call all eight of the neighbors (including diagonal ones) the 8-neighbors of P. There are two versions of many of the definitions in this chapter, depending on what types of neighbors are allowed. It turns out to be important, in order for certain algorithms to work correctly, to use opposite definitions for the set S (of 1's) and for its complement $\bar{S}$.

The pixels P,Q are said to be <u>connected</u> in S if there exists a sequence $P=P_0, P_1, \ldots, P_n=Q$ of pixels in S such that P_i is a neighbor of P_{i-1}, $1 \leq i \leq n$. The set of all pixels connected in S to a given P is called a connected <u>component</u> of S. We assume for convenience that S does not meet the border of the image. Thus one component of $\bar{S}$, called the background of S, contains the image border; any other components of $\bar{S}$ are called <u>holes</u> in S.

In order to count the connected components of S, we must label them, i.e., we must assign a unique label to the pixels belonging to each component; the number of labels needed to do this is the number of components. Labeling is done as follows: We scan the image row by row (moving left to right on each row). Whenever we come to a 1, say P, we check those of its neighbors that have already been visited by this scan. If none of them are 1's, we give P a new label. If some of them are 1's, and they all have the same label, we give P that label; but if they have more than one label, we give P one of their labels, say ℓ, and note that the other labels are equivalent to ℓ. When the scan is complete, we sort out the equivalences and decide on a label to represent each equivalence class; we then rescan and replace the original labels by these representatives. A simple example of how this labeling algorithm works is shown in Figure 1.

3 BORDERS AND CURVES

The <u>border</u> of S is the set of its pixels that are 4-neighbors of pixels in $\bar{S}$. (The 8-neighbor version of this definition is not ordinarily used.) Let C be a component of S and D a component of $\bar{S}$; then the <u>C,D border</u> is the set of pairs of 4-neighboring pixels (P,Q) such that $P \in C$ and $Q \in D$.

Figure 1. Labelling of 8-connected components. (a) Input (b) Result of first scan. On the second scan, the labels all change to a's. If we used 4-connectedness, the result of the first scan would be as shown in (c).

```
  1                        a                          a
1 1 1 1         a   a   b    c              b   c   d   e
  1   1 1 1         a   a   b   (b=a,c=b)       f   g   d   d   (g
       (a)                 (b)                          (c)
```

Digital geometry

Let P,Q be 4-neighboring pixels in C,D, respectively. We shall describe an algorithm that visits all such pairs in sequence and finally returns to the initial pair, so that it "follows" the (C,D)-border. Let A,B be the two 4-neighboring pixels that together with P,Q form a 2-by-2 square, and that we are facing when we stand between P and Q with P on the left (say), e.g., $\begin{smallmatrix}P&Q\\A&B\end{smallmatrix}$ (or a rotation of it by a multiple of 90°). (Keeping P on the left implies that the border defined by C and its background will be followed counterclockwise, while the borders defined by C and its holes, if any, will be followed clockwise.) Then the next pair P',Q' in the sequence is given by the following rules:

a) If we use the 4-neighbor definition for C and the 8-neighbor definition for D:

A	B	P'	Q'
1	0	A	B
1	1	B	Q
0	0,1	P	A

b) If we use the opposite definitions:

A	B	P'	Q'
1	0	A	B
0,1	1	B	Q
0	0	P	A

We can regard this algorithm as following the "cracks" between the pixels of C (regarded as unit squares) and the 4-neighboring pixels of D, while "keeping its left hand on the wall", i.e., always keeping a pixel of C on the left. When following any given crack, one is moving either north, south, east, or west; thus the sequence of moves can be represented by a sequence of 2-bit numbers, denoting the four principal directions by (e.g.) $2\begin{smallmatrix}1\\ \\3\end{smallmatrix}0$. Such a sequence will be called a crack code.

The sequence of first terms of the pairs followed by the algorithm just given contains all the pixels in C that have 4-neighbors in D; but note that the same first term may occur several times in succession (as we follow the cracks on several successive sides of a given pixel), and a given first term may even occur at two places far apart in the sequence (if C is "thin" at that pixel, so that the border first goes through it on one side and then on the other side). In any case, the successive pixels of C obtained in this way are always 8-neighbors of each other, so that the sequence of these pixels can be represented by a sequence of 3-bit numbers, denoting the eight neighbors by (e.g.) $\begin{smallmatrix} 3 & 2 & 1 \\ 4 & & 0 \\ 5 & 6 & 7 \end{smallmatrix}$. Such a sequence is called a <u>chain code</u>. An example is given in Figure 2.

The <u>perimeter</u> of C (with respect to a particular border) can be defined as the number of border pixels, or (better) as the number of moves required to follow the border. In the real plane p^2/a (perimeter squared divided by area) is a measure of the "compactness" of a region, and is least for a circle; but in the digital case (where the area is the total number of pixels in C), for either definition of perimeter, it turns out that p^2/a is smaller for certain octagons than it is for digitized circles.

A <u>curve</u> is a sequence of pixels each of which is a neighbor of the preceding. A curve is called <u>simple</u> if it never crosses or touches itself, i.e., P is a neighbor of Q iff P and Q are consecutive in the sequence. If we allow 8-neighbors, a curve is determined by specifying a starting pixel and a chain code; if we allow only 4-neighbors, we can use a crack code, where the 2-bit numbers represent the four principal directions.

4 CURVATURE AND CONVEXITY

The <u>curvature</u> of a (border or) curve is its rate of change of slope at a given pixel. Since the slope is a multiple of 45°, so is the curvature, unless we perform smoothing on the slope data. The histogram of slopes provides an indication of directional trends in

Figure 2. For the set C shown in Figure 1a, using the 8-neighbor definition, the sequence of points around C's border with its background, starting with C's rightmost pixel, begins as shown in (a), and then returns through 9=5, 10=4,11 and 12=2 to 1. The corresponding chain code is 544353177171. For C's hole border, starting with the uppermost pixel, the sequence is as shown in (b), and the chain code is 7531.

```
           8                                  1
        7    5    11   1                    4    2
           6    4    3    2                    3

           (a)                                 (b)
```

Digital geometry

the curve, and the histogram of curvatures provides an indication of whether the curve is smooth or wiggly. Figure 3 shows a curve, its chain code (starting pixel underlined), and its slope and curvature histograms (without smoothing). Another source of useful information about a curve is its one-dimensional discrete Fourier transform (where the curve is given in terms of parametric equations, $x=f(t)$, $y=g(t)$, or in terms of its chain code, which is essentially its intrinsic equation, specifying slope as a function of arc length); here the

Figure 3. (a) Curve (starting point underlined). (b) Chain code. (c) Slope histogram. (d) Curvature histogram.

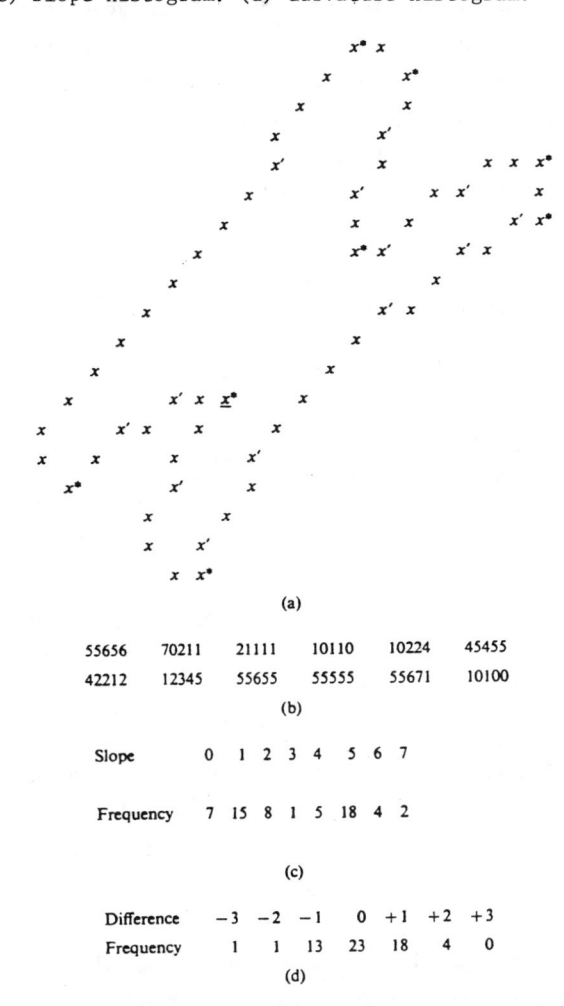

```
                              x• x
                           x         x•
                         x             x
                           x'         x'
                       x'          x         x  x  x•
                   x             x'     x x'          x
                 x             x    x             x'  x•
                   x         x• x'        x'  x
               x                       x
             x                    x'  x
           x                   x
         x              x' x x•       x
       x      x'  x       x       x
     x    x         x       x'
   x•             x'      x
               x      x
             x    x'
           x   x•
                              (a)

      55656    70211    21111    10110    10224    45455
      42212    12345    55655    55555    55671    10100
                              (b)

      Slope        0  1  2  3  4  5  6  7

      Frequency    7  15 8  1  5  18 4  2

                              (c)

      Difference   -3   -2   -1    0   +1   +2   +3
      Frequency     1    1   13   23   18    4    0
                              (d)
```

strength of the high frequencies is an indication of wiggliness, and the occurrence of a very strong coefficient is an indication of periodicity or symmetry.

Pixels at which the curvature is high, or is a local maximum, are good "corners" at which to put the vertices of an approximating polygon; they are starred in Figure 3. Pixels at which the curvature changes sign are (digital) inflections, which separate the curve into convex and concave pieces; they are primed in Figure 3.

In the real plane, set S is convex if for all points P,Q in S, the straight line segment $\overline{PQ}$ is contained in S. Necessary and sufficient conditions have recently been given (Rosenfeld & Kim, 1982) for a subset S of a digital image to be the digitization of a real convex set; these conditions depend on the definition of digitization that is used, and they are also closely related to the conditions for a chain code to be the digitization of a real straight line segment. The convex hull of S is the smallest convex set that contains S; it is the union of S and its concavities.

5 ELONGATEDNESS, SHRINKING, AND THINNING

S is extrinsically elongated if its extent in one direction is much greater than in another; but one really wants to define an intrinsic concept of elongatedness, in which a snake remains elongated even when it is coiled up. To do so, we introduce operations of shrinking and expanding. (These are special cases of operations used in mathematical morphology, which is the topic of another chapter.) Let $S^{(1)}$ denote the result of "expanding" S by changing 0's to 1's when they have 1's as neighbors, and let $S^{(k)}$ denote the result of repeating this k times. Similarly, let $S^{(-1)}$ be the result of "shrinking" S by changing 1's to 0's when they have 0's as neighbors, and let $S^{(-k)}$ be the result of repeating this k times.

Shrinking and expanding do not commute; but for all h,k we have $(S^{(-k)})^{(h)} \subseteq S^{(h-k)} \subseteq (S^{(h)})^{(-k)}$, where $S^{(0)} \equiv S$. To see that the containment relations may be strict, note that if we first shrink S, parts of it may disappear completely, so they cannot reappear when we reexpand since they have nowhere to expand from. Thus small parts of S can be eliminated by shrinking it and then reexpanding it by the same amount. Similarly, if we first expand S, parts of it may "fuse"

Digital geometry

into a large piece which can only get smaller, but not entirely disappear, when we reshrink. These situations are illustrated in Figures 4 and 5.

Figure 4. Elimination of small parts by shrinking and reexpanding.
(a) S. (b) $S^{(-1)}$. (c) $(S^{(-1)})^{(1)}$.

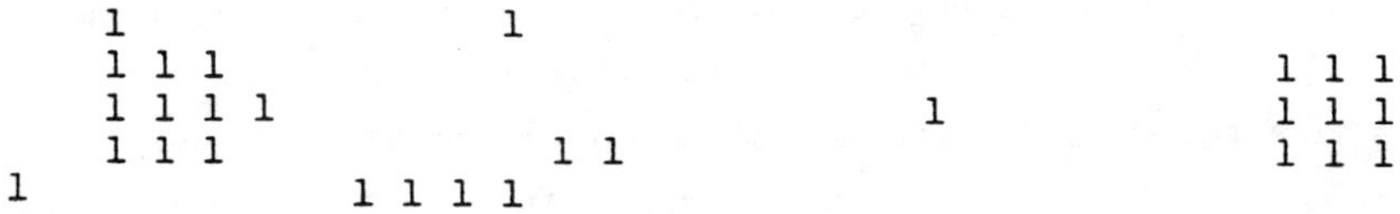

```
            1                                               1 1 1
    1 1 1                                                   1 1 1
    1 1 1 1                              1                  1 1 1
    1 1 1                 1 1
1                 1 1 1 1

        (a)                 (b)                  (c)
```

Figure 5. Fusion of clusters by expanding and reshrinking. (a) S. (b) $S^{(1)}$. (c) $(S^{(1)})^{(-1)}$; points in the original S are underlined.

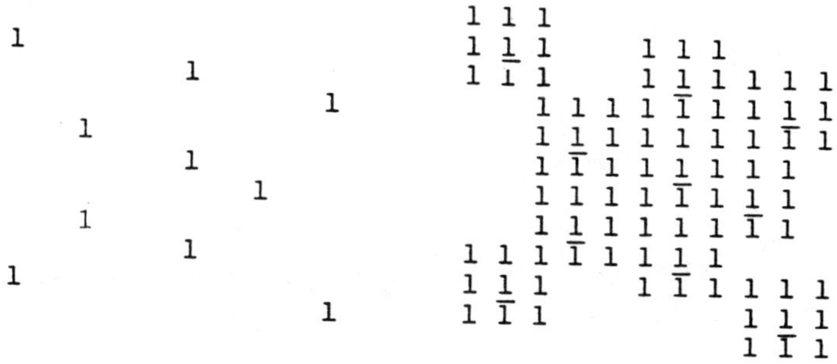

```
                              1 1 1
    1                         1 1 1        1 1 1
            1                 1 1 1            1 1 1 1 1 1
        1            1     1 1 1 1 1 1 1 1 1
                           1 1 1 1 1 1 1 1 1
        1                  1 1 1 1 1 1 1 1
            1        1     1 1 1 1 1 1 1 1
    1                      1 1 1 1 1 1 1 1
            1              1 1 1 1 1 1 1 1
1                          1 1 1
                           1 1 1           1 1 1 1 1 1
                    1      1 1 1                    1 1 1
                                                    1 1 1

        (a)                      (b)
```

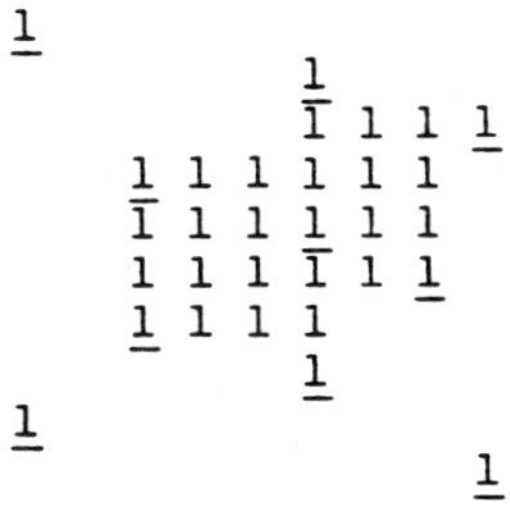

```
            1
                        1
                        1 1 1 1
            1 1 1 1 1 1
            1 1 1 1 1 1
            1 1 1 1 1 1
            1 1 1 1
                        1
        1
                                1

                    (c)
```

Let C be a component of $S-(S^{(-k)})^{(k)}$. Since C entirely disappeared under k steps of shrinking, it is at most 2k "wide". If its area is (say) at least $10k^2$, then its "length" (=area/width) is at least 5k, i.e., at least 2½ times its width, so it is an elongated part of S, as illustrated in Figure 6. By detecting such C's for k=1,2,... we can find elongated parts of S of all possible widths.

If S is elongated, one sometimes wants to approximate it by "thinning" it into a set of curves without changing its connectedness. A thinning algorithm can be defined as follows: A border pixel P∈S is called <u>simple</u> if it has more than one neighboring 1, and if changing it from 1 to 0 does not disconnect its set of neighboring 1's, regarded as a subset of its 3-by-3 neighborhood. To thin S, we simultaneously delete all its simple pixels that are on its north border (i.e., whose north neighbors are 0's); we then repeat this on the south, east, west, north, south, east, west,... until there is no further change. An example of thinning, using the 8-neighbor definition, is shown in Figure 7.

Figure 6. Elongated part detection. (a) S. (b) $S^{(-1)}$. (c) $(S^{(-1)})^{(1)}$. (d) $S-(S^{(-1)})^{(1)}$. The large component is an elongated part of S.

```
1 1 1 1                                              1 1 1 1
1 1 1 1 1 1 1 1 1            1 1                     1 1 1 1
1 1 1 1 1 1 1 1 1            1 1                     1 1 1 1
1 1 1 1                                              1 1 1 1
1

      (a)                    (b)                       (c)
```

```
              1           1 1 1 1
              1           1 1 1 1
        1

              (d)
```

Digital geometry

Figure 7. Thinning. (a) Original. (b-e) Results of thinning from the north, south, east, and west. At the next north step, the uppermost 1 will be deleted; at the next south step, the leftmost lowermost 1; and at the next west step, the leftmost uppermost 1.

```
                    1
                    1 1
                    1 1 1
                    1 1 1 1                    1
                    1 1 1 1 1              1 1
                    1 1 1 1 1 1          1 1 1
                    1 1 1     1 1 1 1
                    1 1 1         1 1 1
            1 1 1 1 1 1 1 1 1 1 1 1
        1 1 1 1 1 1 1 1 1 1     1 1 1
        1 1 1 1 1 1 1 1             1 1 1
                    1 1 1               1 1 1
                    1 1 1
                    1 1 1
                    1 1 1 1 1 1
                    1 1 1 1 1 1
                    1 1 1 1 1 1
                           (a)
```

```
    1
    1 1
    1 1 1
    1 1 1 1                          1
    1 1 1 1 1                      1 1
    1 1 1       1 1           1 1
    1 1 1             1 1 1
    1 1 1             1 1 1
1 1 1 1 1 1 1 1 1       1 1
1 1 1 1 1 1 1 1             1 1
    1 1 1                       1 1
    1 1 1
    1 1 1
    1 1 1
    1 1 1 1 1 1
    1 1 1 1 1 1
          (b)
```

```
    1
    1 1
    1 1 1
    1 1 1 1                         1
    1 1 1 1 1                     1
    1 1 1             1 1     1 1
    1 1 1             1 1 1
    1 1 1             1 1     1 1
1 1 1 1 1 1 1 1                   1
    1 1 1                           1
    1 1 1
    1 1 1
    1 1 1
    1 1 1 1 1 1
          (c)
```

```
    1
    1 1
    1 1 1 1                       1
    1 1       1                 1
    1 1             1 1
    1 1                   1 1
    1 1                   1 1
1 1 1 1 1 1 1 1           1
    1 1                           1
    1 1
    1 1
    1 1
    1 1
    1 1 1 1 1 1
          (d)
```

```
    1
    1 1 1                            1
    1                 1            1
    1                       1 1
    1                             1
    1                       1 1
1 1 1 1 1 1 1 1                    1
    1                                    1
    1
    1
    1
    1
    1 1 1 1 1
          (e)
```

6 3D DIGITAL GEOMETRY

The concepts in this chapter can all be generalized to the three-dimensional case, where we are given a 3D array of 0's and 1's, and S denotes the set of 1's. In 3D, a "voxel" (short for "volume element") has six neighbors in the three principal directions (north/south, east/west/, up/down); if we think of it as a unit cube, these are the neighbors with which it shares a face. It has twelve additional neighbors with which it shares an edge, and eight others with which it shares a corner, for a total of 6+12+8=26. In defining digital connectedness (Rosenfeld, in press), one must specify whether one is using the 6-neighbor, 18 (=6+12)-neighbor, or 26-neighbor definition. For any of these definitions, the concepts of component labeling and counting generalize straightforwardly to 3D. Defining "holes" is more complicated, on the other hand, since there are now two kinds - "cavities" in S (totally surrounded by S), which correspond to components of $\overline{S}$ as in the 2D case, and "holes" in the sense of the hole in a ring, which are much harder to define. (Intuitively, S has no holes if any closed curve in S can be repeatedly shortened, by local changes, until it "shrinks to a point".)

The border of S in 3D is now a set of surfaces rather than a set of curves, and there is no analog of our sequential border-following algorithm. Instead, one can introduce a partial ordering onto the set of border voxels, and traverse a given border surface systematically (with some repetition) using a tree traversal algorithm; see Artzy et al. (1981) for the details. (Simple) curves can be defined in 3D as well as in 2D; a more difficult question is that of defining (simple) surfaces, on which see Morgenthaler & Rosenfeld (in press). The concept of digital convexity also generalizes straightforwardly to 3D, but the criteria for characterizing S's that are digitizations of (real) convex bodies become more complicated (Kim & Rosenfeld, 1981). Shrinking and expanding operations can also be easily extended to 3D, but thinning in 3D is more complicated than in 2D; see Lobregt et al. (1980), Tsao & Fu (1981), and Morgenthaler (in press) on 3D thinning algorithms.

7 CONCLUDING REMARKS

Because of the discrete nature of digital images, care must be taken in applying the definitions of various geometric properties to such images; but useful versions of these defintions can usually be given. The situation becomes more complicated if we want to define geometric properties of 3D arrays; but here too workable definitions can be given, although in some areas, this is still a subject of active research.

REFERENCES

Artzy, E., Frieder, G., & Herman, G.T. (1981). The theory, design, implementation and evaluation of a three-dimensional surface detection algorithm. Computer Graphics Image Processing, 15, 1-24.

Kim, C.E., & Rosenfeld, A. (1981). Convex digital solids. Proc. IEEE Conf. on Pattern Recognition and Image Processing, Dallas, Texas, 175-181.

Lobregt, S., Verbeek, P.W., & Groen, F.C.A. (1980). Three-dimensional skeletonization: principle and algorithm. IEEE Trans. Pattern Analysis Machine Intelligence, 2, 75-77.

Morgenthaler, D.G. (in press). Three-dimensional simple points: serial erosion, parallel thinning, and skeletonization. Information Control.

Morgenthaler, D.G., & Rosenfeld, A. (in press). Surfaces in three-dimensional digital images. Information Control.

Rosenfeld, A. (in press). Three-dimensional digital topology. Information Control.

Rosenfeld, A., & Kak, A.C. (1982). Digital Picture Processing (second edition), Chapter 11. New York: Academic Press.

Rosenfeld, A., & Kim, C.E. (1982). How a digital computer can tell that a line is straight. Amer. Math. Monthly, 89, 230-235.

Tsao, Y.F. & Fu, K.S. (1981). A parallel thinning algorithm for 3-D pictures. Computer Graphics Image Processing, 17, 315-331.

IMAGE SEGMENTATION SURVEY

Robert M. Haralick
Virginia Polytechnic Institute and State University, Dept. of Electrical, Engineering and Dept. of Computer Science, Blacksburg, VA 24061 USA

Abstract. There is no complete theory for image segmentation although there are a variety of techniques for segmenting images. This paper discusses the major ideas behind the measurement space clustering, single linkage, hybrid linkage, region growing, spatial clustering, and split and merge- techniques.

INTRODUCTION

What should a good image segmentation be? Regions of an image segmentation should be uniform and homogeneous with respect to some characteristic such as gray tone or texture. Regions interiors should be simple and without many small holes. Adjacent regions of a segmentation should have significantly different values with respect to the characteristic on which they are uniform. Boundaries of each segment should be simple, not ragged, and must be spatially accurate.

Achieving all these desired properties is difficult because strictly uniform and homogeneous regions are typically full of small holes and have ragged boundaries. Insisting that adjacent regions have large differences in values can cause regions to merge and boundaries to be lost.

There is no theory of image segmentation. Image segmentation techniques are basically ad-hoc and differ precisely in the way they emphasize one or more of the desired properties and in the way they balance and compromise one desired property against another. Image segmentation techniques can be classified as: measurement space clustering, single linkage schemes, hybrid linkage schemes, region growing/centroid linkage schemes, spatial clustering schemes, and split and merge schemes.

The remainder of the paper briefly describes the main ideas behind the major image segmentation techniques. Additional image segmentation surveys can be found in Zucker (1976), Riseman and Arbib (1977), Kanade (1980), and Fu and Mui (1981).

MEASUREMENT SPACE CLUSTERING

This technique for image segmentation uses the measurement space clustering process to define a partition in measurement space. Then each pixel is assigned the label of the cell in the measurement space partition to which it belongs. The segments are defined as the connected components of the pixels having the same label. Because of the large number of pixels in an image, clustering using the pixel as a unit and comparing each pixel value with every other pixel value can require excessively large computation time. Iterative partition rearrangement schemes, such as ISODATA, have to go through the image data set many times and if done so without sampling can also take excessive computation time. Histogram mode seeking, because it requires only one pass through the data, probably involves the least computation time of the measurement space clustering techniques.

Histogram mode seeking is a measurement space clustering process in which it is assumed that homogeneous objects on the image manifest themselves as the clusters in measurement space. Image segmentation is accomplished by mapping the clusters back to the image domain where the maximal connected components of the mapped back clusters constitute the image segments. For images which are single band images, calculation of this histogram in an array is direct. The measurement space clustering can be accomplished by determining the valleys in this histogram and declaring the clusters to be the interval of values between valleys. A pixel whose value is in the ith interval is labeled with index i and the segment it belongs to is one of the connected components of all pixels whose label is i.

Ohlander (1975) recursively uses this idea. He begins by defining a mask selecting all pixels on the image. Given any mask, a histogram of each band of the masked image is computed. Measurement space clustering enables the separation of one mode of the histogram set from another mode. Pixels on the image are then identified with the mode to which they belong. Then each connected component of all pixels with the same mode is, in turn, used to generate a mask which during successive iterations selects pixels in the histogram computation process.

For ordinary color images, Ohta, Kanade, and Sakai (1980) suggest that histograms not be computed with the red, green, and blue (R,G, and B), but with a set of variables closer to what the Karhunen Loeve transform would suggest. They suggest $(R + G + B)/3$, $(R - B)/2$ and $(2G - R - B)/4$.

Weszka and Rosenfeld (1978) describe one way for segmenting white blobs against a dark background by a threshold selection based on busyness. Panda and Rosenfeld (1978) suggest another approach for segmenting the white blob against a dark background. Pixels having high gradients and gray levels in the valley between the two histogram modes are likely to be edge pixels. Non edge pixels are those with low gradient values and either high or low gray levels. The segments are the connected components of the non edge pixels.

Watanabe (1974) suggests choosing a threshold value which maximizes the sum of gradients taken over all pixels whose gray level equals the threshold value. A survey of threshold techniques can be found in Weszka (1978)

For multiband images such as LANDSAT, determining the histogram in a multi-dimensional array is not feasible. For example, in a six band image where each band has intensities between 0 and 99, the array would have to have $100^6 = 10^{12}$ locations. A large image might be 10,000 pixels per row by 10,000 rows. This only constitutes 10^8 pixels, a sample too small to estimate probabilities in a space of 10^{12} values were it not for some constraints of reality: (1) there is typically a high correlation between the band to band pixel values and (2) there is a large amount of spatial redundancy in image data. Both these factors create a situation in which the 10^8 pixels can be expected to contain only between 10^4 and 10^5 distinct 6-tuples. Based on this fact, the counting required for the histogram is easily done by hashing the 6-tuple into an array.

Clustering using the multidimensional histogram is more difficult than univariate histogram clustering. Goldberg and Shlien (1977, 1978) threshold the multidimensional histogram to select all N-tuples situated on the most prominent modes. Then they perform a measurement space connected components on these N-tuples to collect together all the N-tuples in the top of the most prominent modes. These measurement space connected sets form the cluster cores. The clusters are defined as the set of all N-tuples closest to each cluster core.

An alternate possibility (Narendra and Goldberg, 1977) is to locate peaks in the multi-dimensional measurement space and region grow around it constantly descending from the peak. The region growing includes all successive neighboring N-tuples whose probability is no higher than the N-tuple from which it is growing. Adjacent mountains meet in their common valleys.

SINGLE LINKAGE IMAGE SEGMENTATION

Single linkage image segmentation schemes regard each pixel as a node in a graph. Neighboring pixels whose properties are similar enough are joined by an arc. The image segments are maximal sets of pixels all belonging to the same connected component. Single linkage image segmentation schemes are attractive for their simplicity. They do, however, have a problem with chaining, because it takes only one arc leaking from one region to a neighboring one to cause the regions to merge.

The simplest single linkage scheme defines similar enough by pixel difference. Two neighboring pixels are similar enough if the absolute value of the difference between their gray tone intensity values is small enough. Bryant (1979) defines similar enough by reference to the quantity (square root of 2) times the root mean square value of neighboring pixel distances taken over the entire image.

For pixels having vector values, the obvious generalization is to use a vector norm of the pixel difference vector. Instead of using a Euclidean distance, Asano and Yokoya (1981) suggest that two pixels be joined together if this absolute value of their difference is small enough compared to the average absolute value of the center pixel minus neighbor pixel for each of the neighborhoods the pixels belong to. Haralick and Dinstein (1975), however, do report some success using the simpler Euclidean distance on LANDSAT data. They, as did Perkins (1980), region grew the edge pixels in order to close gaps. The ease with which unwanted region chaining can occur with this technique limits its potential on complex or noisy data.

Hybrid single linkage techniques are more powerful than the simple single linkage technique. The hybrid techniques seek to assign a property vector to each pixel where the property vector depends on the KxK neighborhood of the pixel. Pixels which are similar, are similar because their neighborhoods in some special sense are similar. Similarity is thus established as a function of neighboring pixel values and this makes the technique better behaved on noisy data.

One hybrid single linkage scheme relies on an edge operator to establish whether two pixels are joined with an arc. Here an edge operator is applied to the image labeling each pixel as edge or non-edge. Neighboring pixels, neither of which are edges, are joined by an arc. The initial segments are the connected components of the non-edge labeled pixels. The edge pixels can either be left assigned edges and be considered as background or they can be assigned to the spatially nearest region having a label.

The quality of this technique is highly dependent on the edge operator used. Simple operators such as the Roberts and Sobel operator may provide too much region linkage, for a region cannot be declared as a segment unless it is completely surrounded by edge pixels. Yakimovsky (1976) uses a maximum likelihood test to determine edges. Edges are declared to exist between pairs of regions if the hypothesis that their means are equal and their variances are equal has to be rejected.

Haralick (1980) suggests fitting a plane to the neighborhood around the pixel, and testing the hypothesis that the slope of the plane is zero. To determine roof or V-shaped edge, Haralick suggests fitting plane to the neighborhoods on either side of the pixel and testing th hypothesis that the coefficients of fit are identical. Haralick (1982) discusses a very sensitive zero-crossing of second directional derivative edge operator. In this technique, each neighborhood is fitted by least squares with a cubic polynomial in two variables. The first and second partial derivatives are easily determined from the polynomial. The first partial derivatives at the center pixel determine the gradient direction. With the direction fixed to be the gradient direction, the second partials determine the second directional derivative. If in the gradient direction, the second directional derivative has a zero-crossing inside the pixel, then an edge is declared in the neighborhood's center pixel.

Another hybrid technique first used by Levine and Leemet (1976) is based on the Jarvis and Patrick (1973) shared nearest neighbor idea. Using any kind of reasonable notion for similarity, each pixel examines its KxK neighborhood and makes a list of the N pixels in the neighborhood most similar to it. Call this list the similar neighbor list, where we understand neighbor to be any pixel in the KxK neighborhood. An arc joins any pair of immediately neighboring pixels if each is in each other's shared neighbor list and if there are enough pixels common to their shared neighbor lists; that is, if the number of shared neighbors is high enough.

REGION GROWING / CENTROID LINKAGE

In region growing, as contrasted to single linkage, pairs of neighboring pixels are not compared for similarity. Rather, the image is scanned in some predetermined manner such as left-right top bottom. A pixel's value is compared to the mean of an already existing but not necessarily completed segment. If the values are close enough, then the pixel is added to the segment and the segment's mean is updated. If no neighboring region has its mean close enough, then a new segment is established having the given pixel's value as its first member.

Pavlidis (1972) suggests a more general version of this idea. Given an initial segmentation where the regions are approximated by some functional fit guarranteed to have a small enough error, pairs of neighboring regions can be merged if for each region the sum of the squares of the differences between the fitted coefficients for each region and the corresponding averaged coefficients, averaged over both regions, is small enough. Pavlidis gets his initial segmentation by finding the best way to divide each row of the image into segments with a sufficiently good fit. He also describes a combinatorial tree search algorithm to accomplish the merging which guarantees a best result.

Gupta, Kettig, Landgrebe, and Wintz (1973) suggest using a t-test based on the absolute value of the difference between the pixel and the region near as the measure of dis-similarity. Kettig and Landgrebe (1975) discuss the multi-band situation leading to the F-test and report good success with LANDSAT data.

Nagy (1972) just examines $|y-\bar{X}|$. If this distance is small enough pixel y is added to the region. If there is more than one region, then y is added to that region with smallest distance.

The Levine and Shaheen scheme (1981) is similar. The difference is that Levine and Shaheen attempt to keep regions more homogeneous and try to keep the region scatter from getting too high. They do this by requiring the differences to be more significant before a merge takes place if the region scatter is high.

Brice and Fennema (1970) accomplish the region growing by partitioning the image into initial segments of pixels having identical intensity. They then sequentially merge all pairs of adjacent regions if a significant fraction of their common border has a small enough intensity difference across it. Muerle and Allen (1968) suggest a Kolmogorov-Smirnov test for merging one region with another.

Simple single pass approaches which scan the image in a left right top down manner are, of course, unable to make the left and right sides of a V-shaped region belong to the same segment. To be more effective, the single pass must be followed by some kind of connected components algorithm in which pairs of neighboring regions having means which are close enough are combined into the same segment.

One minor problem with region growing schemes is their inherent dependence on the order in which pixels and regions are examined. A left right top down scan does not yield the same initial regions as a right left bottom up scan or for that matter a column major scan. Usually, however, differences caused by scan order are minor.

SPATIAL CLUSTERING

It is possible to determine the image segments by combining clustering in measurement space with a spatial region growing. Such techniques are called spatial clustering. In essence, spatial clustering schemes combine the histogram mode seeking technique with a region growing or a spatial linkage technique.

Haralick and Kelly (1969) suggest it be done by locating, in turn, all the peaks in measurement space. Then determine all pixel locations having a measurement on the peak. Beginning with a pixel corresponding to the highest peak not yet processed, simultaneously perform a spatial and measurement space region growing in the following manner. Initially, each segment is the pixel whose value is on the current peak. Consider for possible inclusion into this segment the neighbors of this pixel (in general, the neighbors of the pixel we are growing from) if the neighbor's N-tuple value is close enough in measurement space to the pixel's N-tuple value and if its probability is not larger than the probability of the pixel's value we are growing from. Matsumoto, Naka, and Yamamoto (1981) discuss a variation on this idea.

Milgram (1979) defines a segment for a single band image to be any connected component of pixels all of whose values lie in some interval I and whose border has a higher coincidence with the border created by an edge operator than for any other interval I. The technique has the advantage over the Haralick and Kelly technique in that it does not require the difficult measurement space exploring of climbing down a mountain. However, it does have to try many different intervals for each segment. Extending it to efficient computation in multiband images appears difficult. However, Milgram does report good results of segmenting white blobs against a black background. Milgram and Kahl (1979) discuss embedding this technique into the Ohlander (1975) recursive control structure.

Minor and Sklansky (1981) make more active use of the gradient edge image than Milgram but restrict themselves to the more constrained situation of small convex-like segments. They begin with an edge image in which each edge pixel contains the direction of the edge. The orientation is so that the higher valued gray tone is to the right of the edge. Then each edge sends out for a limited distance a message to nearby pixels and in a direction orthogonal to the edge direction. The message indicates what is the sender's edge direction. Pixels which pick up these messages from enough different directions must be interior to a segment.

The spoke filter of Minor and Sklansky counts the number of distinct directions appearing in each 3x3 neighborhood. If the count is high enough they mark the center pixel as belonging to an interior of a region. Then the connected components of all marked pixels is obtained. The gradient guided segmentation is then completed by performing a region growing of the components. The region growing must stop at the high gradient pixels, thereby assuring that no undesired boundary placements are made.

Burt, Hong, and Rosenfeld (1981) describe a spatial clustering scheme which is a spatial pyramid constrained ISODATA kind of clustering. The bottom layer of the pyramid is the original image. Each successive high layer of the pyramid is an image having half the number of pixels per row and half the number of rows of the image below it. Initial links between layers are established by linking each father pixel to the spatially corresponding 4x4 block of son pixels. Each pair of adjacent father pixels has 8 son pixels in common. Each son pixel is linked to a 2x2 block of father pixels. The iterations proceed by assigning to each father pixel the average of his son pixels. Then each son pixel compares his value with each of his father's values and links himself to his closest father. Each father's new value is the average of the sons to which he is linked etc. The iterations converge reasonably quickly and for the same reason the ISODATA iterations converge. If the top layer of the pyramid is a 2x2 block of great grandfathers, then these are at most 4 segments which are the respective great grandson of these 4 great grandfathers. Pietikainen and Rosenfeld (1981) extend this technique to segment an image using textural features.

SPLIT AND MERGE

The split method for segmentation begins with the entire image as the initial segment. Then it successively splits each of its current segments into quarters if the segment is not homogeneous enough. Homogeneity can be easily established by determining if the difference between the largest and smallest gray tone intensities is small enough. Algorithms of this type were first suggested by Roberston (1973) and Klinger (1973). Kettig and Landgrebe (1975) try to split all non-uniform 2x2 neighborhoods before beginning the region merging. Fukada (1980) suggests successively splitting a region into quarters until the sample variance is small enough.

Because segments are successively divided into quarters, the boundaries produced by the split technique tend to be squareish and slightly artificial. Sometimes adjacent quarters coming from adjacent split segments need to be joined rather than remain separate. Horowitz and Pavlidis (1976) suggest a split and merge strategy to take care of this problem.

Chen and Pavlidis (1980) suggest using statistical tests for uniformity rather than a simple examination of the difference between the largest and smallest gray tone intensities in the region under consideration for splitting. The uniformity test requires that there be no significant difference between the mean of the region and each of its quarters. The Chen and Pavlidis tests assume that the variances are equal and known.

The data structures required to do a split and merge on images larger than 512x512 are extremely large. Execution of the algorithm on virtual memory computers results in so much paging that the dominant activity is paging rather than segmentation. Browning and

Tanimoto (1982) give a description of a split and merge scheme where the split and merge is first accomplished on mutually exclusive subimage blocks and the resulting segments are then merged between adjacent blocks to take care of the artificial block boundaries.

REFERENCES

T. Asano and N. Yokoya, "Image Segmentation Schema For Low-Level Computer Vision," Pattern Recognition, Vol. 14, No. 1, 1981, pp.267-273.

C. Brice and C. Fennema, "Scene Analysis Using Regions," Artificial Intelligence, Vol. 1, 1970, p205-226.

J.D. Browning and S.L. Tanimoto, "Segmentation of Pictures into Regions With a Tile by Tile Method," Pattern Recognition, Vol. 15, No 1, 1982, pp.1-10.

P.J. Burt, T.H. Hong, and A. Rosenfeld, "Segmentation and Estimation of Image Region Properties Through Cooperative Hierarchial Computation" IEEE Transactions on Systems, Man, and Cybernetics, Vol. SMC-11, No. 12. December 1981

J. Bryant, "On the Clustering of Multidimensional Pictorial Data", Pattern Recognition, Vol. 11, 1979, p115-125.

P.C. Chen, and T. Pavlidis, "Image Segmentation as an Estimation Problem," Computer Graphics and Image Processing, Vol. 12, 1980, pp. 153-172.

K.S. Fu and J.K. Mui, "A Survey On Image Segmentation", Pattern Recognition, Vol 13, 1981, p3-16.

Y. Fukada, "Spatial Clustering Procedures For Region Analysis", Pattern Recognition, Vol 12, 1980, p395-403.

M. Goldberg and S. Shlien, "A Clustering Scheme For Multispectral Image," IEEE Systems, Man, and Cybernetics, Vol. SMC-8, No 2, 1978, pp.86-92.

M. Goldberg and S. Shlien, "A Four-Dimensional Histogram Approach to the Clustering of LANDSAT Data," Machine Processing of Remotely Sensed Data, IEEE CH 1218-7 MPRSD, Purdue University, West Lafayette, Indiana, June 21-23, 1977, pp.250-259.

J.N. Gupta, R.L. Kettig, D.A. Landgrebe, and P.A. Wintz, "Machine Boundary Finding and Sample Classification of Remotely Sensed Agricultural Data," Machine Processing of Remotely Sensed Data, IEEE 73 CHO 834-2GE, Purdue University, West Lafayette Indiana, October 16-18, 1973, p4B-25 to 4B-35.

R.M. Haralick and G.L. Kelly, "Pattern Recognition with Measurement Space and Spatial Clustering for Multiple Image," Proceedings of IEEE, Vol. 57, No. 4, April 1969, pp.654-665.

R.M. Haralick and I. Dinstein, "A Spatial Clustering Procedure for Multi-Image Data," IEEE Circuits and Systems, Vol. CAS 22, No. 2, May 1975, pp.440-450.

R.M. Haralick, "Edge and Region Analysis For Digital Image Data", Computer Graphics and Images Processing, Vol. 12, 1980, p60-73. Also reprinted in Image Modeling a. Rosenfeld (Ed.), Academic Press, 1981 New York, p171-184.

R.M. Haralick, "The Digital Step Edge," to be published in 1982.

S.L. Horowitz and T. Pavlidis, "Picture Segmentation by a Tree Traveral Algorithm", *Journal of the Association for Computing Machinery*, Vol 23, No 2, April 1976, pp.368-388.

R.A. Jarvis and E.A. Patrick, "Clustering Using A Similarity Measure Based on Shared Near Neighbors," *IEEE Transactions on Computers*, Vol. C-22, No. 11, 1973, pp.1025-1034.

T. Kanade, "Region Segmentation: Signal vs Semantics" *Computer Graphics and Image Processing*, Vol. 13, 1980, p279-297.

R.L. Kettig and D.A. Landgrebe, "Computer Classification of Multispectral Image Data by Extraction and Classification of Homogeneous Objects". The Laboratory for Application of Remote Sensing LARS Information Note 050975, Purdue University, West Lafayette, Indiana 1975.

A. Klinger, "Data Structures and Pattern Recognition," 1st *International Joint Conference on Pattern Recognition*, Washington D.C., Oct 1973 pp.497-498.

M.D. Levine and J. Leemet, "A Method For Non-Purposive Picture Segmentation," 3rd *International Joint Conference on Pattern Recognition*, 1976.

M.D. Levine and S.I. Shaheen, "A Modular Computer Vision System For Picture Segmentation and Interpretation", *IEEE Transactions on Pattern Analysis and Machine Intelligence*, Vol. PAMI-3, No.5, Sept. 1981.

K. Matsumoto, M. Naka, and H. Yamanoto, "A New Clustering Method For LANDSAT Images Using Local Maximums of a Multi-Dimensional Histogram," *Machine Processing of Remotely Sensed Data*, IEEE CH 1637-8 MPRSD, Purdue University, West Lafayette, Indiana, June 23-26, 1981, p321-325.

D.L. Milgram "Region Extraction Using Convergent Evidence", *Computer Graphics and Image Processing*, Vol. 11, 1979, p1-12.

D.L. Milgram and D.J. Kahl "Recursive Region Extraction", *Computer Graphic and Image Processing*, Vol 9, 1979, p82-88.

L.G. Minor and J. Sklansky "The Detection and Segmentation of Blobs in Infrared Images" *IEEE Transactions on Systems Man and Cybernetics*, Vol. SMC-11, No 3, March 1981, p194-201.

J. Muerle and D. Allen, "Experimental Evaluation of Techniques for Automatic Segmentation of Objects In a Complex Scene", *Pictorial Pattern Recognition*, G. Cheng et. al Eds., Thompson Book Co., Washington D.C., 1968, p3-13.

G. Nagy and J. Tolaba, "Nonsupervised Crop Classification Through Airborne Multispectral Observations", *IBM Journal of Research and Development*, Vol 16, No.2, March 1972, P138-153.

P.M. Narendra and M. Goldberg, "A Non-parametric Clustering Scheme for LANDSAT", *Pattern Recogniton*, Vol. 9, 1977, p207-215.

P.M. Narendra and M. Goldberg "Image Segmentation with Directed Trees", *IEEE Transactions on Pattern Analysis and Machine Intelligence*, Vol. PAMI-2, No 2, March 1980, p185-191.

R. Ohlander, *Analysis of Natural Scenes*, Ph.D. dissertation, Carnegie-Mellon University, Pittsburgh, Penn, 1975.

Y. Ohta, T. Kanade, and T. Sakai, "Color Information For Region Segmentation", *Computer Graphics and Image Processing*, Vol. 13, 1980, p222-241.

D.P. Panda and A. Rosenfeld, "Image Segmentation by Pixel Classification in (Gray Level, Edge Value) Space", *IEEE Transactions on Computers*, Vol. C-27, No. 9, September 1978, p875-879.

T. Pavlidis, "Segmentation of Pictures and Maps Through Functional Approximation", Computer Graphics and Image Processing Vol. 1, 1972, p360-372.

W.A. Perkins, "Area Segmentation of Images Using Edge Points", IEEE Transactions on Pattern Analysis and Machine Intelligence, Vol. PAMI-2, No. 1, January 1, 1980, p8-15.

M. Pietikainen and A. Rosenfeld, "Image Segmentation by Texture Using Pyramid Node Linking" IEEE Transactions on Systems, Man, Cybernetics, Vol. SMC-11, No. 12, December 1981.

E. Riseman and M. Arbib, "Segmentation of Static Scenes", Computer Graphics and Image Processing, Vol. 6, 1977, p221-276.

T.V. Robertson, "Extraction and Classification of Objects In Multispectral Images," Machine Processing of Remotely Sensed Data, IEEE 73 CHO 837-2GE, Purdue University, West Lafayette Indiana, October 16-18, 1973, p3B-27 to 3B-34.

S. Watanabe and the CYBEST group, "An Automated Apparatus for Cancer Prescreening: CYBEST", Computer Graphics and Image Processing Vol. 3, 1974, p350-358.

J.S. Weszka, "A Survey of Threshold Selection Techniques", Computer Graphics and Image Processing Vol. 7, 1978, p259-265.

J.S. Weszka and A. Rosenfeld, "Threshold Evaluation Techniques", IEEE Transactions on Systems, Man, and Cybernetics, Vol. SMC-8, No.8, August 1978, p622-629.

Y. Yakimovsky, "Boundary and Object Detection in Real World Image", Journal of the Association for computing Machinery., Vol 23, No. 4, October 1976, p599-618.

S. Zucker, "Region Growing: Childhood and Adolescence" Computer Graphics and Image Processing, Vol 5, 1976, p382-399.

SEGMENTATION: PIXEL-BASED METHODS

A. Rosenfeld
Computer Vision Laboratory, Computer Science Center, University of Maryland, College Park, MD 20742

1 INTRODUCTION

Segmentation of an image into parts is usually the first step in the process of image analysis, as discussed earlier. There are two principal approaches to segmentation. One approach uses pattern classification techniques (defined in an earlier chapter) to assign the pixels to classes; this results in a segmentation in which the parts are just the sets of pixels that belong to the various classes. The second approach constructs a partition of the image into maximal homogeneous regions, typically by starting with an initial partition and performing splitting and merging operations (an inhomogeneous region is split; two regions are merged if their union is still homogeneous) until a partition into maximal homogeneous regions is obtained; this is then the desired segmentation. The region partitioning approach to segmentation will be discussed in a later chapter; in this chapter we treat the pixel classification approach.

In a grayscale image, pixels can be classified on the basis of their gray levels (Section 2), e.g., into "dark" and "light" classes. In a multispectral or color image, more refined classifications are possible based on the pixels' spectral signatures (Section 3). Even in a grayscale image, we can obtain more refined classifications using properties computed in the neighborhood of each pixel; this approach can be used to segment the image into differently textured regions (Section 4). In all of these cases we assume that the probability densities of the property values for the various classes are known, so that Bayesian methods can be used for classification; or else that the classes constitute substantial subpopulations of pixels, so that clustering methods can be used to discriminate them. Pixel classification is also used to detect pixels of "rare" types (not constituting a large subpopulation) in an image, e.g., pixels belonging to special types of local image features such as edges or curves; this subject is treated in a separate chapter.

A general treatment of segmentation techniques, including pixel-based methods, can be found in Rosenfeld & Kak (1982). The specific approaches described in this section are all relatively standard methods; individual references on these techniques will not be given here (see Rosenfeld & Kak (1982) for historical and bibliographical notes).

2 SEGMENTATION BASED ON GRAY LEVEL

Many types of images are composed of parts which differ significantly in gray level. Printed or written characters are (usually) darker than the paper; the chromosomes in a mitotic cell are darker than their background; in a white blood cell, the nucleus is darker than the cell body, which is in turn darker than the background; while cloud cover seen from a satellite is brighter than most terrain backgrounds. In such cases, pixel classification based on gray level can be used, at least in principle, to segment the images. Since the classification is being done on the basis of a single feature (gray level), we are working in a one-dimensional feature space, and the decision surfaces needed to partition the space are simply (sets of) points, i.e., thresholds; we assign a pixel to a given class if its gray level is above or below a given threshold, or between two given thresholds, as appropriate. Pixel classification on the basis of gray level is therefore known as thresholding.

As mentioned in Section 1, if we know the gray level probability densities for the classes, we can use Bayes' theorem to determine the most likely class for each pixel; but if we do not know these densities, we must use clustering methods to distinguish the classes. In a one-dimensional feature space, clusters give rise to peaks on the histogram of feature values; thus we can determine thresholds for segmentation based on gray level by detecting peaks on the image's gray level histogram, and choosing thresholds that separate these peaks. This process is illustrated in Figure 1.

Segmentation: Pixel-based methods 227

Often, due to variations in illumination, one cannot use the same threshold(s) to segment the entire image correctly; for example, the background gray level on one side of the image may be higher than the object gray level on the other side. In such a situation one can break the image into arbitrary blocks; determine a threshold for each block, if possible, based on histogram peak analysis.

Figure 1. Segmentation by thresholding. (a) Image (blood cells). (b) Gray level histogram. (c) Result of choosing two thresholds that separate the peaks on the histogram; the three ranges are displayed as black, gray, and white.

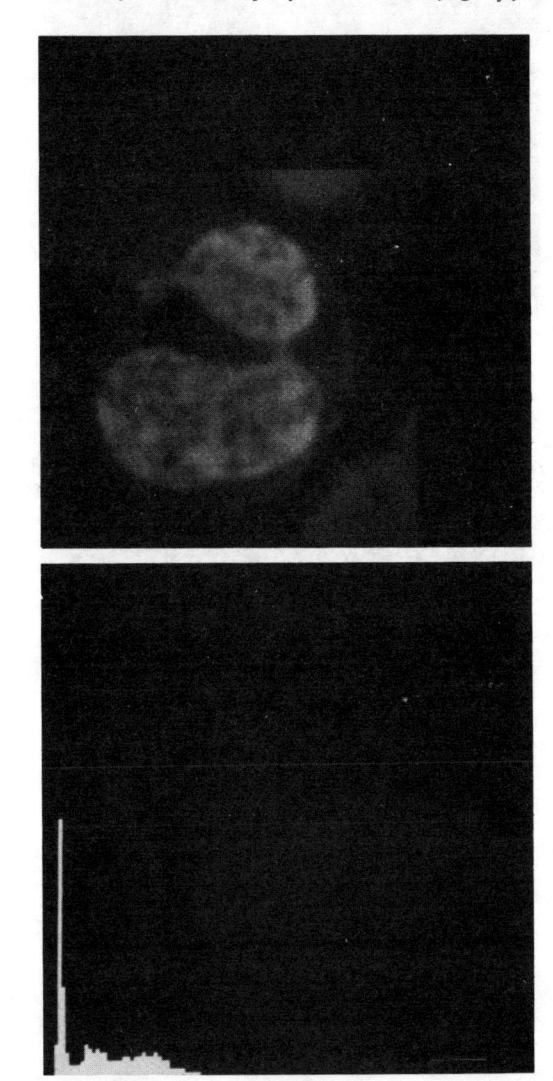

(a)

(b)

and interpolate on the values of these thresholds to define a smoothly varying threshold for the entire image. An example of this approach is shown in Figure 2.

Figure 1 (cont'd.)

(c)

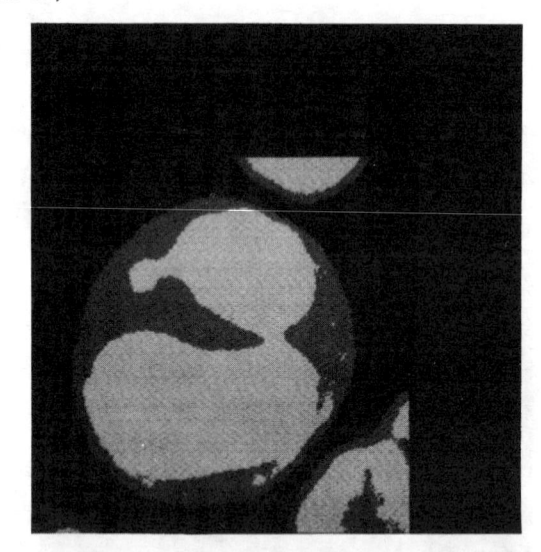

Figure 2. Variable thresholding. (a) Image of mechanical parts. (b) Result of global thresholding. (c) Local histograms. (d) Varying threshold, obtained by interpolating thresholds derived from the local thresholds. (e) Result of variable thresholding.

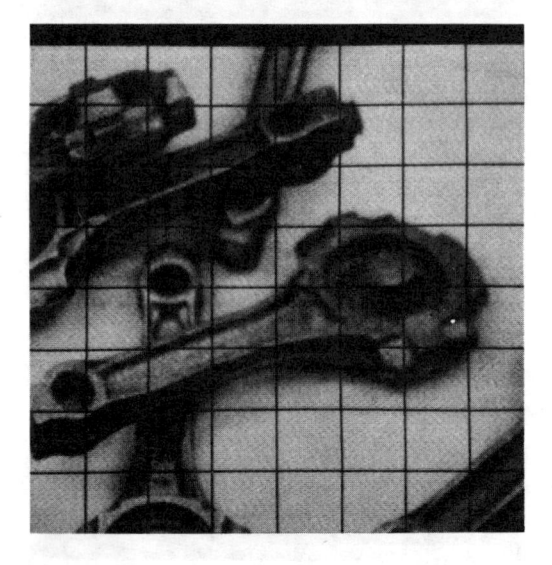

(a)

Segmentation: Pixel-based methods

3 SEGMENTATION BASED ON COLOR

Images usually cannot be segmented into more than a few classes of pixels based on gray level alone; if there are too many classes, they overlap and become impossible to discriminate. For example, in the house scene shown in Figure 3, there are five major types of regions: sky, brick, grass, bushes, and shadow, but the histogram of this image has fewer than five significant peaks.

Figure 2 (cont'd.)

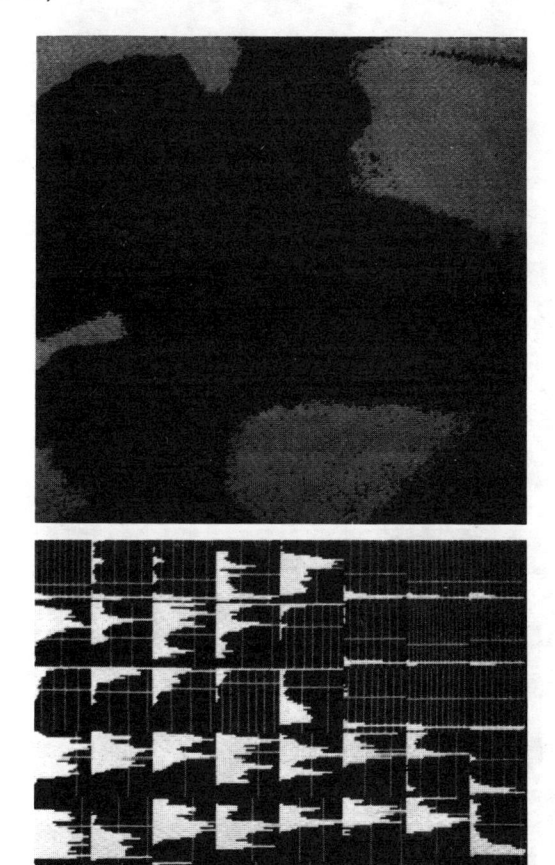

(b)

(c)

More refined segmentations can be obtained if we have more than a single feature (gray level) for each pixel. In a multispectral or color image, for example, each pixel is characterized by a set of color components, or a set of brightnesses in the various spectral bands. Since this is a multidimensional feature space, there is room to discriminate among a larger set of classes. To illustrate this, Figure 4 shows the red, green, and blue components

Figure 2 (cont'd.)

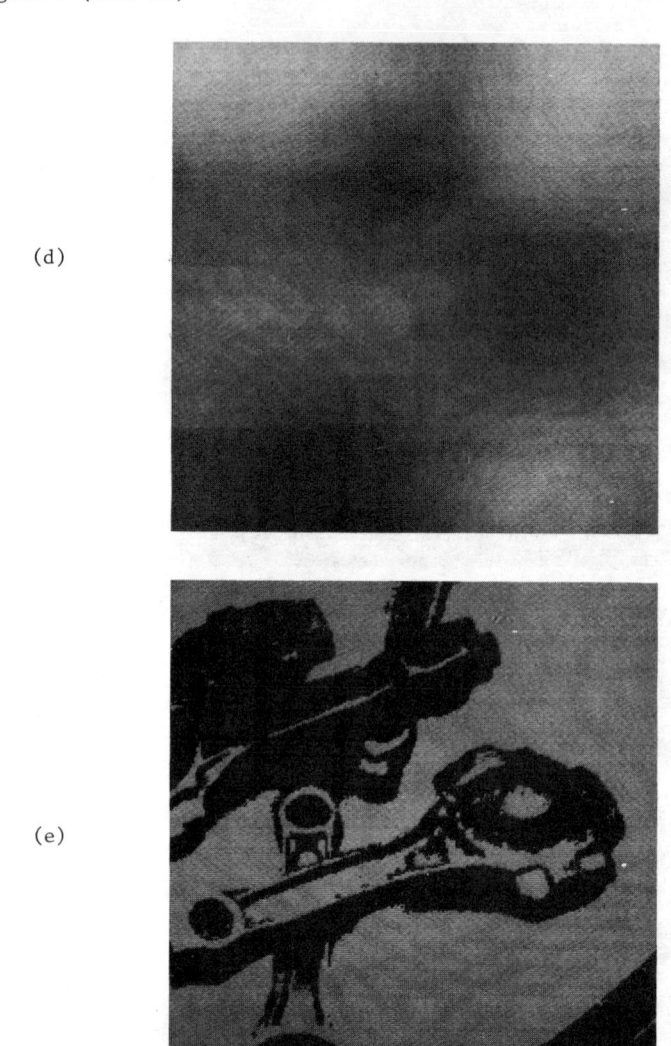

(d)

(e)

Segmentation: Pixel-based methods   231

of a color image of the house scene in Figure 3. The scatter plot of color values contains at least five clusters, and if we define decision

Figure 3. House image and its gray level histogram.

Figure 4. (a) Color components of the house image. (b) Scatter plot of color values, projected on the (red, green), (green, blue), and (blue, red) planes. (c) Pixels belonging to each of five major clusters in the scatter plot, displayed as white.

(a)

surfaces that separate these clusters, we obtain a good segmentation into the five types of regions.

4 SEGMENTATION BASED ON LOCAL PROPERTY VALUES

Even for grayscale images, we can obtain more refined segmentations by using additional features derived from the neighborhood of each pixel. For example, in Figure 3 the shadow and bush regions

Figure 4 (cont'd.)

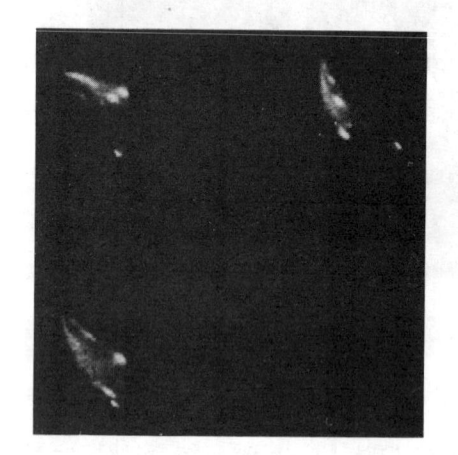

(b)

(c)

have similar gray levels, but the bush region is "busier", i.e., it has higher local variability in gray level. Thus we should be able to distinguish between these regions by using some measure of local "busyness" as a pixel property. An example of such a measure is the "total variation" of gray level in the neighborhood of the pixel,

Figure 4 (cont'd.)

(c')

Figure 5. (a) "Busyness" values for the house image. (b) Scatter plot of (gray level, busyness) values. (c) Pixels belonging to each of five major clusters in the scatter plot.

(a)

(b)

i.e., the sum of absolute differences between all the adjacent pairs of pixels in the neighborhood. Figure 5 shows the values of such a measure for the house scene; we see that these values indeed tend to be higher in the bush region. Note that low values also occur even in the bush region; but we can reduce the variability of the values by local smoothing. When this is done, a scatter plot of (gray level, busyness) values (Figure 5) shows distinct clusters representing the bush and shadow regions. If we define decision

Figure 5 (cont'd.)

(c)

(c')

Segmentation: Pixel-based methods

surfaces that separate the clusters in this scatter plot, we once again obtain a reasonable segmentation of the image into the five types of regions. This approach can be used to segment images into various types of differently textured regions; see a later chapter for further details on texture analysis.

Local property values can also be used as an aid in segmentation based (primarily) on gray level. As an example, suppose that the image contains an object on a background, but that the contrast between object and background is low, so that the peaks corresponding to the object and background subpopulations overlap, and it is hard to choose a threshold that separates these peaks, as illustrated in Figure 6. Suppose that for each pixel, in addition to its gray level, we compute its gradient magnitude (see the chapter on edge detection). In the object interior or background interior, this magnitude will be relatively low, while on the object/background border it will be relatively high. Thus if we exlude pixels with high gradient magnitudes from the gray level histogram, we are primarily eliminating pixels that lie on the border and so have intermediate gray levels. These pixels fell between the peaks on the

Figure 6. Use of gradient magnitude for histogram improvement. (a) Infrared image of a tank. (b) Histogram. (c) Scatter plot of gray level vs. gradient magnitude. (d) Result of suppressing high-gradient pixels from histogram.

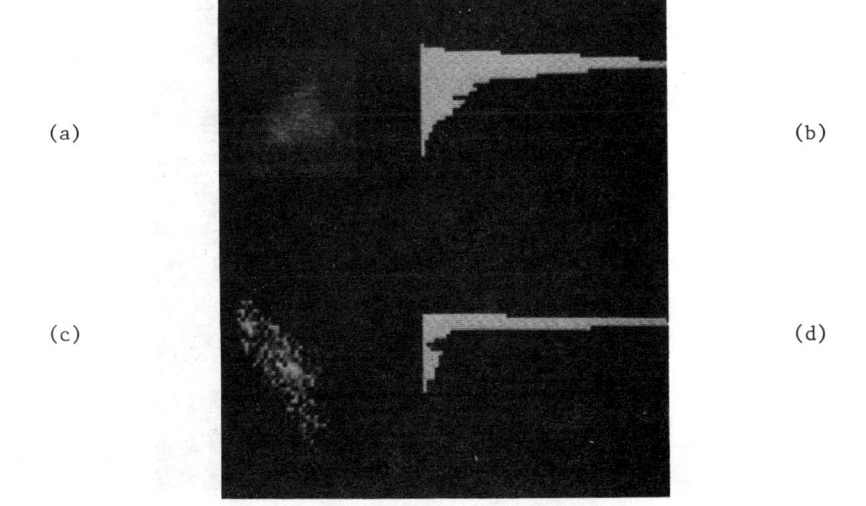

histogram, so that by eliminating them we are deepening the valley between the peaks, and making the threshold easier to select. This valley deepening effect is illustrated in Figure 6. In effect, in this method we are using a two-dimensional feature space - namely, (gray level, gradient magnitude) space - as an aid in selecting a one-dimensional decision criterion (i.e., a threshold).

Another way of using local property values in connection with gray level thresholding is to try a range of thresholds, and choose the one that yields the least "busy" result, on the assumption that the correctly extracted objects are not noisy. Still another possibility is to choose the threshold for which the object/background borders best coincide with edges in the original image. An example of threshold selection based on border/edge coincidence is shown in Figure 7.

Figure 7. Segmentation using border/edge coincidence. (a) Image and results of thresholding at several levels. (b) Edges (raw values and maxima). (c) Results of overlaying edge maxima on thresholded images. (d) Original image and thresholded image for which borders coincide with edge maxima.

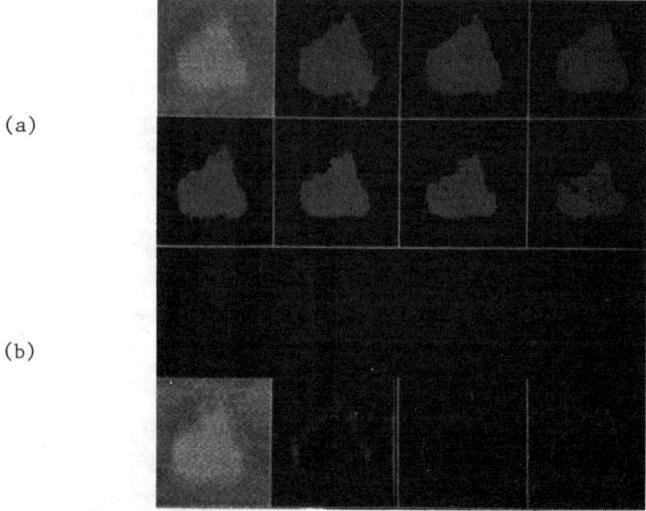

(a)

(b)

5 CONCLUDING REMARKS

Segmentation based on pixel classification is the standard approach in many application areas (e.g., in multispectral remote sensing), and many specific techniques for segmenting images in this way have been developed. In this section we have reviewed only a few basic approaches; for further details and examples, the reader is referred to the extensive literature on this topic (see Rosenfeld & Kak (1982)).

REFRERENCES

Rosenfeld, A. , & Kak, A.C. (1982). Digital Picture Processing (second edition), Chapter 10. New York: Academic Press.

Figure 7, cont'd.

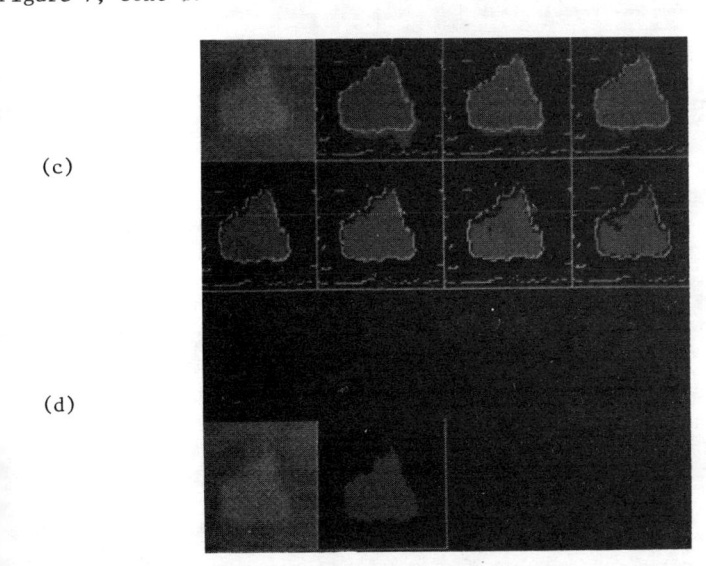

(c)

(d)

BASIC IDEAS FOR IMAGE SEGMENTATION

S. Levialdi

Institute for Information Sciences, University of Bari, Italy

1. Introduction

Image processing is the activity (science or art?) which deals with the development and refinement of techniques for using digital computers to perform image analysis, manipulation, evaluation and representation according to a specific task. Sometimes this task considers biomedical images or remote sensing ones, other times the class of images is immaterial and the goal is the creation of a pictorial data base, still other times the main purpose is the display of different projections of the stored images with different light conditions taking into account the occluding lines (or not) etc. Although in this last case, computer graphics are also invoked, broadly speaking, image processing deals with all kinds of operations on digital images: at the input of the system, in storage and at the output (display or hard copy unit).

It is common to distinguish two different areas of processing: a first one in which general techniques are applied to enhance and restore the image so as to improve (artificially or naturally) its original contents and a second one which is more task-oriented and may be of a more complex nature since it is generally designed for feature extraction, parameter evaluation, classification, etc. The first area is known under the name of pre-processing and the second one as analysis.

In practice, the second area strongly benefits from the results obtained in the first area, namely after preprocessing. It is common to say that the classification part of a procedure becomes very simple if the preliminary processing is "adequately" performed; just what adequately means is still an open question but establishes a common background from which algorithms for preprocessing may start operating on the image. For instance, once the concept of a regular contour is established (a path of connected elements where each one has exactly two neighbors) then a contour-extraction algorithm will be effective only on regular contours in the sense that by means of preprocessing, all the contour elements of an image are firstly transformed into regular ones and then the contour algorithm will run on the image.

It often happens that an image contains a number of "objects" (generally not well defined) which correspond to real physical objects in nature, for instance, cells, letters, trees, etc.

In order to process the image, one has to extract these objects, one by one, so that the analysis may concentrate on each single object. To do this, a segmentation procedure must be employed so as to separate the objects by means of boundaries which will appear between regions. The definition of boundary, of region and, above all, of the segmentation procedure are the subject of this paper. To end this introduction we may simply consider segmentation as one preprocessing technique of general purpose nature, and refer the reader to (1,2) which contain both an introduction to this problem as well as a number of algorithms for performing region segmentation.

2. Segmentation approaches

We may consider two main approaches as the starting points for classes of segmentation algorithms: the pixel one and the region one. The first one uses the very local information contained in a pixel whilst the second one considers some uniformity property which is shared by a group of pixels (a region). This second class of algorithms (to which we will refer in this paper) provides a "convenient basis for semantic analysis by reducing both the amount of detail and the ambiguities of interpretation found at the picture-element level"(3). For this reason a number of authors have tried to consider different logical predicates for establishing the homogeneity condition, for breaking regions into subregions(known as splitting) and for obtaining convergence after a small number of iterations of a "split-and-merge" segmentation technique"(4).

Originally(5) considered firstly a partition (of the image) and then a region growing process so as to generate the "segments" of the image. Since region analysis is time consuming it is better to decrease the amount of pixels which must be processed.

2.1 A segmentation program

Let us now briefly review(3) where the grouping of adjacent pixels is made on the basis of the fact that they have similar attributes. In their example, the picture was sampled with 40 x 40 pixels and the brightness of the pixels was considered the attribute on which the pixels were compared. Different sampling methods were employed: modal, mean and straight sampling. The first one considers the brightness value most frequently found within a given neighborhood, as the significant grey level value. This reduces noise but also reduces fine detail (which is the same result obtained when a low pass filter is used for enhancement).

The mean sampling of the image (which performs the arithmetical mean of a given neighborhood) is not adequate because it smooths discontinuities and therefore normal sampling was finally chosen to give the best results. Another possibility is to use color information (in contrast to only the brightness information used above). Grouping regions having quantization intervals corresponding to characteristic colors of objects was tried but the authors report worse results if compared to grouping obtained only with brightness values.

As a complementary method for extracting some objects in multi sensory data, some distinguishing features may be used. In the specific case of office scenes, the first five criteria of the table I (from(3)) give a simplification of the partition process. Unfortunately, for outdoor scenes the main features for region segmentation appear to be shape and texture so that image point features are not helpful. Conversely, in office scenes, a factor of 2 (half the regions) can be obtained when using sequential criteria as well as brightness criteria. The next step in the process is to merge regions (which generally turn out to be too many) on the basis of some similarity measurement, generally defined ad-hoc.

The experiments described in (3) consider the first partition on the sampled brightness and then using one, out of six different criteria, for merging regions until about 250 regions remained in the scene. The final judgement was left to the operator who compared this result with the real organization of the scene in terms of its components. In order to introduce the similarity measurements, the factors which were considered must be defined:

br_{ai} is the brightness through a neutral density filter of the ith pixel on the boundary of region a

br_{bi} the brightness of adjacent pixels in region b

r_{ai}, g_{ai}, b_{ai} the brightnesses of the ith boundary pixel from region a as seen from the red, green and blue filters respectively and

r_{bi}, g_{bi}, b_{bi} the corresponding brighthnesses from region b

$\bar{r}_a, \bar{g}_a, \bar{b}_a$ the average brighthnesses over region a seen trhough the usual filters and

$\bar{r}_b, \bar{g}_b, \bar{b}_b$ the corresponding averages over region b.

Two useful measurements of similarity are the Boundary Color Contrast and the Region Color Contrast:

BCC similarity criterion: $(\sum_{i=1}^{N} |r_{ai} - r_{bi}| + |g_{ai} - g_{bi}| + |b_{ai} - b_{bi}|)/N$

RCC   "   "   : $|\bar{r}_{ai} - \bar{r}_{bi}| + |\bar{g}_{ai} - \bar{g}_{bi}| + |\bar{b}_{ai} - \bar{b}_{bi}|$

Since these two criteria gave the best results, their combined use was chosen with a conservative rule, namely merging only if both gave the maximum value (for given pairs of pixels coming from different regions).

In order to avoid erroneous merging, semantic information must be introduced so that two regions should not merge if they correspond to interpretations of different objects. This condition may be slightly relaxed by allowing merging of two regions in which one of them has not yet been interpreted. The new regions inherit the interpretations of their parents (or parent); the interpretations are given interactively to the system by the operator (cooperative approach) for the fast segmentation of complex scenes (having many details) and the results obtained may help in providing the semantic information to the system by means of the automatic analysis of local attributes and contextual constraints.

As has been seen from this example, there are many levels at which the information should be used in order to produce a meaningful segmentation. The main difficulty in obtaining the correct regions lies in the definition of the levels (from the pixel level to the more abstract one of the semantic value) and in the organization of their interaction during the whole process.

Before entering into the description of other approaches and experimental results obtained by different groups working in this area (see 6), it seems adequate to formalize the notion of uniformity of a region, according to (2).

2.2 Segmentation definitions

Let us assume, that the image is made of N by N elements which are obtained after a spatial and intensity sampling of the original picture, scene, etc. The result of this process (acquisition and digitalization) is a function $f(i,j)$ which is non-negative and integer valued for every digital element (pixel) in the N by N rectangular array. The output from the process of segmentation will be another array of the same number of elements (N x N) in which the elements will have a common label(value) if they belong to the same region (part, object) in the image.

The regions will be defined in terms of a certain property which must either be constant or, at least, contained within a range of values. For simplicity we will start by considering a uniformity predicate which is defined as follows:

Uniformity predicate:

Let X denote the grid of sample points of a picture, i.e. the set of pairs.

(i,j) $i = 1,2,3,\ldots,N$, $j = 1,2,3,\ldots,N$

Let Y be a subset of X containing at least two points, a <u>uniformity predicate</u> $p(Y)$ is one which assigns the value true or false to Y depending only on properties of the brightness matrix $f(i,j)$ for the points of Y. (Moreover, if Z is a nonempty subset of Y then $P(Y)$ = true implies always $P(Z)$ = true).

We now give some examples:

$P_1(Y)$ = true, if the brightness value at any two points of Y is the same.

$P_2(Y)$ = true if the brightness value of elements of Y does not exceed a certain maximum. On the other hand, $P_3(Y)$ = true if Y has less than ten points, is <u>not</u> a uniformity predicate since the true/false value of the predicate must only depend on the values of $f(i,j)$ (brightness function).

We now give a formal definition of segmentation according to (7):

A segmentation of a grid X, for a uniformity predicate P, is a partition of X into n disjoint nonempty subsets X_1, X_2, ..., X_n such that

1) $\bigcup_{i=1}^{n} X_i = X$

   $X_1 + X_2 + \ldots + X_n = X$

2) X_i is directly connected for $i = 1, 2, \ldots, n$

   $X_i \cap X_j = \emptyset$ for $\forall\ i$

3) On each X_i the uniformity predicate $P(X_i)$ is true

   $P(X_i)$ = true for $\forall\ i$

4) P is false on the union of any number of adjacent members of the partition

   $P(X_i \cup X_j)$ = false for $\forall\ i \neq j$ (X_i, X_j are adjacent).

This definition does not imply that only one segmentation exists nor that n is the smallest possible value for a partition.

Since segmentation algorithms require the inspection of each point and of neighbouring points ("near" in some fashion to a central point) we must consider the "data structure" which constitutes the original image we wish to segment. For this reason we require the notion of a RAG (Region Adjacency Graph) which is homomorphic to the more well known adjacency matrix. (Refer to figure 1).

In the figure, a number of solid lines separates regions which have some common property; to each one of these regions a node is associated and a branch will join two nodes if the regions represented by the nodes are adjacent. The _degree_ of a node corresponds to the number of branches that are associated to this node, for instance nodes of degree one correspond to holes (regions not connected to the external border of the image). The image frame will be represented by a node in much the same way as nodes represent loops in Circuit Theory and common circuit elements are shown by branches on the graph.

An image may be processed on a line by line basis(7) and each line may be segmented into a number of intervals so that a graph may be built by associating to each line interval a node and two nodes will be connected by a branch if the corresponding intervals lie on adjacent lines and their projections on a single line overlap. Refer to fig. 2. This particular graph may be called Line Adjacency Graph or LAG. It is important to note that adjacent intervals on the same line are not joined by a branch.

In order to show the inclusion of regions instead of the adjacency another graph may be built, namely the Picture Tree (PT).

The root of this tree corresponds to the whole picture and the leaves to single pixels (refer to fig. 3). If the grid has dimensions N x N, with $N = 2^L$, then the picture tree has L levels. Nodes at a distance j from the root correspond to squares of size 2^{L-j}; since every node in the tree corresponds to four pixels, the picture tree is called quartic picture tree (QPT). The importance of the QPT is due to its use in image segmentation where particular branches must be found in order to "cut" the graph according to some properties of the regions of the image.

A cut of a tree is a subset of its nodes such that no two nodes belong to the same path from the root to a leaf and no more nodes can be added to it without its loosing the previous property.

Finally, a segmentation of an image always corresponds to a cut of the QPT as may be seen on the fig. 4, where the nodes belonging to the cut set are 1,2,5,6,b,9,10,13,14,d. The reason for introducing the graphs and the cut sets is not only formal but also practical since if an algorithm may be developed for extracting the cut set in a graph this algorithm will also segment the image provided this image has been mapped into a picture tree.

Many authors have devised representations of images which have a number of levels (in a cross section of a cone(8) in a pyramid(9)) and each level corresponds to a degree of coarseness (or definition) where some information may be better extracted. This, in turn, bears a connection with the computational structure of the machine on which the images are to be processed. In fact, data representation (and retrieval), processing elements and their interconnection generally have a trong influence on the algorithms (for segmentation or analysis) which will be better matched to such computational structure.

2.3 Split-and-merge segmentation

It is still an open question whether it is better to introduce semantics at an early stage of the segmentation task and therefore get involved with a larger computational burden or if it is perhaps better to use it at a later stage for substantially improving the results obtained only on the basis of the segmentation algorithm. Lastly, it may sometimes be unnecessary to use semantics at all if the problem is not too involved and, or, if the procedure for segmentation is powerful enough.

Splitting- and-merging tries to make up for the disadvantages of only merging or only splitting and essentially relabels an initial set of regions on an image until some (formal) segmentation properties are satisfied.

Since an image may be decomposed in "quarter" (see the Quartic Picture Tree mentioned above) every single pixel may be grouped together with other three adjacent pixels (the four of them will be the leaves of a QPT) so as to become one region, this process is referred to as merging. On the other hand if coming from the root of the tree we subdivide a region into four elements, this is referred to as splitting.

In both cases we will merge or split provided a certain predicate is true and when all the merging and splitting is accomplished (the convergence of the method may be proved) there is still need for a final procedure to eliminate small regions which are generally created in the process. The new predicate will not only contain values of $f(i,j)$, but also values of the areas in which $f(i,j)$ is computed, in this way, a smaller number of regions is obtained and the final segmented image is easier for human interpretation.

Returning now to the formal definition of segmentation, which was given above; we should note that merging starts with a partition satisfying condition 3) and proceeds to fulfill condition 4) whilst splitting starts from a partition satisfying condition 4) and proceeds to fulfill condition 3). A split-and-merge procedure starts with an arbitrary partition which satisfies neither 3) nor 4) and ends satisfying both.

In order to achieve the segmentation, the PAT (picture adjacency tree of the image) is considered so as to extract the cutnode (see definition above) which is the only part of the graph which requires to be stored in memory and all the new arrangement of the nodes will modify the cutset until the last obtained cutset provides the final segmentation.

Let us refer to fig. 5 (from (4)), where an artificial image is shown together with a hypothetical segmentation into regions labeled with numbers of one, two or three digits according to the level in the tree to which they belong in their node representation.

The root of the tree represents the "external world" or the frame (sometimes labeled node 0), node 4 has no descendents, nodes 1, 2 and 3 have a number of descendents shown on the figure. In this tree the leaves are blocks of one element (pixels) and the other nodes at level L, named b, represent a block of elements having sides of length $n/2^L$. The element is positioned on the top left corner of such a square and will have coordinates i,j. The brightness function $f(i,j)$ will be associated to each node with its maximum and minimum value for the corresponding block.

Let us now see how is the merging performed between adjacent blocks. As mentioned earlier, two regions should be merged if they share some common property, for instance if the difference between the maximum of $f(i,j)$ and the minimum of $f(i,j)$ is below a certain threshold, then th components of the block (which might be other blocks) should be merged, this will be done on all the levels of the tree, level by level for all blocks. On the other hand splitting will be performed inside a block if the difference between the maximum and minimum value of brightness exceeds the given threshold. Merging may proceed until only the root is left (and therefore it will terminate) and splitting may only proceed until all the leafs are obtained and, again, will terminate; furthermore the two processes are mutually exclusive since once merging has been done and followed by splitting no other legal partition might be achieved (no merged blocks may be splitted and no splitted blocks may be merged again).

2.4. A segmentation paradigm.

As a general approach to the problem of segmentation it is convenient to mention the one in (10) where, after a critical review of the pitfalls and mishappenings of image segmentation based on picture features, a new model of the interaction between different domains (signal, physical and semantic) is presented to introduce his own methodological approach for solving this task in the most general way. Let us start from the beginning and use Kanade's simple definition of segmentation: the location of regions where pixels share some consistent characteristics. The main point to stress is that whilst this definition does not attach any specific value to the regions, we strive for a general purpose program which could, in principle, give names to the regions: i.e. find the semantic values of each one of the regions. It is in this sense that segmentation may also be seen as a method to extract meaningful objects from a picture. These "atoms" are, for instance: cells, cars, sky, house, table etc. In order to do this it is very important to have at our disposal the knowledge of the class of images we want to segment as well as the clear definition of the task we are going to perform.

Kanade suggests a general knowledge model which includes three different levels: a _signal_ level where the measurements on the image might be performed, a _physical_ level where the picture cues may be extracted both independently or as a function of the scene cues (scene is 3D) and a _semantic_ level where the abstract model of the world of images we are going to process is contained (see fig. 6).

The cues from the picture should not be confused with the cues from the scene; in fact the first ones are line segments, homogeneous regions, intensity gradients whilst the second ones (relative to the same objects) are edges, surfaces and reflectance.

Furthermore the model, which is made with general information about the objects in the scene and their relationships, after being matched with the scene cues, will produce an _instantiated model_ which contains the specific information of the objects present in the scene and their sizes. This instantiated model gives rise to a _view sketch_ which is an abstract description of the anticipated picture (sometimes organized in a graph) and this sketch is tested with the image where new measurements may be made to confirm or discard some components of the sketch. New cues are obtained from the picture and are checked with cues from the scene domain so closing the loop with the model from the semantic domain. The consistency in the image features (which constitutes a preprocessing stage of the whole method) leads to the extraction of picture domain cues, the consistency in the scene characteristics leads to the extraction of the scene domain cues and the consistency in the semantic association will produce the view sketch which includes information from all three levels: signal, physical and semantic.

The most well known methods for segmentation use and operate only within the signal domain and make large use of spectral information in one of the three following possible ways: 1) by considering local spatial criteria for merging regions, 2) by considering the distributions in the image feature space for region splitting and 3) a combination of 1) and 2). Typically region merging is accomplished whenever a "distance" between regions is under a certain threshold or, as defined in the literature, a boundary is "weak" and therefore it will be erased and a new boundary comprising both regions is defined (phagocyte).

In the splitting techniques use is made of histograms for valley detection but many different features may be used, and the more the better.

The combined methods post process the first segmented region on the basis of the histogram analysis by merging regions with pixels having similar values or surrounded by pixels of a given value, etc.

Relaxation techniques generalize this approach and a probability of belonging to a region is assigned at the beginning of the process and then this probability is updated iteratively using the previous values of its own and neighboring pixels (error removal is 5 times better than without relaxation). Segmentation may also start at the center of regions, pyramidal data structures (with a compaction factor of 1:4 for each new plane) are a good example. If many redundant image features are used, 1-D histograms can be employed for valley detection better than multidimensional clusters. Another very important point is the representation of the segmentation results (partial or final) so as to compare different segmentation of the same image. As shown in the previous example the segmentations tree has a node for each region and a branch to indicate containment of children regions; in top-down methods (splitting) good retrieval facilities are required in order to extract, for instance, "regions which are vertically long and have a darker region on their left". But the real problem is the one of giving names to regions and one possibility (11,12) is to use a graph in which a node is a region and the branches represent relationships between regions (to the right, above, etc.). The process of interpretation is performed by matching the graph or the subgraph with the "best" assignments of object names (to the graph nodes). Unfortunately two problems arise: a) occlusions and rotations change the graph structure and b) perfect regions (homogeneous) are not always extracted in real images.

A way to represent semantic information is to use the probability that a certain global interpretation, on the basis of context and a set of measuerements, is maximal for a certain region with respect to a relational role which is defined between two adjacent regions. For the computation (which may be iterative) tables are used having object-object relationship and object-property relationship. In (13) a top down approach is presented having a relational data base with a production system as a computational procedure; in (14) use is made of a semantic network with a collection of knowledge blocks.

A common feature of this approach is that object-dependent control is possible by the use of attached procedures in the knowledge of the scene domain, if not correctly used the corespondence with the image is not always adequate. The true meaning of the relations is dependent on depth values which are sometimes unknown. A short cut is performed on the circular information flow of the paradigm by parsing the model and assuming that there is an equivalence with the longer route (see fig.7).

Sometimes this may still be true, for instance in outdoor scenes where the observer is far away and the whole scene might be considered to be flat: 2D. In practice the picture domain cues generate the view sketch without passing through the scene domain cues, the model,etc. The other negative issue about this approach is that the shape, if known, has to be checked on the image but the boundaries of this shape are still to be found and therefore we run in circles: this is called the shape dilemma(10).

We therefore need to know about the mapping of the scene into the picture and this knowledge refers to the physical world.

Basic ideas for image segmentation

The ARGOS system(15) uses alla the knowledge about Pittsburgh downtown to precompile the expected views from the information contained on maps and on 3D data sources, so that whenever a picture is analyzed (from a range of expected angles) a search is initiated on a semantic network. On this network a path is located which corresponds to a set of assignments of semantic labels to pixels which best satisfy both local image feature properties and the expected constraints. This approach is different from the one storing descriptions of relations between objects (Model in the scene domain) where a good recovery of the scene domain cues is required. ARGOS stores signal level evidences which are anticipated and precompiled from a given task, it does not require the recovery of scene domain cues, this holds for a knowledge rich task world if the input image belongs to a given class. In this instance the picture and scene domains are interchangeable.

The physical level knowledge includes lighting conditions, angles, etc. for instance a plane surface painted uniformly is not always a region of homogeneous intensity in the image, but has homogeneous reflectance. Edges may have different profiles (step, peak, roof) according to physical conditions and, again, lines may have different scene domain interpretations. Little is known on the physical level of knowledge, shape-from-shade theory constitutes a first step in this direction. Surface orientation may be represented by the gradient along x and y called p and q; slopes (relative to the viewer) along the x and y directions. The mapping is $f:(p,q) \rightarrow I$, the intensity I is a function of p and q.

If the light source and the viewer coincide and the object reflectance is proportional to the cosine of the incident angle (Lambert's law) of light:

$$f_1 : I(p,q) = \frac{1}{\sqrt{p^2+q^2+1}}$$

the inverse mapping ($f_1^{-1} : I \rightarrow (p,q)$) is generally not unique, in fact, points having a given intensity I may be on any point of a circle having radius $(1/I^2 - 1)$.

2.5 Segmentation by texture

In (16) an algorithm is proposed, called MITES (model driven, iterative texture segmentation algorithm) which starts by considering the existence of texture models in an image (of statistical nature). A pixel on the image can be given one of the three possible descriptions based on the analysis of its kxk neighborhood: 1) the pixel is interior to a region described by M_i, or 2) on an edge between an M_i region and an M_j region and being in the M_i region or 3) on an edge between an M_i region and an M_j region in the M_j region. According to which of the conditions hold for the pixel, a non-linear smoothing (originally named E^k algorithm(17))will be performed choosing a subset of the neighboring pixels which are contiguous. For instance, if the pixel is interior to the region, all the neighbors will be considered in the smoothing process. The process is iterated a number of times (in the examples of (16) good results appeared at the 5th iteration).

MITES may also run in an unsupervised mode in which the models are computed by a clustering procedure and the cluster definitions are modified in time. For computational economy, special pixels may be considered (instead of all the pixels on the image) either on the base of some shape constratins: edge maps of texture fields illustrated in (16) or using a hierarchical representation of texture models.

3. Conclusions

The problem of image segmentation has been considered from the different angles which have originated the image partition strategies; in itself this problem is of a very general nature and constitutes the first

Basic ideas for image segmentation

stage of any image analysis system. Due to the difficulties underlying the formal definitions of "uniformity" for establishing the necessary conditions for region growing the techniques employed so far have tried to involve semantic information to improve their performance. When and how this information must be used during the process are relevant questions that must still be answered to give a general and efficient solution to this problem area.

REFERENCES

(1) A. Rosenfeld, A. Kak, "Digital Image Processing", Academic Press, New York, 1976.

(2) T. Pavlidis, "Structural Pattern Recognition",Springer-Verlag, Berlin, 1977.

(3) J.M. Tenenbaum, S. Weyl, "A Region-Analysis subsystem for interactive scene analysis", Proc IJCAI, 1975, 682-687.

(4) S.L.Horowitz, T.Pavlidis, "Picture segmentation by a directed split-and-merge procedure", 2nd IJCPR, Copenhagen, 1974, 424-433.

(5) C.R. Brice, C.L. Fennema, "Scene analysis using regions", Artificial Intelligence, vol. 1, n° 3, 1970, 205-226.

(6) S.Levialdi, "Segmentation techniques and parallel computation for image processing", NATO ASI on Map Data Processing,edits., H.Freeman, G.Pieroni, Academic Press, New York, 1981, 279-307.

(7) S.M. Selkow, "One-pass complexity of digital picture properties", Journal of ACM, 19, 1972, 283.

(8) L. Uhr, "Layered recognition cone networks that preprocess, classify and describe", IEEE Trans. on Comp. C-21, 1972, 758-768).

(9) M.D. Levine, J. Leemet, "A method for non-purposive picture segmentation" 3rd IJCPR, Coronado, 1976, 494-498.

(10) T. Kanade, "Region segmentation: signal vs semantics", 4th IJPCR, Kyoto, 1978, 95-105.

(11) M.G.Barrow, J.M.Tenenbaum, "Recovering intrinsic scene characteristics from images", Technical Note, 157, SRI, 1978.

(12) M.G. Barrow, R.J. Popplestone, "Relational descriptions in picture processing", in <u>Machine Intelligence</u>, vol. 6, edits. B. Meltzer, D. Michie, American Elsevier, 19.

(13) M.D. Levine, "A knowledge-based computer vision system", R 77-3 El. Eng., McGill University, 1977.

(14) T. Kanade, "Model representation and control structures in image understanding", Proc. IJCAI, vol. 2, 1977, 1074-1082.

(15) S. Rubini, "The ARGOS image understanding system", Ph.D. Thesis, Carnegie-Mellon Univ., 1978.

(16) L.S. Davis, A. Mitiche, "MITES: a model driven iterative texture segmentation algorithm", Univ. of Texas, TR 81-1.

(17) L.S. Davis, A. Rosenfeld, "Noise cleaning by iterated averaging", IEEE Trans. on Systems, Man and Cybernetics, 9, 1978, 705-710.

TABLE 1

SEQUENTIAL CLASSIFICATION CRITERIA

(1) Extract floor samples by height (0.1 foot)

(2) Extract chairseat samples by characteristic height and horizontal orientation

(3) Extract tabletop samples by characteristic height and horizontal orientation

(4) Extract picture samples in two passes:

  (a) By characteristic height and saturation greater than maximum saturation for wall

  (b) By characteristic height and hue outside the hue range of wall

(5) Extract chairback samples by characteristic height, vertical orientation, and saturation

(6) Partition remaining samples by brightness.

Basic ideas for image segmentation

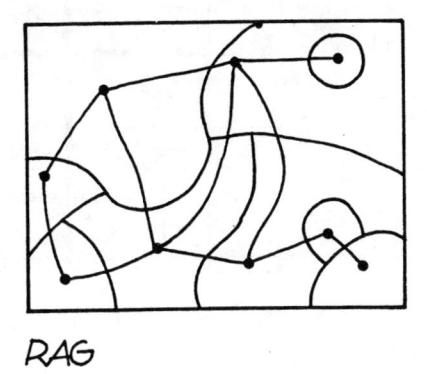

Fig. 1 - Region adjacency graph

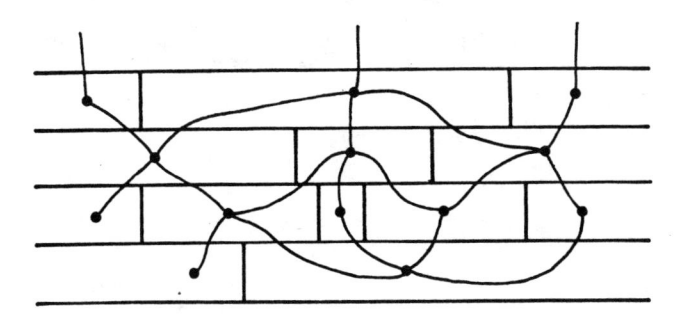

Fig. 2 - Line adjacency graph

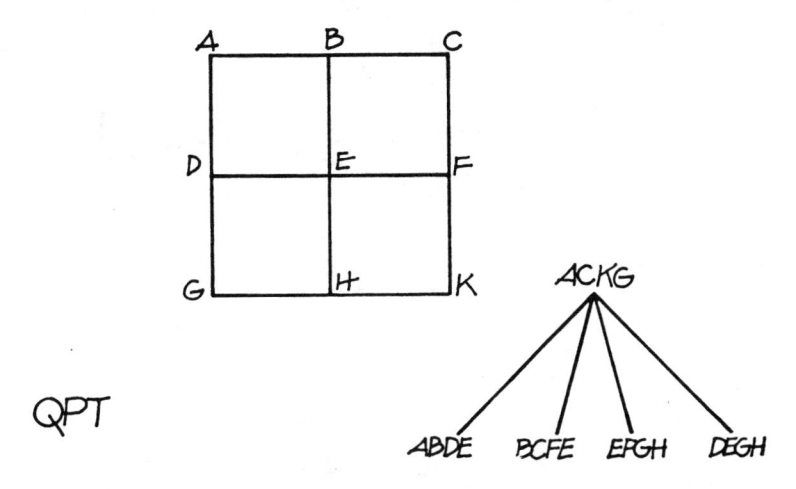

Fig. 3 - Quartic picture tree

TREE CUT

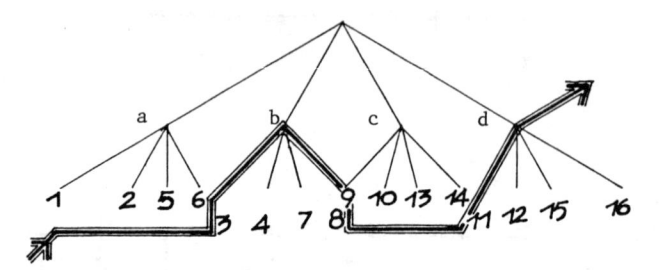

Fig. 4 - Tree cut

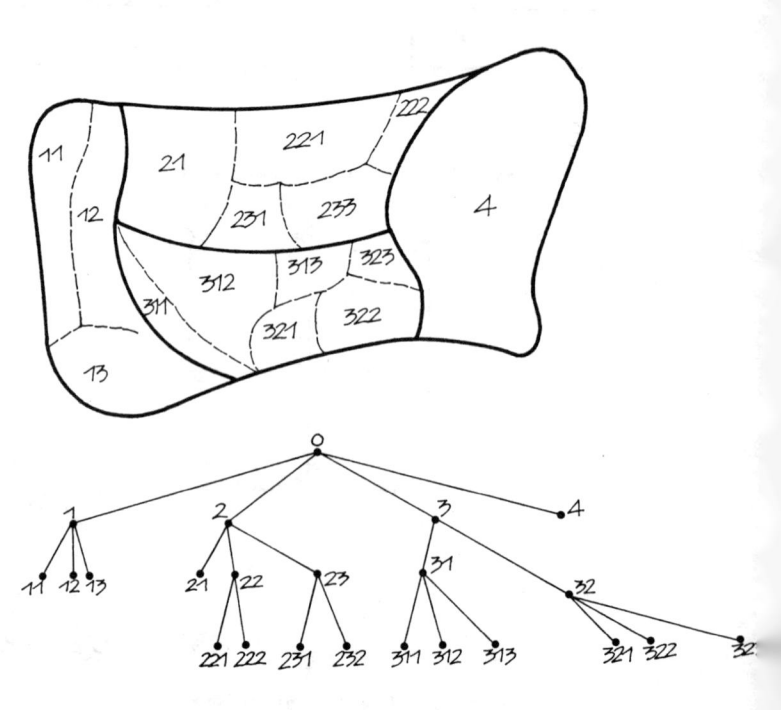

Fig. 5 - Split-and-merge example

Basic ideas for image segmentation

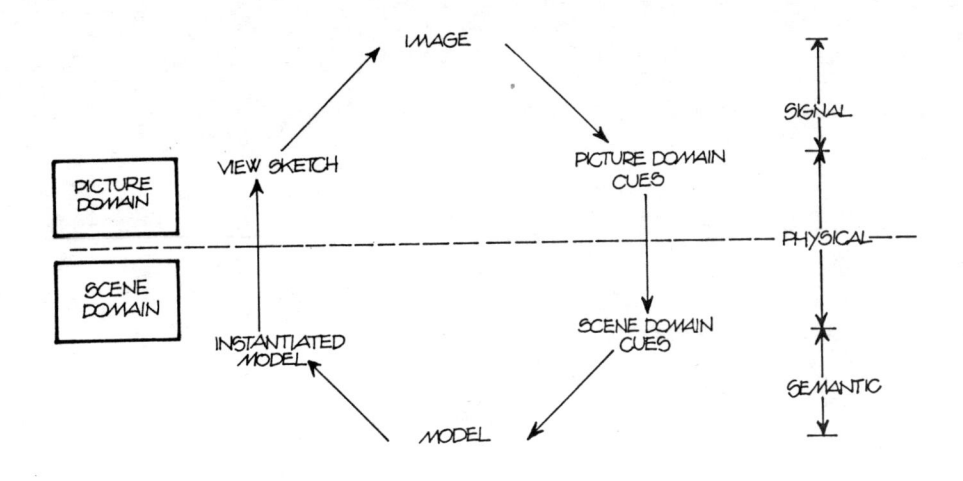

Fig. 6 - Signal, physical and semantic spaces.

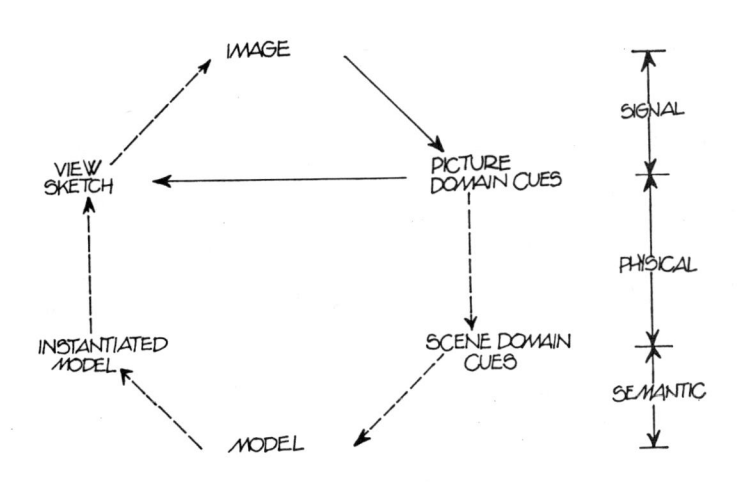

Fig. 7 - As with short cut.

FEATURE-BASED 2-D SHAPE MODELS

T. C. Henderson
Department of Computer Science
The University of Utah
Salt Lake City, Utah 84112

The solution to a complex shape analysis problem generally requires the design of shape modeling techniques and procedures for organizing unknown shapes according to such models. Once a shape modeling mechanism has been chosen, specific models for the classes of shapes to be analyzed can be constructed. This might involve simply determining values for specific shape features or detailing the spatial organization of a shape. The process of choosing a modeling mechanism and then constructing shape models is complicated by a variety of factors that influence both the appearance of specific shapes in images and the segmentation of shapes of individual objects from images. These factors include: geometric transformations, obstruction, agglomeration and noise. In this chapter, we consider shape models based on features of the area, boundary or special axes of the objects to be modeled.

AREA METHODS

Area methods are based on the knowledge of all points belonging to the shape, that is, both interior and boundary points. One commonly used area model is the method of moments. The theory of 2-D moment invariants for planar geometric figures is described by Hu (1962). The method provides for recognition of shapes independent of position, size, and orientation. The (p,q)th moment of a shape is defined as:

$$m(p,q) = \iint x^p y^q r(x,y) \, dx \, dy \qquad p,q = 0,1,2\ldots$$

where $r(x,y)$ is the characteristic equation for a bounded shape, i.e., $r(x,y) = 1$ if (x,y) is in the shape and 0 otherwise. The sequence $\{m(p,q)\}$ is uniquely determined by $r(x,y)$, and vice versa. Definitions and properties of moment invariants under translation, scaling, orthogonal transformations and general linear transformations are developed.

These moment invariants rather than the actual moments can be used for shape modeling.

Thus, the method of moments presents many desirable features. However, the method has several drawbacks. A significant amount of computation is required to compute them. Also, although the first few moments may suffice to model a simple shape, more complicated shapes require many more terms of the sequence $\{m(p,q)\}$. Moreover, distortion of a shape by noise and poor segmentation is not easily modeled in terms of transformations of the moments. Finally, if parts of the shape are obscured, or if agglomeration occurs, then the moments of the resulting shapes are radically different from those of the original shape.

In a digital image, $f(x,y)$, the projection of that image (see Rosenfeld (1976)) may be defined as a function:

$$p(i) = \sum_{x,y \in I_i} f(x,y)$$

where I_i describes a family of curves, e.g., a set of lines or circles. Some standard projections include the x-projection of f, and the y-projection of f, which are just the column sums and the row sums, respectively. These projections can be used to detect blobs by merely looking for plateaus in the projection function. Linear objects can be detected if they run perpendicular to the projection axis. Thus, if their orientation is unknown, several projection axes might be tried. If several objects must be recognized, and the objects have a known orientation, then a projection function might suffice to distinguish between them (e.g., character identification).

A different approach to shape decomposition is to segment the area of a shape into convex subregions. These area-based shape models use convex polygons as the primitive elements of the representation. Given a set of points, S, the usual definitions for convexity include:

-For every two points p and q in S, the line segment from p to q lies entirely in S, or

-For every two points p and q in S, the midpoint of the line segment from p to q lies in S.

However, these definitions require some care if they are to be
implemented for a digital image. Namely, line segments must be
digital line segments, and the midpoint must be compared to points
within half a unit of the midpoint. The definition of convex objects in
a digital setting remains a problem for the application of these
methods.

Once a suitable definition of convexity has been chosen, a
set of points must be broken into its convex subsets. In general,
there is no unique convex segmentation. One approach to solve this
problem is to start with several (say 3) close points, and "grow" a
convex face containing those parts, i.e., points of S are added until no
point can be added and still satisfy the definition of convexity. A
structural approach to this problem is described by Pavlidis (977).
Such an approach is complicated by the fact that a polygon cannot, in
general, be expressed uniquely as the finite union of its convex
subsets. Moreover, decompositions into convex subsets may not
necessarily correspond to a natural organization of the shape.

BOUNDARY METHODS

Several classes of shape models based on the boundary of a
shape have been proposed, including, polygonal approximations, Fourier
descriptors, B-splines, Hough transforms and shape numbers. All of
these methods depend on extracting the object in terms of the boundary
between the object and the background.

Chain codes, and more generally, boundary segments
determined by piecewise functional approximations provide a slope
intrinsic representation of shape. A grid is superimposed on the
shape, and one of two methods is used to encode the boundary of the
shape. Some starting point on the boundary is arbitrarily chosen; then
for each point of intersection of the shape boundary with a grid line,
the nearest point of the grid is included in the encoding of the shape
boundary. If the elements of the encoding are joined together, they
define a polygonal arc which is an approximation to the shape boundary.
This arc is called a chain and can be completely specified by giving a
starting point and the successive directions necessary to follow the
chain. For an eight neighbor grid, these directions can be efficiently
encoded with 3 bits corresponding to directions 0 degrees to 315 degrees

in 45 degree increments. Such a list is called a chain code, and each element is a chainlet.

Various properties of chain codes can be easily derived and used in defining shapes (see Freeman (1974)). For example, the Euclidean length of a chain is the sum of the number of even chainlets and the /2 times the number of odd chainlets. Some other properties which can be easily computed include the maximum cross section length in any of the eight given orientations and the first moments in any of these orientations. This is quite attractive in many applications.

When the approximating functions are line segments, then this is a generalization of the chain code which allows for arbitrary lengths and directions of chain elements (also see Freeman (1979) on generalized chain codes). In general, the shape boundary is segmented into pieces described by arbitrary functions. It is usually sufficient for the functions to be restricted to low-order polynomials (of degree less than three).

Perhaps the most frequently used shape boundary representation is the piecewise linear approximation of the boundary. Many algorithms have been proposed for computing various piecewise linear approximations - see, e.g., Pavlidis (1973a, 1974, 1975), Ramer (1972), Rosenfeld & Johnston (1973) and Davis (1977). These procedures can be categorized as one of two types: (1) those that attempt to find the lines directly, and (2) those that attempt to find the break points directly. The first class of procedures search for boundary segments which are well fit by lines, while the second class search for boundary points with locally high curvature, that is, they are angle detectors. Of all these algorithms, the split-and-merge algorithm proposed by Pavlidis (1974) seems to be the most robust (i.e., it has very slight sensitivity to small changes in the underlying shape) yet is at the same time computationally efficient.

Piecewise linear approximation can be used not only for feature extraction, but also for noise filtering and data compaction (see Pavlidis (1973b)). Let $S = \{(x_i, y_i)\}$, $i = 1, N$, be a set of points; the problem is to find a minimum partition of S into n subsets $s_1, \ldots, s_n$ where $s_i = \{(x_j, y_j)\}$ $j = K_i, l_i$ and s_i is approximated by a straight line with an error norm less than a prespecified threshold, E_{max}. Given an error norm E and a segmentation $s_1, \ldots, s_r$, let E_i be the value of the norm evaluated on s_i, and let $E = E_i$, $i = 1,r$. Then the split-and-merge algorithm is:

1. For $i = 1,r$, if $E_i > E_{max}$, then split s_i into two subsets, set $r = r + 1$ and calculate the error norms of the two subsets. The breakpoint for obtaining the two new sets can be chosen in one of several ways. Pavlidis proposes using the point which contributes the most to the error value or the midpoint.

2. For $i = 1,r$, if E_i and E_{i+1} can be merged and the error norm of the new segment is less than E_{max}, then merge E_i and E_{i+1} and reduce r by one. Compute the error norm of the new segment.

3. Adjust the segment end points to minimize E. If no changes were made in steps (2) or (3), then terminate, else go to (1).

There are many reasonable choices for the error norm. Pavlidis describes the Euclidean distance between $\{x_i, y_i\}$ and the approximating curve, while Horowitz (1977) demonstrates the use of the mean square error norm. To minimize the latter norm means to find:

$$\min_{a,b} \{ \Sigma [y_i - (ax + b)]^2 \}$$

There exist unique closed forms for a and b:

$$a = (n\Sigma x_i y_i - \Sigma x_i \cdot \Sigma y_i)/(n\Sigma x_i^2 - (\Sigma x_i)^2)$$

$$b = (\Sigma y_i - a \cdot \Sigma x_i)/n.$$

An important practical advantage of this error norm is that the various sums used in defining a and b need not be recalculated completely for updating purposes; they can be directly added or subtracted as sums.

One can obtain higher order approximations using the split-and-merge algorithm with higher degree polynomials, but as Pavlidis discusses (1973a), the computational cost increases dramatically as one raises the degree of the approximating curves. Furthermore, the algorithms become numerically less stable. Pavlidis suggests using the results of the piecewise linear approximations to selectively guide the application of higher order approximation procedures to pieces of the shape boundary.

A description of 2-dimensional non-intersecting closed curves proposed by Bribiesca & Guzman (1979) called the shape number is associated with the boundary of simply connected regions. This shape number is obtained by laying a grid over a shape and encoding the border around the grid squares that fall (at least 50% of the square) within the shape. In particular, the shape number has several chain code representations, namely, one for each starting point on the boundary. If the derivative of the chain code is used, i.e., replace each convex corner of the chain by a 1, each straight corner by a 2, and each concave corner by a 3, then the chain with the minimum value when viewed as an integer can be used to represent the shape. In order to normalize the shape, a grid is chosen after the shape is surrounded by its basic rectangle (i.e., its orientation coincides with the major axis of the shape). Then the shape can be represented to any level of detail by refining the grid. In this way, a shape description is obtained that is independent of size, orientation and position.

The boundary of a 2-D shape can be expressed in term of slope of the boundary as a function of arc length, Zahn (1972), or as a complex parametric function, Granlund (1972). In either case, the function is periodic and can be expanded in a Fourier series. The shape can be approximated to any desired degree of accuracy by retaining a sufficient number of terms of the series. For example, suppose that the boundary of a shape is expressed in parametric form:

$$Z(t) = (x(t), y(t)), \quad 0 \le t < L,$$

where L is the length of the boundary. Let $T(t)$ be the angular direction at point t, and let $T(0) = C$. Define $P(t)$ to be the amount of angular bend between the starting point and t. Then, $P(t) + C = T(t)$. The function $P(t)$ can be normalized over the interval [0,2] as follows:

$$P^*(t) = P(Lt/2) + t.$$

Then, $P^*(0) = P^*(2) = 0$. There exists a sequence $\{A_k, a_k\}$, k=1 to ∞, and u such that

$$P^*(t) = u + \Sigma A_k \cos(kt - a_k).$$

The A_k, a_k k = 1 to ∞ are the Fourier descriptors.

Transforms other than the Fourier transform can be used, e.g., the Walsh transform. However, one of the main advantages of the Fourier transform is the well-developed theory and software to implement it. Fourier shape models can be made independent of position, orientation and scale. The major disadvantages of the method are that local features of the shape are difficult to describe without taking many coefficients and that obstruction or agglomeration produce coefficients unrelated to those of the original shape.

An approach to boundary representation that has received much attention in the areas of computer graphics and CAD/CAM is the use of B-splines. As described by Gordon & Riesenfeld (1974), a polynomial spline is a generalized polynomial with specified points of derivative discontinuity. Usually, a spline function of degree m-1 is defined over a sequence of intervals and is a polynomial of degree m-1 on each subinterval, and its derivatives of orders 1,2,...,m-2 are everywhere continuous. Various bases exist for the space of all such spline functions, however, the B-spline basis functions are a commonly used basis. Arbitrary primitive functions can be represented by a B-spline approximation which consists of a weighted sum of the basis functions. B-spline approximations are variation diminishing representations and provide local approximations. Although these are convenient properties of general splines, only first order splines have had much success in shape analysis since the theoretical and computational problems are much more complex for higher-order splines; moreover, in shape perception, linear approximations capture most relevant information and continuity conditions are usually not too important (see Pavlidis (1977) for a discussion of the use of splines in shape analysis).

Another approach to 2-D shape modeling is to transform the elements of the shape into a parameter (or transform) space. The Hough transform is perhaps the most important example of this approach. Originally, the Hough transform was used to detect simple curves, e.g., lines or circles. Usually this method is applied to an edge image, i.e., a binary or thresholded description of the edges in the original image. The (x,y) location of an edge response restricts the set of possible lines that this edge could lie on. This set of lines can be represented by a couple of quantized parameters, e.g., slope and intercept. The complete set of lines possible for the whole image can be represented by an accumulator array whose axes are slope and intercept.

Then, for every (x,y) location of an edge response the accumulator array is incremented for every possible line through (x,y).
If many edge responses lie on the same line then this results in a high value at the position in the accumulator array corresponding to that line.

The Hough transform has been generalized for arbitrary shapes in the plane by Merlin & Farber (1975). Given a list $(x_i,y_i) i=1,n$ of locations designating a shape, then the shape is modeled by choosing a reference point, (x_o,y_o), and keeping list of displacement vectors $D=(\delta x_i, \delta y_i)$ i = 1 to n where $\delta x_i = x_o - x_i$ and $\delta y_i = y_o - y_i$. This list constitutes the model for that shape. In order to detect the shape in an image, an accumulator array H is initialized to zero and for each edge location (x,y) detected in the image, $H(x+\delta x_i, y+\delta y_i)$ is incremented by one. Maxima in H should represent the location of the reference point (x_o,y_o). This algorithm is actually an efficient binary convolution of the shape model and the edge image. Moreover, this approach can be extended to account for orientation and scale changes (see Davis & Yam (1981)). If one is willing to pay the overhead of a complete edge description, e.g., edge likelihood and orientation, then one can use even more efficient generalization of the Hough transform has been proposed by Ballard (1981).

SPECIAL AXES METHODS

Sweep representations describe shape in terms of special axes which can be given as a set of points or as a function. Associated with each point of the axis is either a geometric object, e.g., a circle, or some deformation of that object. The two major examples of this approach are the medial axis transform and ribbons: the 2-D version of generalized by cylinders.

The medial axis transform proposed by Blum (1964 differs from the previously discussed methods in that a new object is derived from the given one. This is one of the earliest proposed shape modeling techniques and has been widely studied. Let R be the set of points defining an object and let B be the set of boundary points of R. Let N_x be the set of points that are in B and whose distance from x is less than or equal to the distance of x from any other point of B. Then medial axis transform consists if two parts:

- The set $S = \{x : N_x \text{ has more than one member}\}$, and

- The radius of the largest disk contained in R and centered at x for each x in S.

Thus, a spatial decomposition of R is given in terms of S, also known as the skeleton of R. The medial axis transform has many desirable properties; for example, to determine whether or not a point, p, is in the interior or R, one need only compute the distance of p from each point s of the skeleton and see if that distance is less than the radius of the disk associated with s.

The medial axis transform has several disadvantages. The skeleton of an arbitrary object is not as economical a representation as the boundary of the object. Moreover, digital approximations to the skeleton may not be connected and are very sensitive to noise. Finally, there is no obvious way to compute properties of the original shape directly from the skeleton.

Another successful sweep representation is that of ribbons. A ribbon is a 2-D restriction of the 3-D shape modeling method of generalized cylinders. Basically, a ribbon is a means of describing the projection of a generalized cylinder. 3-D objects are described by a basic 2-D shape, an axis along which the shape is moved and a description of the transformation of the 2-D shape as a function of position along this axis. A ribbon describes the relations between lines in an image in the same way, except that the special axis is restricted to stay in the plane, as is the 2-D shape (see Brook (1981)).

REFERENCES

Ballard, D.H. (1981). Generalizing the Hough Transform to Detect Arbitrary Shapes. Pattern Recognition, $\underline{13}$, no.2, 111-122.

Blum, H. (1964). A Transformation for Extracting New Descriptions of Shape. Symp. on Models for the Perception of Speech and Visual Form, MIT Press.

Bribiesca, E. & Guzman, A. (1979). How To Describe Pure Form and How To Measure Differences in Shape Using Shape Numbers. Proc. PRIP, Chicago, 427-436.

Brooks, R. (1981) Symbolic Reasoning Among 3-D Models and 2-D Images. Rept. No. STAN-CS-81-861, June.

Davis, L. (1977). Understanding Shape, I: Angles and Sides. IEEE T. on Computers, $\underline{C-26}$, 236-242.

Davis, L. & Yam, S. (1980). A Generalized Hough-like Transformation for Shape Recognition. TR-134, Univ. of Texas.

Freeman, H. (1974). Computer Processing of Line-Drawing Images. Computing Surveys, March, 57-97.

Freeman, H. (1979). Use of Incremental Curvature for describing and Analyzing 2-D Shape. Proc. PRIP, Chicago, 437-444.

Gordon, W.J. & Riesenfeld, R.F. (1974). B-Spline Curves and Surfaces. In Computer Aided Geometric Design, ed. R.E. Barnhill & R.F. Riesenfeld, pp. 95-126. New York: Academic Press.

Granlund, G.H. (1972). Fourier Preprocessing for Hand Print Character Recognition. IEEE T. on Computers, $C-21$, 195-201.

Horowitz, S. (1977). Peak Recognition in Waveforms. In Syntactic Pattern Recognition Applications, ed. K.S. Fu, pp. 1-31. Berlin: Springer-Verlag.

Hu, M. (1962). Visual Pattern Recognition by Moment Invariants. IRE T. on Inf. Theory, $IT-8$, 179-187.

Merlin, P.M. & Farber, D.J. (1975). A Parrallel Mechanism for Detecting Curves in Pictures. IEEE T. on Computers, $C-24$, 96-98.

Pavlidis, T. (1973a). Waveform Approximation Through Functional Approximation. IEEE T. on Computers, $C-22$, 689-697.

Pavlidis, T. & Horowitz, S. (1973b). Piecewise Approximation of Plane Curves. Pattern Recognition, 346-405.

Pavlidis, T. & Horowitz, S. (1974). Segmentation of Plane Curves. IEEE T. on Computers, $C-23$, 860-870.

Pavlidis, T. & Ali, F. (1975). Computer Recognition of Handwritten Numerals Using Polygonal Approximations. IEEE T. on Systems, man, and Cybernetics, $SMC-5$, 610-614.

Pavlidis, T. (1977). Structural Pattern Recognition. Berlin: Springer-Verlag.

Ramer, U. (1972). An Iterative Procedure for Polygonal Approximation of Plane Curves. CGIP, 1, 244-256.

Rosenfeld, A. & Johnston, E. (1973). Angle Detection in Digital Curves. IEEE T. on Computers, $C-22$, 874-878.

Rosenfeld, A. & Kak, A. (1976). Digital Picture Processing. New York: Academic Press.

Zahn, C.T. & Roskies, R.Z. (1972). Fourier Descriptors for Plane Closed Curves. IEEE T. on Computers, $C-21$, 269-281.

SYNTACTIC AND STRUCTURAL METHODS I

Thomas C. Henderson
Department of Computer Science
The University of Utah
Salt Lake City, Utah 84112

The syntactic and structural methods developed for 2-D shape analysis derive from their more traditional counterparts in formal language theory. The notions of alphabet, string, grammar and language have been extended to account for the description of shapes in the plane. However, in applying syntactic techniques to the analysis of 2-D shapes, several major problems must be overcome which do not arise in classical language theory:

- On a theoretical level, concatenation is rarely the only relation possible between symbols; e.g., one piece of a shape can be above, below, left of or right of another piece of the shape.

- On the practical side, the quality of an actual image may require a preprocessing step, e.g., enhancement, filtering, etc., and for most structural approaches a pattern representation must be determined from the image before any syntax analysis can be performed.

The approaches described here attempt to provide solutions to these problems (see Gonzales & Thomason (1978) for an introduction to syntactic pattern recognition).

Some familiarity with formal language theory, e.g., Hopcroft & Ullmann (1969) is assumed, however, a brief review of the basic notions is in order. An <u>alphabet</u>, A, is a finite set of symbols or characters. A <u>string</u>, x, over A is defined as either the empty string, e, or as the concatenation of a non-empty symbol of A with a string s over A. The number of symbols in a string is denoted by $|x|$, and $|e| = 0$. The <u>closure</u> of A, called A^*, is the set of all sentences over A, including e. The <u>positive closure</u> of A, called A^+, is $A^* - \{e\}$. A <u>language</u> is a set of sentences over an alphabet, i.e., a subset of A^*. We define a^n as the empty string if n = 0, otherwise, a concatenated with a^{n-1}. A <u>grammar</u> for a language is a finite set of rules for generating exactly the set of strings in the language, and is defined formally as a 4-tuple, $G = (N,T,P,S)$, where N is a finite set of nonterminals, T is a finite set of terminals, P is a finite set of productions, and S in N is the start symbol. Let $V = N \cup T$, $N \cap T = 0$, and P consists of rewrite rules of the form a -> b, where a in V^*NV^* and b in V^*, meaning that the string a may be rewritten as the string b. Let r and s be strings in V^* and a -> b be in P. By ras => rbs is meant that rbs is derivable from ras by the application of a single rewrite rule. The symbol $=>^*$ indicates the application of zero or more rewrite rules, and the symbol $=>^+$ means one or more applications. If for every production a -> b in P, a is in N and b is of the form cd, where c is in T and d is in N, then the grammar is <u>regular</u>. If a is in N and b is in V^+, then the grammar is <u>context-free</u>. If a = tcs and b = tds, where t and s are in V^* and d is in V^+ and c is in N, then the grammar is <u>context-sensitive</u>. An unrestricted grammar is one with any kind of productions. The subset of T^* obtained from S by using a finite number of productions is denoted by L(G) and is called the language generated by G:

$$L(G) = \{x | x \text{ in } T^* \text{ and } S =>^* x\}.$$

A <u>sentential form</u> of a grammar G is a string over V derivable from S.

Once a grammar has been defined, some type of recognition device is required. The application of such a recognizer is called <u>parsing</u>. The goal is to produce a description of the test string in terms of the syntactic model. String parsing techniques can also be used in shape analysis, and such diverse methods as Earley's algorithm and precedence parsers have been used. Finally, syntax-directed translation has been used to create structural descriptions of shape (see Fu (1974)).

Syntactic pattern recognition analysis consists of three major steps:

1. <u>preprocessing</u>-improving the quality of an image containing the shape, e.g., filtering, enhancement, etc.

2. <u>pattern representation</u>-segmenting the shape and assigning the segments to the parts in the structural model, and

3. <u>syntax analysis</u>-organizing the primitive shape according to the syntactic model.

The preprocessing step should make the primitive extraction easier and more reliable. The primitive parts produced by the segmentation step will in turn be used by the syntax analysis component. The syntax analysis step consists of determining if the set of primitives (or some subset) is syntactically well-formed. As a result of the syntax analysis, a parse tree is produced, and the parse tree can be used not only for recognition purposes, but also as a description of the shape.

A structural shape model describes the spatial decomposition of a shape, and consequently, must describe the primitive parts composing the shape. There are no established guidelines for choosing shape primitives; however, there are several desirable characteristics. Primitives should provide a compact description of the shape with little or no loss of information, and the extraction of shape primitives from an image should be relatively simple using existing non-syntactic techniques.

Once the primitives have been obtained, relations between the primitives are computed. For piecewise linear approximations, relations such as adjacency, collinearity and symmetry may be computed. The exact relations computed are determined by the form of the structural model. There are, essentially, two kinds of structural model: a single-level grammar or relational network, and a multi-level grammar. If relational networks are used, then shape analysis involves graph matching. Grammatical models require a parsing procedure. In this chapter, the application of string grammar techniques to shape analysis will be studied.

STRING GRAMMARS

Perhaps the greatest attraction of the syntactic method is the convenience with which a hierarchical description of a shape can be specified. For example, an airplane can be described as composed of a head section and a tail section. These in turn can be described in terms of the nose, wings, fuselage, etc. Thus, not only is a global description of the shape available, but also very local descriptions.

There are two main disadvantages in the use of the syntactic method for analyzing the boundaries of planar objects according to a string grammar:

1. Noise tends to complicate the process of computing the appropriate string structure, and

2. Context sensitive grammars are ordinarily necessary to analyze the complete boundary of a closed curve.

If piecewise linear approximations are first fit to the boundary of a shape, then using these primitives as input for syntax analysis will suppress the noise in the boundary proportionately to the degree that the linear segments fit the boundary. If the syntax analysis is restricted to only a certain level, then context sensitive grammars are not necessary.

When the boundary of a shape is modeled by a string grammar, the formal theory of automata can be used to parse strings of shape primitives, and a class of shapes can be characterized as a set of strings. A major difficulty is the contruction of a simple grammar (preferably regular or context free) which generates only these strings. Another approach is to regard combinations of the lowest level symbols, i.e., terminal symbols, as features. Then a shape can be characterized by these features.

String grammars have been used to detect peaks in waveforms by Horowitz (1977). Selected waveform features are extracted and compared with predetermined standards to monitor and classify a wide range of phenomena. The technique is composed of two parts:

1. The waveform is fit with piecewise linear approximations and encoded as a string of symbols, and

2. The string is parsed to recognize peaks.

The symbols of the grammar correspond to positive, negative, and zero slope, and each linear segment of the piecewise linear approximation is assigned one of these symbols according to its slope. The set of peaks is described by a set of regular expressions, and a deterministic context free grammar is derived which recognizes peaks (if any) in the waveform if the non-terminal symbol corresponding to a peak is derived for some part of the waveform.

PICTURE DESCRIPTION LANGUAGE

Major theoretical impetus was given to the area of syntactic shape analysis when Shaw (1969) designed the Picture Description Language (PDL). PDL was intended not only for interpretation of pictures, but also for picture generation. In PDL, picture primitives are described as named objects, and objects have attributes. Two special attributes are the head and the tail of an object, denoted head(a) and tail(a), respectively, where a is an object. Six different concatenation operations are defined as follows:

1. a+b: Join head(a) to tail(b),
2. axb: Join tail(a) to tail(b),
3. a-b: Join head(a) to head(b),
4. a*b: Join tail(a) to tail(b) and Join head(a) to head(b),
5. a&b: Join tail(a) to head(b) and Join head(a) to tail(b), and
6. /a: Reverse the head and tail of a.

Objects can be defined using these six operators. Any arbitrary graph of connected primitives can be recognized or generated using these six operators. For describing string languages, only + is needed. A grammar can be defined to express a given class of pictures.

The PDL description of a simple house shape is given by:

T = {+,x,*,-,&,/,0,1,2,3}, where 0,1,2,3 represent the unit vector oriented at 0, 45, 90, and 135 degrees, respectively and whose heads are at the right, upper right, top, and upper left, respectively.

N = {<house>, <top>, <bottom>, <roof>, <right>, <down>},

P = { <house> -> <top>*<bottom>,
       <top>   -> <roof>*0,
       <roof>   -> 1+<right>,
       <right> -> /3,
       <bottom> -> <down>+0+2,
       <down>   -> /2 }, and

S = <house>.

Although a successful experiment was described concerning spark chamber events, there are several problems with PDL. It is a very unnatural representation for plane objects. Moreover, spatial relationships such as inside, to the right of, and above are difficult, if not impossible, to represent. Finally, PDL allows the description of self-intersecting patterns, e.g., (0+3)x1 with the symbols given above, and it is not easy to determine if a grammar allows such patterns.

A generalization due to Feder (1971) allows an arbitrary number of attachment points for syntactic objects. Such grammars are called plex grammars, and they can be used to describe general graphs.

Such grammars, however, are even more cumbersome to use than PDL, and a set of productions may easily have unforeseen side effects.

TREE GRAMMARS

In using higher-dimensional grammars for describing structured patterns, one way to limit the complexity of the grammar is to use trees instead of arbitrary graphs. These grammars are called tree grammars, and their use in pattern recognition has been demonstrated in fingerprint classification by Moayer & Fu (1974), bubble chamber event analysis by Bhargava (1972) and LANDSAT data interpretation by Li & Fu (1976).

Our development of tree grammars follows that of Fu & Bhargava (1973). Let U be the free monoid generated by N^+, the set of positive integers. Let + be the operator and 0 the identity element of U. Let d(a) denote the depth of a in U, where d(0) = 0 and d(a+i) = d(a) + 1, for i in N^+. We say a < b iff there exists x in U such that a+x = b. A tree domain D is a finite subset of U satisfying: 1) if b in D and a < b, then a in D, and 2) if a+j in D and i < j in N^+, then a+i in D. The first condition of a tree domain ensures a path to the root, while the second condition ensures that if any son of a node is in the domain, then all sons less than that son are in the domain.

A stratified alphabet is a pair <T,r>, where T is a finite set of symbols and r : T -> N, where N = N^+ {0}.

For x in T, r(x) is called the stratification of x. A tree is defined as a mapping from a tree domain to a set of symbols. A tree over <T,r> is a function a : D -> T such that D is a tree domain and r[a(x)] = max{i|x+i in D}. Let D(x) denote the domain of a tree, and let Tr(T) be the set of all trees over T.

A regular tree grammar over <T,r> is a system G = (V,r',P,S), where

1. <V,r'> is a finite stratified alphabet with T in V and r' restricted to T = r,

2. P is a finite set of production pairs of the form (a,b), where a and b are trees over <V,r'>, and

3. S is a finite subset of Tr(V), the set of trees over <V,r>.

We must now show how to represent patterns using trees. Let

the root of a tree be the special symbol $ which corresponds to a physical point. Just as in PDL, symbols in the grammar may correspond to geometric objects. Given a tree, the corresponding pattern can be drawn by starting at the root and drawing line segments according to the descendent of each node. The tail of the descendent is connected to the head of the current node until only nodes with no descendents remain. For example, to describe the same house as was shown in PDL, let a, b, c and d be the same as 0, 1, 2 and 3 in the PDL grammar. Then the house can be described as:

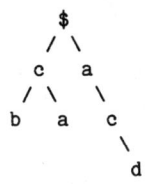

```
       $
      / \
     c   a
    / \   \
   b   a   c
            \
             d
```

Fu & Bhargava give examples of grammars for directed triangles, chemical structures, electrical circuits, the English alphabet and bubble chamber events.

A major difficulty with using tree automata for pattern recognition is the problem of noisy or distorted data. Error-correcting tree automata have been proposed by Lu & Fu (1978) to deal with these problems; in particular, a distance between two trees can be defined in terms of the least number of error transformations (e.g., substitution or deletion) required to obtain one from the other. The similarity of trees can be defined in terms of the given tree metric.

ARRAY AND WEB GRAMMARS

Pictures are not very easily described as a concatenation of strings. For picture processing, the 2-D analog of the string is the array. Generalizations from phrase structure grammars to grammars capable of dealing with pictures have been proposed by several investigators (see Rosenfeld (1979)). The organization of a picture is in terms of an array, each of whose elements is an intensity in the picture. A picture can then be viewed as composed of subarrays which are themselves composed of subarrays. This gives rise to the following definition:

An <u>array</u> <u>grammar</u>, G, is a quadruple (N,T,P,S), where

- N is a finite set of non-terminals,
- T is a finite set of terminals,
- P is a finite set of pairs (a,b) called productions in which a and b are geometrically identical arrays; a production is applied to an array s by finding a and replacing it with b,
- S in N is the start symbol.

A special symbol, $, is used to allow the definition of geometrically identical arrays; this end marker symbol is sometimes added as the fifth part of the grammar. The language of G is the set of arrays derivable from the initial array (an infinite array composed of the symbol S surrounded by an infinite number of $'s) and consisting of $'s and terminal symbols, where all non-$'s are connected.

An array grammar is <u>monotonic</u> if the productions do not allow for the creation of $. Monotonicity guarantees connectedness of the non-$'s if the non-$'s in all rewrite rules are connected. An array grammar is called <u>isotonic</u> if for every production pair (a,b), a and b are geometrically identical arrays. For us then, array grammars are by defintion isotonic. When replacing one string by another in a string grammar, the result is a new string. However, when replacing an arbitrary subarray b, a problem is encountered in making the rows and columns of b match the size of the array left vacant by a. One solution to this problem is to impose the isotonic property on the productions and use the special symbol $ as filler.

Thus, we see that array grammars provide a reasonable notation for describing pictures when pictures are thought of as arrays. On the other hand, defining more general relations between object parts and specifying attributes of these parts is not a straight-forward process. A generalization of this notion will allow us to accomplish these things.

Web grammars generalize the notion of phrase structure grammars from strings to labeled graphs (see Pfaltz & Rosenfeld (1969)). Both the nodes and the arcs of the graph can be labeled, and the arcs can be directed. A web grammar is still a quadruple, G=(T,N,P,S), but the vocabulary symbols now correspond to webs (labeled graphs). The start symbol, S, is a one-node graph. A production is a pair (a,b), where a and b are webs, and every production must now specify how b is attached to the neighborhood of a. This specification is called an **embedding**. A **normal embedding** has been defined to be one that calls for the same number of nodes in a as in b. Other approaches give embeddings in terms of the neighbor labels or in terms of homomorphisms.

In describing an embedding, it has been found useful to use negative context as well as positive context. Positive context includes neighbor labels that can appear, while negative context rules out possible labels for neighbor nodes. This does not arise in string and array grammars since only a finite number of neighbors are possible, and therefore, negative context can be written as positive context. Since webs allow for an arbitrary number of connections between nodes, negative context becomes necessary.

Much of the hierarchy (in the Chomsky sense) of string grammars carries over to web grammars, and finite-state, context-free, context-sensitive and monotonic web grammars can be defined in a similar way. Brayer & Fu (1974) have shown context-sensitive and monotonic web grammars to be equivalent; they also provide a wide range of other results. For a discussion of inclusion relations between classes of web grammars, see Abe et al (1973).

Brayer et al (1977) have described an application of web grammars in pixel classification in LANDSAT images. Grammars are developed for clouds, shadows, downtown areas and highways, and these provide guidance in recognizing complexly structured land use classes in the imagery in a way that is not possible with spectral properties alone. It was shown that the use of spatial information was of some help in complex natural scenes, and that the amount of structure present allows a significant application of this syntactic technique to the land use problem.

REFERENCES

Abe, N., Mizumoto, M., Toyoda, J. & Tanaka, K. (1973). Web Grammars and Several Graphs. J. Comp. and Sys. Sci., 7, 37-65.

Bhargava, B.K. (1972). Application of Tree Systems Approach to Classification of Bubble Chamber Photographs. TR-EE 72-30, Purdue U.

Brayer, J.M. & Fu, K.S. (1974). Some Problems of Web Grammars. TR-EE 74-19, Purdue U.

Brayer, J.M., Swain, P.H. & Fu, K.S. (1977). Modeling of Earth Resources Satellite Data. In Syntactic Pattern Recognition, Applications, ed. K.S. Fu, pp. 215-242, Berlin: Springer-Verlag.

Feder, J. (1971). Plex Languages. Inf. Sci., 3, 225-241.

Fu, K.S. & Bhargava, B.K. (1973). Tree Systems for Syntactic Pattern Recognition. IEEE T. on Computers, C-22, 1087-1099.

Fu, K.S. (1974). Syntactic Methods in Pattern Recognition. New York: Academic Press.

Gonzalez, R.C. & Thomason, M.G. (1978). Syntactic Pattern Recognition. Reading, Ma.: Addison-Wesley.

Hopcroft, J.E. & Ullman, J.D. (1969). Formal Languages and Their Relation to Automata. Reading, Ma: Addison-Wesley.

Horowitz, S. (1977). Peak Recognition in Waveforms. In Syntactic Pattern Recognition, Applications, ed. K.S. Fu, pp. 1-31, Berlin: Springer-Verlag.

Li, R.Y. & Fu, K.S. (1976). Tree System Approach for LANDSAT Data Interpretation. Purdue-LARS Symp. on Mach. Proc.. of Remotely Sensed Data, West Lafayette, In.

Lu, S. & Fu, K.S. (1978). Error-Correcting Tree Automata for Syntactic Pattern Recognition. IEEE T. on Computers, C-27, no. 11, 1040-1053.

Moayer, B. & Fu, K.S. (1974). Syntactic Pattern Recognition of Fingerprints. TR-EE 74-36, Purdue U.

Pfaltz, J.L. & Rosenfeld, A. (1969). Web Grammars. Proc. 1st IJCAI, 609-619.

Rosenfeld, A. (1979). Picture Languages. New York: Academic Press.

Shaw, A.C. (1969). A Formal Picture Description Scheme as a Basis for Picture Processing Systems. Inf. and Con., 14, 9-52.

SYNTACTIC AND STRUCTURAL METHODS II

Thomas C. Henderson
Department of Computer Science
The University of Utah
Salt Lake City, Utah   84112

The emphasis up to now has been on the use of string grammar techniques for the description of shape. However, it is sometimes preferable to trade parsing efficiency for descriptive power, and to this end a much richer class of geometrical grammars and parsing techniques for those grammars have been developed. Alternatively, one can forego grammars altogether and describe shape as a relational structure; this implies a graph matching approach to shape analysis. A side effect of this approach is the construction of a structural description of the shape being analyzed; in the grammatical approach, this description depends directly on the grammar, whereas for relational models, the description depends on the defined relations.

GEOMETRICAL SHAPE GRAMMARS

The notions of formal language theory can be modified so as to allow for a more direct description of the geometrical relations between the pieces of a shape. Thus, geometrical shape grammars must not only describe how the primitve pieces of the shape are joined together (syntactic coincidence), but also how well various relations between geometrical properties of the primitives hold (semantic consistency).

An example of this approach is the <u>stratified context-free shape grammar</u>, or SCFSG, of Henderson & Davis (1981c), which is a quadruple (N,T,P,S) just like for a string grammar, but with a much richer structure. Let V = N U T, then for every v in V,

v = <name part> {attachment part} ⌊semantic part⌋

where <name part> is the unique name of the symbol, {attachment part} is a set of places where the symbol can be attached to other symbols, and ⌊semantic part⌋ is a set of properties, usually geometric, of the symbol.

Each vocabulary symbol represents a piece of the boundary of a shape.

This formulation differs significantly from grammars such as PDL or tree grammars in that a terminal symbol no longer represents an oriented line segment or physically defined piece of the shape, but rather a logical piece of the shape. For example, the terminal symbol of a tree grammar might represent a horizontal line segment of a specific length, while for a SCFSG the terminal symbols would represent wall, floor, roof, etc.

Productions in the grammar are also quite different from ordinary productions. Every p in P is of the form $(v::=a,A,C,G_a,G_s)$, where the rewrite part $v::=a$ means that the symbol v in N is composed of the string of symbols $a = v_1 v_2 ... v_k$ in V^+. Stratification of the grammar is achieved by assigning a level number to each symbol, indicated by $l(v)$; $l(S) = n$, for t in T, $l(t) = 0$, and in any rewrite rule $v::=a$, if $l(v) = k$, then for every v_i in a, $l(v_i) = k-1$. Furthermore, in order to apply a rewrite rule, it is necessary to check for appropriate syntactic coincidence and semantic consistency which are described by the A and C parts, respectively, of the production. Finally, if the rewrite rule is to be applied, then the G_a and G_s parts of the production describe how to form the attachment part and semantic part, respectively, of the new symbol.

Such a grammar allows an anthropomorphic layout of the semantics of the shape, i.e., productions can be described graphically, and relations such as parallel, equal length, etc. can be used in describing shape. Other approaches to the use of attributes and geometrical shape grammars are given in Vamos & Vassy (1973), Fu (1974) and Gonzalez and Thomason (1978).

PARSING TECHNIQUES

Parsing techniques for shape grammars are usually nondeterministic in nature, and classical top-down and bottom-up techniques have been applied to shape analysis. However, a start has been made toward a theory of shape grammar compilers, see Henderson (1981a) and Henderson & Davis (1981b). The approach advocated there involves producing a parsing mechanism based on both syntactic and semantic constraints between the pieces of the shape. A similar method has been proposed by Masini and Mohr (1978).

A constraint between two symbols can be represented in complete generality by a relation between two sets, see Montanari (1974). If X and Y are sets, then the characteristic function F can be used to represent relation R:

$$F : X \times Y \to \{0,1\}; \quad F(x,y) = 1 \quad \text{iff} \quad (x,y) \in R.$$

If X has m elements and Y has n elements, then 2^{mn} different relations exist between X and Y. An m by n matrix whose entries are just $F(x,y)$ can be used to represent a relation.

Negation, union, intersection and a partial ordering relation of set inclusion can be defined in a straightforward way. Relations form a complete lattice with greatest element the matrix of all ones and least element the zero matrix. Union and intersection act as sup and inf, respectively. A composition of relations is defined as:

$$R_{13} = R_{12} R_{23}$$
$$\text{iff} \quad R_{13}(r,s) = V[R_{12}(r,t) \wedge R_{23}(t,s)]$$

This is just Boolean matrix multiplication when relations are represented as matrices. A relation is total if every element of X and Y are in relation to some other element.

We would like to consider constraints between more than two variables at a time, but the amount of information grows exponentially with the number of variables. This poses practical difficulties since an n-ary relation R is any subset of $X = X_1 \times X_2 \times \ldots \times X_n$, and if each X_i has m elements, then there are a total of 2^p n-ary relations, where $p = m^n$. However, projections of n-ary relations to networks of binary relations present a useful alternative. A <u>network of binary relations</u> is a set of sets $X = \{X_1, X_2, \ldots, X_n\}$, and a relation R_{ij} from every set X_i to every set X_j, for $i,j = 1,n$. The nodes of the network are the elements of X, and the relations R_{ij} can be viewed as labeling the edges between nodes. Given X as above, there are 2^p networks of binary relations, assuming $R_{ij} = R_{ji}$ and $p = m^2 * [n*(n+1)/2]$.

N-ary relations can be defined in terms of networks of binary relations. The projection formula proposed by Montanari is:

$$R'_{ij}(i_1,i_2) = V\ a(i_1,i_2,i_3,\ldots,i_n)$$

where $i_3, i_4, \ldots, i_n = 1, m$.

Not all n-ary relations are representable by a network of binary constraints. Moreover, an n-ary relation may have many distinct network representations. Montanari shows that the projection formula provides a minimal (with respect to inclusion) network to represent an n-ary relation.

The central problem in using networks of constraints to represent an n-ary relation is to find the minimal network equivalent to the given network. No algorithm is known other than complete enumeration.

We now discuss the procedures for deriving local constraints from the shape grammar. Two types of constraints, syntactic and semantic, are described. The semantic attributes of a vocabulary symbol are computed from the attributes of the symbols which produce it (see Knuth (1968) for a discussion of defining semantics for context-free languages using both synthesized and inherited attributes; we use only synthesized attributes here). Consider a vocabulary symbol as representing a piece of the boundary of a shape. If a hypothesized vocabulary symbol is part of a complete shape, then it is adjacent to pieces of the shape which can combine with it to produce a higher level vocabulary symbol. Therefore, if the set of all possible neighbors of a vocabulary symbol is known, and at one of its attachment points no hypothesis for any of these symbols exists, then that hypothesis can be eliminated. This type of constraint is called a syntactic constraint. The other type of constraint involves some geometric relation between the semantic features of two vocabulary symbols, e.g., the main axis of an airplane is parallel to the axis of its engines.

Let $G = (N,T,P,S)$ be a SCFSG, let v,w,x be in V, let $at(v)$ denote the attachment points of v, and let av be in $at(v)$. We define:

1. v $\underline{Ancestor}_{av,aw}$ w iff there exists p in P such that the rewrite rule of p is $v::=\ldots w\ldots$ and there exists an aw in $at(w)$ such that aw is identified with av in G_a of p. That is, the attachment point of the left hand side symbol, v, is associated with endpoint aw of the right hand side symbol w.

2. w $\underline{Descendent}_{aw,av}$ v iff v $Ancestor_{av,aw}$ w.

3. v $\underline{\text{Neighbor}}_{av,aw}$ w iff there exists p in P such that the rewrite rule of p is x:=...v...w... and aw is specified as being joined to av in A of p, or there exists x in V with ax in at(x) and y in V with ay in at(y) such that x $\text{Ancestor}_{ax,av}$ v, and y $\text{Neighbor}_{ay,ax}$ x, and w $\text{Descendent}_{aw,ay}$ y.

Using matrix representations for these relations, the descendents and neighbors of a symbol at a particular attachment point can be computed, see Henderson & Davis (1981c) or Davis & Henderson (1981) for details.

Semantic constraints can be generated in exactly the same way as syntactic constraints, i.e., by defining binary relations and compiling their transitive closure. For example, the axes of two symbols are parallel if a production states this explicitly, or if each symbol has an ancestor parallel to itself and these ancestors are parallel. Such constraints also permit global constraints to be accounted for, e.g., the orientation of the main axis of an airplane could be fixed, and this certain information propagated throughout the system.

The result of such a parse of an unknown shape is a description in terms of the grammar used. This approach to shape analysis can be viewed as a type of pattern-directed translation in that the graph structure of relations on the primitives is mapped into a structural description based on the underlying grammar.

AMBIGUITY AND NOISE

One of the major problems facing any shape analysis method is that of the noise in the data. For example, extraneous primitives may be generated by spurious edges, or actual primitives of a shape may go undetected due to the poor quality of the image. Another major concern is the ambiguity of the primitives that are extracted to represent the patterns. For example, if the terminal symbols of a grammar represent the line segments oriented at 0, 45, 90 and 135 degrees, then what symbol is assigned an edge oriented at 22.5 degrees? How can this ambiguity be accounted for in the model or the analysis?

The problem of noise can be overcome to a great extent by smoothing the data. Many syntactic shape modeling methods use line segments as the primitive shape elements. Piecewise linear approximations provide an efficient means of data compression and noise suppression. However, the noise elimination techniques are usually directly related to the type of image analyzed and the type of shape model. e.g., string, tree or geometrical.

Elimination of ambiguity in the data is also closely related to the specific shape model used. A common method is to quantize the "primitive space" fine enough so that ambiguous primtives have little effect or are easily accounted for in the model. Alternatively, ambiguity can be accounted for in the grammar by allowing some leeway in the application of productions.

STOCHASTIC GRAMMARS

Stochastic grammars provide a mechanism for describing the effect of random events on a pattern. As such, stochastic grammars are interesting in their own right and need not be restricted to shape grammars. When noise and ambiguity affect the terminal symbols of a grammar, more terminal symbols may be introduced, i.e., new symbols account for deformed versions or perhaps parts of the original terminal symbols. These new symbols have a lower a priori likelihood of being in the data. We consider how this can be accounted for in a string grammar.

Given a language L, every string x in L can have a probability p(x) from (0,1] associated with it such that the sum of the probabilities of all the strings in the language is one. A <u>stochastic grammar</u> is a quadruple (N,T,P,S), where N and T are the non-terminals and terminals, respectively, S in N is the start symbol, and P is a finite set of stochastic production triples (a_i, b_{ij}, p_{ij}), $j = 1,\ldots,n_i$, and $i = 1,\ldots,k$, where a_i is in $(N \cup T)^* N (N \cup T)^*$, b_{ij} is in $(N \cup T)^*$, and p_{ij} is the probability associated with the application of the production, p_{ij} in (0,1] and the sum of the p_{ij} is one for $j = 1$ to n_i. Let (a_i, b_{ij}, p_{ij}) be in P. Then $c = da_i e$ may be replaced by $f = db_{ij} e$ with probability p_{ij}, that is, $c \Rightarrow f$ with probability p_{ij}. If a sequence of strings $w_1, w_2, \ldots, w_{n+1}$ exist such that $c = w_1$, $f = w_{n+1}$ and $w_i \Rightarrow w_{i+1}$ for $i = 1$ to n, then c generates f with probability $p = \prod p_i$, and $c \Rightarrow^* f$.

Let L(G) denote the stachastic language generated by G:

$$L(G) = \{(x, p(x) | x \text{ in } T^*, S \Rightarrow^* x,$$

$$\text{and } p(x) = \sum p_j, j = 1, k\}$$

where k is the total number of distinct derivations of x from S, and p_j is the jth derivation of x. See Fu (1974) for a complete description of stochastic grammars and recognizers.

GRAMMATICAL INFERENCE

A major inconvenience of grammatical methods is the necessity of designing a shape grammar to model the desired class of shapes. Some effort has been expended to directly infer the grammar from a set of examples. Grammars generated in such a way can be assigned a measure of goodness in terms of their complexity, that is, the number of non-terminal symbols and productions. The two major approaches to grammatical inference are enumeration and induction. Both of these methods will be briefly presented here, and their application to shape grammars discussed; see also Fu (1974) and Fu and Booth (1975a, 1975b).

Gold (1967) has formulated the theoretical study of grammatical inference by enumeration. Basically, given a set of strings in the target language and a set of strings not in the target language, then a grammar must be found. Different types of languages are produced depending on the method of presentation of information: text or informant presentation. The former provides both the set of strings in the grammar and those not in the grammar, while the latter gives only the set of strings in the grammar. An enumeration method to infer finite-state grammars has been developed by Pao (1969). A finite-state grammar is constructed for the set of strings; then, with the help of the informant, the grammar is generalized. Crespi-Reghizzi (1971) developed a method for inferring an operator precedence grammar. The search for a grammar can be made more efficient if the form of the grammar is known; this allows the elimination of a large class of grammars. The notion of one grammar covering another also allows elimination of all those grammars covered by an unsuccessful grammar.

Inductive inference methods proceed by discovering the recursive structure of a grammar. Given a valid string, delete substrings from it and determine if the resulting string is acceptable. If so, then substitute repetitions of the deleted substring and determine if the resulting string is acceptable. If it is, then a recursive structure of the language has been found.

Grammatical inference of shape is, like parsing, somewhat more complicated than its string grammar counterpart. Given a set of patterns, the problem is first to describe the patterns in terms of the chosen shape primitives and relations between them, and then to determine a shape grammar that generates that set. Thus, three things must be determined:

the primitives, the relations and the productions. Evans (1971) has produced a heuristic inference method. After generating a set of possible grammars, a measure of goodness determines which one is chosen. For example, such a measure could be proportional to the number of primitives and relations in the grammar and inversely proportional to the number of productions squared. Lee & Fu (1972) have extended string grammar inference techniques to an interactive shape grammar inference system.

A method for inferring tree grammars is described in Gonzalez & Thomason (1978). Given a sample set of trees, productions are formed for each tree separately. Productions are checked for embedding. Next, equivalent non-terminals are merged. After this step, non-terminals of the appropriate degree are combined. Finally, productions are added so that all the samples derive from the start symbol, and a tree grammar is formed. Examples of 2-D pattern grammar inference are given; however, the number of non-terminals and productions seems inordinately high. Techniques also exist for stochastic grammar inference, and Gonzalez and Thomason (1978) and Fu (1974) dicuss these.

RELATIONAL MODELS

Several high-level relational models have been described in the literature, see Fischler & Elschlager (1973), Davis (1979), Shapiro & Haralick (1979) and Shapiro (1980). Such models use convex parts or line segments as shape primitives. Once a decomposition of the shape in terms of the primitives is achieved, a relational description of the shape is constructed in terms of the relations. For example, Shapiro (1980) uses ternary relations between primitives which form intrusions and protrusions. A search is then performed to find mappings from a prototype to a test shape. This is equivalent to a subgraph isomorphism problem, and consequently is NP-complete. However, special look-ahead operators are used to reduce the amount of search required, see Haralick & Shapiro (1979). This approach to shape modeling allows for simple, yet sufficient models, which permit inexact and partial matches. However, the construction of the relations is quite expensive, being of order n complexity.

Semantic networks offer a similar approach to shape description, e.g., see Ballard et al (1978). Each node of the net represents a primitive, and nodes are connected by one or more arcs which indicate the relation between those nodes. Operations defined on a semantic net include: pattern-directed access, updating (deletion, addition and modification of nodes or arcs) and inference. Semantic nets are very limitied in scope unless they allow quantification and the use of logical connectives. Moreover, in most systems, encoding and decoding are done by hand.

REFERENCES

Ballard, D., Brown, C. & Feldman, J. (1978). An Approach to Knowledge-Directed Image Analysis. In Computer Vision Systems, eds. A. Hanson and E. Riseman, New York: Academic Press.

Crespi-Reghizzi, S. (1971). An Effective Model for Grammar Inference. IFIP Congress, Yugoslavia.

Davis, L. (1979). Shape Matching Using Relaxation Techniques. IEEE T. on Patt. Anal. Mach. Intell., PAMI-1, no. 1, 60-72.

Davis, L. & Henderson, T. (1981). Hierarchical Constraint Processes for Shape Analysis. IEEE T. on Patt. Anal. Mach. Intell., PAMI-3, no. 3, 265-277.

Evans, T.G. (1971). Grammatical Inference Techniques in Pattern Analysis, In Software Eng., ed. J.T. Tou, Vol. 2, New York: Academic Press.

Fischler, M. & Elschlager, R. (1973). The Representation and Matching of Pictorial Structures. IEEE T. on Computers, C-22, 67-92.

Fu, K.S. (1974). Syntactic Methods in Pattern Recognition. New York: Academic Press.

Fu, K.S. & Booth, T.L. (1975a). Grammatical Inference: Introduction and Survey - Part I. IEEE T. on Systems, Man, and Cybernetics, SMC-5, no. 1, 95-111.

Fu, K.S. & Booth, T.L. (1975b). Grammatical Inference: Introduction and Survey - Part II. IEEE T. on Systems, Man, and Cybernetics, SMC-5, no. 4, 409-423.

Gold, E.M. (1967). Language Identification in the Limit. Inf. and Control, 10, 447-474.

Gonzalez, R.C. & Thomason, M.G. (1978). Syntactic Pattern Recognition. Reading, Ma: Addison-Wesley.

Haralick, R.M. & Shapiro, L. (1979). The Consistent Labeling Problem. IEEE T. on Patt. Anal. Mach. Intell., PAMI-1, no. 2.

Henderson, T.C. (1981a). Shape Grammar Compilers. In Digital Image Processing, eds. J.C. Simon and R.M. Haralick, pp. 327-336, Dordrecht, Holland: D. Reidel.

Henderson, T.C. & Davis, L. (1981b). Compilateurs de Grammaires de Formes. Proc. 3rd AFCET Conf. on Patt. Rec. and Art. Intell., Nancy, France, 415-424.

Henderson, T.C. & Davis, L. (1981c). Hierarchical Models and Analysis of Shape. Pattern Recognition, 14, nos. 1-6, 197-204.

Knuth, D. (1968). Semantics of Context-Free Languages. Math. Systems Theory, 2, no. 2, 127-145.

Lee, H.C. & Fu, K.S. (1972). A Syntactic Pattern Recognition System with Learning Capability. Proc. Int. Symp. Comput. and Inf. Sci. (COINS-72), Miami Beach, Fl, 14-16.
Masini, G. & Mohr, R. (1978). Un Systeme de Reconnaissance des Formes. Proc. AFCET Symp. on Patt. Rec., 107-114.
Montanari, U. (1974). Networks of Constraints: Fundamental Properties and Applications to Picture Processing. Inf. Sci., 7, 95-132.
Pao, T.W. (1969). A Solution of the Syntactical Induction-Inference Problem for a Non-Trivial Subset of Context-Free Languages. Interim TR 69-19, U. of Pennsylvania.
Vamos, T. & Vassy, Z. (1973). Industrial Pattern Recognition Experiment - A Syntax Aided Approach. IJCPR, Wash., 445-452.

3-D SHAPE REPRESENTATION

O.D. FAUGERAS
I.N.R.I.A. Domaine de Voluceau - Rocquencourt
B.P. 105, 78153 - LE CHESNAY Cédex France

1. - INTRODUCTION

Computer Vision is an Information Processing problem in which it is useful to distinguish several levels of description [1].

1) The computational level : that is the formulation of a problem along with the mathematics and the physics relevant to its solution.

2) the algorithm level : that is a set of abstract solutions to the problem specified at the computional level. These are specifications of how a description in a particular input representation is to be manipulated to produce a desired output representation. Of course there can be many distinct algorithms for the same problem which differ in the kind of representations they employ and the kind of manipulations they use.

3) the mechanism level : the level of the physical entities which implement particular algorithms and explanations of how this is done.

Information processing problems, when viewed at the computational theory and algorithm levels involve the manipulation of abstract descriptions to produce other abstract descriptions. This means that the development of computational theories and algorithms for them requires an understanding of what these abstract descriptions are and how they differ. We will use the term representation to indicate a particular way of describing a given subject. A description in a representation will be an instance of its use. The characterization of representations is an important part of the development of computational theories and their algorithms. Because there are many possible ways to represent a given subject the clarity with which a problem and its solution can be formulated depends largely on identifying the representations that make explicit just that information upon which the problem depends.

In this chapter we are concerned with the shape of 3-D Objects which can be defined as the geometry of an objects physical surface. As we will see in the following, it is convenient to discuss shape representations in terms of their <u>primitive elements</u>, i.e. the elemental pieces of information out of which a description in the representation is constructed.

2. - ISSUES IN 3-D SHAPE REPRESENTATION

It should be clear from the above discussion that some representations are better than others for a given task. Badler and Bajcsy [2] and Marr and Nishihara [3] have stated some of the important issues that are raised when judging the suitability of shape representations for three-dimensional recognition problems and computer graphics :

1) what are the primitive elements in the representation ?

2) the coordinate system those primitives are specified in.

3) the organization of those specifications in a description.

4) what is the storage cost associated with the representation ?

5) what operations on the representation are natural, which one difficult ?

6) how can one representation be converted into another ?

7) can the desired description be computed from an image and at which cost ? (accessibility)

8) what class of shapes is the representation designed for . In particular does there exist a well-defined description for each shape in the class of shapes we are interested ? (scope and uniqueness)

9) do differences between descriptions in the representation reflect the relative importance of differences between the shapes described with respect to the task at and ? (stability and sensitivity).

2.1. - Coordinate system

The coordinate system used by a representation can be viewer-centered or object-centered. In the first case locations are indicated relative to the viewer whereas in the second case they are indicated relative to the object described. In terms of accessibility, a viewer-centered representation is preferable to an object-centered one since it does not require a three-dimensional coordinate transformation. However viewer-centered descriptions depend as much on the orientation of a shape as they do on the shape itself. As a result a viewer-centered approach to shape recognition gains accessibility in exchange for an enormous set of possible descriptions even when the class of shapes to be recognized is small. It is interesting to note that all computer graphics representations are object-centered and most of the work is spent on going from this representation to a viewer-centered representation in order to display an image of the shape. In computer vision most of the work is spent doing the reverse that is on going from a viewer-centered representation obtained from the output of the various sensors that look at the scene to an object-centered representation.

2.2. - Primitives

A representation's primitive elements can be surface-based or volumetric. A 3-D object can be modelled by storing its surface as :

1) a set of surface points.
2) a polygonal network.
3) curved surface patches.

Surface points are represented by a list of coordinate triples and can be obtained from digitizers [5, 6], from stereo pairs [7, 8] or derived from serial section [9, 10]. A polygonal network is a decomposition into planar polygon patches. They have been used extensively for computer graphics and have also been proposed for scene analysis purpose [11, 12]. Curved surface patches are normally used in the context of computer aided design [13, 14] and computer graphics. Alternatively, the object volume or its decomposition can be stored as :

  4) cellular spaces.
  5) convex polyhedra.
  6) sweep representations such as spheres and cylinders.

Volume representations based on an extension of two-dimensional digitized images are called cellular spaces since each cell represents some identical cubical volume. Arranging these cells in a tree structure yields the 3-D analog of a quadtree (see chapter IV.5) which is an octtree. Polyhedral decompositions are best exemplified by Robert's scene analysis system [15] and the following works on the Blocks world [16]. Blum [17] and Serra [18] proposed the so-called skeleton or medial axis representation. Skeleton descriptions are defined as the set of centers fo maximal spheres contained within the shape. The skeleton indicates orientations in space and the sphere radii indicate size. A similar but less general representation called the generalized cylinder was proposed by Binford [19], Agin [20] and Nevatia [21] as an efficient means for describing the surfaces in terms of their variation along the length of continuous axes.

2.3. - Stability and Sensitivity

Surface descriptions tend to be large and cumbersome in the sense that they are more sensitive to local properties of the shape's surface than they are to its larger features. For example, it would be difficult to quickly determine the disposition of an arm specified by hundreds of little surface patches. Furthermore, a description in terms of many surface elements does not measure up well to the stability and sensitivity criterion.

The resolution at which the volumetric primitives are derived has an important effect on the stability and sensitivity characteristics of the representation. For cellular spaces, changing the resolution is related to dilation which is awkward : increasing the resolution of the original data requires an estimation of boundary points in the new cells. The sensitivity of the skeleton representation is inversely proportional to size. That is smaller features of a shape tend to override larger ones in a description. The same is true of generalized cylinders. Thus these representation are most sensitive to details at the finest resolution at which the respective primitive elements can be derived properly.

2.4. - Organization of descriptions

In the simplest case, the representation is homogeneous and all primitive elements are at the same level of detail. That is, the shape is segmented so that each primitive corresponds to a distinct part of the shape as is the case for the skeleton and generalized cylinders representations and the above mentioned surface representations. It is also possible to represent a shape at several levels of details simultaneously, maintaining information about the general orientation and structure of the shape at the coarset level while also making information about the structure of smaller components explicit at finer levels of detail.

2.5. - Operations

The operations of interest which can be performed on the various representations can be classified as :

a) measurements : topology, surface area, surface derivatives, volume, center of gravity, inertia axis, stable positions, grasp positions.

b) combination : various set operations such as union, intersection, difference.

c) transformation : translation, rotation, dilation, interpolation (to change resolution).

Table I, reproduced in part from [2], summarizes each group of operations with respect to each representation type.

Finding the topology consists of counting the number of components, cavities and holes ("handles"). For a cellular space there are linear algorithms ([22] and the chapter on digital geometry), for polyhedra we can use the Euler formula :

$$V - E + F - 2-2G = 0$$

vertexes   edges   faces   genus

[11, 23, 24], for 3-D surface points the computation is of course impossible since the notion of point-neighbors is absent and for all other types of representations one has to convert either to the cellular space or the polyhedral representations (see chapter IV.5).

Surface area is easier to estimate for surface-based representations. The case of 3-D surface points is tackled by converting to the polygonal network representation using for example the technique described in Chapter IV.5. For volumetric representations things are slightly more difficult and solutions depend mostly upon the capability of computing the unions of the instance primitives. This can be done efficiently for convex polyhedra [23] but solutions are unknown for generalized cylinders or skeleton-based representations. The easiest case is that of the cellular space representation where the surface area can be estimated by counting cells bounding non-object cells.

Computing surface derivatives (usually to find the surface normal or curvature) can be done in principle for all mentioned representations.

Volume measurements are generally easier for surface-based than for volume-based representations. For the latter the problem is to be able to compute the intersections of the instance primitives. Again this can be done efficiently for convex polyhedra [23] but solutions are unknown for generalized cylinders or skeleton-based representations. Just as for the surface area measurement, volume is easiest to measure in a cellular space (just the number of points). Polyhedra and curved surface representations admit piecewise integration techniques for volume measurements.

Combining instances of primitives by the algebraic operations of union, intersection and difference are boolean point operations for cel-

3-D shape representation

lular space representations and geometric computations for polygonal networks and polyhedral representations. Curved patches and volumetric representations are more difficult to manipulate because for the former no fast algorithms are known for the algebraic operations and for the latter the results of these operations are not generally primitives. A solution to this last problem is found in Constructive Solid Geometry (CSG) [25] where primitives are r-sets that is topological polyhedra and are closed under the so-called regularized set intersection, union and difference.

All representations are transformable by translations, rotations and dilations, except for cellular spaces where dilations and rotations are awkward [26]. Changing the resolution of the representation is related to dilation, causing a similar problem to arise for cellular spaces. All other representations can be refined by interpolation, sub-division, or parametric substitution (altering the step fo the parameter controlling the bounding curves).

Operation Representation	Topology	Surface area	Surface derivative	Volume	Manipulation	Transformation	Interpolation
1) 3-D surface points	NO	via 2	Y	via 2	N	Y	via 2
2) Polygonal network	via 5	YES	Y	Y	Y	Y	Y
3) Curved surface patches	via 4	YES	Y	Y	N	Y	Y by subdivision
4) Cellular spaces	[2], [22]	YES	Y	Y	Y	P	P
5) Polyhedra	[11]	YES	Y	Y volume of union	Y	Y	Y
6) Sweep representations	via 4 POOR	POOR *	Y	P **	***	Y	Y
7) CSG	via 4 POOR	POOR	P	P	Y	Y	Y

*   depends on method of computing union
** depends on method of computing intersection
*** result may not be primitive

TABLE 1

2.6. - Conversion

Converting an object description in one representation into a description in another is motivated if a desired operation is easier to perform in another representation. The various conversion possibilities are summarized in Table 2 reproduced from [2].

Conversion to from	3-D surface points	Polygonal Network	Curved Surfaces	Cellular Spaces	Polyhedra	Sweep	CSG
1) 3-D surface points		Triangulation	[27]	via 2	via 2	[5], [9]	?
2) Polygonal network	Sampling		Numerical fit	fill algorithm	Computational geometry	via 1 ?	?
3) Curved surfaces	Sampling	Interpolation		fill algorithm	via 2 ?	via 1 ?	?
4) Cellular spaces	Pts not inside	via 1 ?	via 1 ?		via 1 ?	via 1 ? or [18]	?
5) Polyhedra	Union + Sampling	Union	via 2 ?	fill algorithm ?		via 1 ?	?
6) Sweep representations	Union + Sampling or via 2	Union + interpolation	via 1 or via 2 ?	fill algorithm ?	via 2 ?		?
7) CSG	via 2 ?	Sampling of the boundary Rep.	Approx. of the boundary Rep.	fill algorithm ?	via 2	?	

TABLE 2

Checking through the columns of Table 2, it can be seen that 3-D surface points and cellular spaces are sorts of "universal representations" into which a description in any other system can be easily converted. The polygonal network representation is also such an example.

3. - CONCLUSION

The problem of 3-D shape representation for computer Vision is far from being completely understood. We have stressed the importance of understanding the characterization of representations for the development of computational theories and their algorithms. We have raised several key issues to judge the suitability of shape representations for three-dimensional recognition problems. The main conclusion is that none of the known representations is completely "universal" and that conversion algorithms must be investigated. This is the subject of the next chapter.

REFERENCES

[1] D. MARR and T. POGGIO, "From Understanding Computation to Understanding Neural Circuity", Neuronal Mechanisms in Visual Perception, Neuro-sciences Research Program Bulletin (Eds E. Poppel et al.), 15, 470-488, 1977.

[2] N. BADLER and R. BAJCSY, "Three-dimensional representations for computer graphics and computer vision", ACM Comput. Gr. 12, 3 (Aug. 1978), 153-160.

[3] D. MARR and H.K. NISHIHARA, "Representation and Recognition of the Spatial Organization of Three-Dimensional Shapes", Proc. R. Soc. Lond. B. 200, 269-294, 1978.

[4] Proc. of the Workshop on the Representation of Three-Dimensional Objects, Ed. Ruzena Bajcsy, University of Pennsylvania, May 1-2, 1979.

[5] G.J. AGIN and T.O. BINFORD, "Computer description of curved objects", Proc. IJCAI (1973), 629-640.

[6] J.D. BOISSONNAT and F. GERMAIN, "A new approach to the problem of acquiring randomly oriented workpieces out of a lim", Proc. IJCAI (1981), 796-802.

[7] D. MARR and T. POGGIO, "Cooperative computation of stereo disparity", Science 194, 283-287 (1976).

[8]   D.B. GENNERY, "A stereo vision system for an autonomous vehicle", Proc. IJCAI (1977), 576-582.

[9]   C. LEVINTHAL and R. WARE, "Three-dimensional reconstruction from serial sections", Nature 236, 207-210, March 1972.

[10] H. FUCHS, Z.M. KEDEM and S.P. USELTON, "Optimal surface reconstruction from planar contours", CACM 20, 693-702, October 1977.

[11] B.G. BAUMGART, "Geometric modelling for computer vision", Stanford University, Department of Computer Science, Technical Report, October 1974.

[12] B. BHANU, "Shape matching and image segmentation using stochastic labeling", Ph. D. Dissertation, University of Southern, California, Los Angeles, 1981.

[13] R.E. BARNHILL and R.F. RIESENFELD, Computer Aided Geometric Design, Academic Press, New-York, 1974.

[14] D.F. ROGERS and J.A. ADAMS, Mathematical Elements for Computer Graphics, Mac Graw-Hill, New-York, 1976.

[15] L.G. ROBERTS, "Machine perception of three-dimensional solids", in Optical and Electro-Optical Information Processing, J.T. Tippell et al. (Eds), MIT Press, Cambridge, Mass., 159-197, 1965.

[16] The Psychology of Computer Vision, P. Winston (Ed.), Mc Graw-Hill, New-York, 1975.

[17] H. BLUM, "A transformation for extracting new descriptors of shape", in Models for the Perception of Speech and Visual Form, Walthen Dunn (Ed.), MIT Press, 1967.

[18] J. SERRA, "Image Analysis and Mathematical Morphology", Academic Press, 1982.

[19]   T.O. BINFORD, "Visual Perception by Computer", IEEE Systems Science and Cybernetics Conference, Miami, December 1971.

[20]   J.G. AGIN, "Representation and Description of Curved Objects", Stanford Artificial Intelligence Project, Memo AIM-173, Stanford University.

[21]   R. NEVATIA, "Structured Descriptions of Complex Curved Objects for Recognition and Visual Memory", Stanford Artificial Intelligence Project, Memo AIM-250, Stanford University.

[22]   S.B. GRAY, "Local properties of binary images in two and three-dimensions", Information International Report (1970).

[23]   B.M. CHAZELLE, "Computational Geometry and Convexity", Ph. D. Dissertation, Yale University, Computer Science Department, 1980.

[24]   D. HILBERT and S. COHEN VOSSEN, Geometry and Imagination, Chelsea.

[25]   A.G. REQUICHA, "Representations for rigid solids : Theory, Methods and Systems", Computing Surveys, Vol. 12, N° 4, 437-464, Dec. 1980.

[26]   S.L. JACKSIN and S.L. TANIMOTO, "Oct-trees and their use in representing three-dimensional objects", CGIP 14, 249-270, 1980.

[27]   S. WU, J.F. ABEL and D. GREENBERG, "An interactive computer graphics approach to surface representation", CACM 20, 703-712, Oct. 1977.

CONVERSION ALGORITHMS BETWEEN 3-D SHAPE REPRESENTATIONS

O.D. FAUGERAS

I.N.R.I.A. Domaine de Voluceau - Rocquencourt
B.P. 105, 78153 - LE CHESNAY

I) INTRODUCTION :

The purpose of this chapter is not to explore all possible conversion algorithms between all possible shape representations. This would be outside the scope of this book. We rather want to highlight some problems commonly encountered when attempting to go from a raw 3-D point representation of objects obtained directly from a sensor to some higher level representation. We will concentrate here on surface representations. Volume representations can be obtained from the latter. We first describe briefly an acquisition system for the set of 3-D surface points and then explain how various planar and nonplanar approximations can be built from them.

II) THE ACQUISITION SYSTEM :

The system which has been developed [1] is composed of a laser rangefinder, a system of cameras, a Computer controlled table and a microprocessor system.
II.1) General principle :
The sensor used provides the z coordinate of a point on the surface of an object as a function of the x coordinate (Fig. 1). The basic principle is that of active stereoscopy. Stereoscopy because it uses at least two cameras to yield images from different viewpoints. Active because the difficult problem in Stereo Vision of matching the two views is avoided by overilluminating one point of the surface to be observed.
This strong lighting is provided by a laser beam which can be moved in the x direction. The deflection system is Computer controlled and this makes the sensor random access. The laser beam creates a small spot on the surface which forms by diffusion a secondary source of light which yields one image on each camera. From the position of these images on the retinas, and the geometrical parameters of the cameras, it is possible to compute the x and z coordinates of the brightly lit point.
The platform on which the object rests is equiped with step by step motors. These motors enable us to raise and lower the table (this is the y-axis) and also to rotate it (around the y-axis). We can digitize objects of sizes up to 750*750*600 mm. The redundancy introduced by this rotation with respect to the sweeping of the laser beam allows us to accurately analyze concave objects.

II.2) The existing System :

As we have just seen, the principle of the sensor is very simple. The realisation is more tedious and requires the design and coordination of three distinct systems : the cameras, the laser and the logical unit.

The optics of the laser has two functions. First the beam must be swerved in the x direction. This can be done by means of an opto electronic device, but a system composed of a mirror steered by a galvanometer has been preferred. This is the only moving part of the whole apparatus. Second, the beam must be focussed. The optics are arranged so as to maintain the diameter of the laser beam constant and minimal in the whole spatial range of measurements. This arrangement is also a function of the laser physical parameters.

Each camera is associated with a microprocessor which determines the significantly lighted diodes and performs an interpolation to obtain a position accurate to $1/8^{th}$ of a diode width. A second microprocessor controls the whole system, in particular the galvanometer, and performs the triangulation from the data provided by each camera.

This sensor is therefore relatively autonomous. The connection to a Computer is done through the second microprocessor. Exchanges between the two are very limited. The Computer sends the sensor the direction of the beam and receives the two coordinates of the measured point. In our current implementation, the measurement time is approximately 1ms.

II.3) Gathering 3-D data :

This data is in the observer centered coordinate system. While creating a 3-D model of the object, an object centered representation is required. This is computed by marking the zero position for the x- and y-axis and obtaining a reference value for the z-axis on the platform on which the object rests (Fig. 1). Thus all the points can be transformed to the same coordinate system. While acquiring the data related to an unknown view of the object, the actual position or orientation of the object on the platform does not matter.

As an example, we show the complicated casting of an automobile piece in Fig. 2. In order to create a 3-D model of the object, a range data image was produced for every $30°$ rotation of the object around the y-axis. Finally, top and bottom views of the object were taken. These two views were put in correspondence with the other views by using several control points on the object to compute the transformation. A view obtained with the range data acquisition system described above is shown in Fig. 3.

III) PLANAR APPROXIMATIONS :

This Section deals with the problem of constructing a planar surface representation from a set of 3-D surface points obtained, for example, by the acquisition system of Section II.

III.1) 3-point seed [2] :

this first technique approximates objects by a set of planar faces. It is a two-step process. In the first step we find the set of points that belong to various faces of the object using a three point seed algorithm and in the second step, approximate the face points obtained by polygons. The three point seed method for the extraction of planar faces from range data is a sequential region growing algorithm.

In a well-sampled 3-D object, any three points lying within the sampling distance of each other (called a 3-point seed) form a plane (called the seed plane) which :
    a) coincides with that of the object face containing the points, or
    b) cuts any object face containing any of the three points.
A seed plane satisfying a) results in a plane from which a face should be extracted, while a seed plane satisfying b) should be rejected. Two simple conditions that suffice to determine if a plane falls into category b) are convexity and narrowness.

The algorithm involves the following steps.
    1. From the list of surface points select 3 points which are noncolinear and close relative to the sampling distances.
    2. Obtain the equation of the plane passing through the 3 points chosen in step 1.
    3. Find the set of points P which are very close to this plane.
    4. Apply the convexity condition to the set P to obtain a reduced convex set P'. This separates faces lying in the same plane.
    5. Check the set P' obtained in step 4 for narrowness.
    6. If the face is obtained correctly (i.e. The convexity and narrowness conditions are satisfied), remove the set of points belonging to this face from the list of points and proceed to step 1 with the reduced number of points in the list.

After the surface points belonging to a face have been obtained using the 3-point seed algorithm, two checks are made.
    1. All the points which have been previously associated with various faces are checked for possible inclusion in the present face.
    2. The set of points in the present face is checked for possible inclusion in previous faces.

The application of the above two tests provides the points which belong to more than one face. This information in turn provides knowledge about the neighbors of a face and relations among them which are in turn useful for object recognition tasks.

Now the surface points have been associated with various planar faces. Although some edge points and vertices will be known, an independent step is required to obtain polygonal faces. The polygonal approximation of a face involves the following steps.
    1. Get the points belonging to a face.
    2. Obtain the binary image of the face points.
    3. Trace the boundary of the image obtained in step 2 using a boundary follower.
    4. Perform a polygonal approximation of the boundary of the face.

The complexity of the 3-point seed algorithm is $O(n^4)$ where n is the number of data points.

The 3-point seed method was applied to the individual view shown in Fig. 3. Fig. 4 shows the faces found. In this figure various faces are shown in different shades of grey. As can be seen, most of the faces found are reasonable. The object has major curved surfaces that were split into different faces.

III.2) Triangulation [3] :
We show how to construct a polyhedral approximation of an object whose faces are triangles. We first introduce the spherical representation of a surface [4].

Suppose that we know, apart from the coordinates of the points, the normals of the surface and the sign of the Gaussian curvature at these points. In that case, we can reduce the general case to several convex problems by means of the spherical representation of surfaces.

This representation associates to a point M of a surface, the end-point of the normal vector to the surface at M. So the image of a surface lies on the unit sphere (the Gaussian sphere). The important fact is that the spherical representation of sufficiently small regions that are everywhere cup-shaped (called surfaces of positive curvature) or everywhere saddle-shaped (called surfaces of negative curvature) is one-to-one (see fig.5); and so a triangulation on the unit sphere corresponds to triangulation on the surface.

It is to be noticed that, under the spherical representation the points of positive or negative curvature play the same role. The triangulation of saddle-shaped regions is as easy as the triangulation of the cup-shaped regions. Moreover such a triangulation does not depend on the initialization for it is unique, preserves the Gaussian curvature which is an important invariant of the surface, and, by the way, produces nice computer graphics displays.

The regions where the curvature is zero can not be studied in the same manner. This occurs when the surface is developable which means that it can be obtained by bending some planar region. Such a surface is covered entirely by straight lines (called the generators of the surface). Moreover the surface has one and the same tangent plane at all the points of the generator. In that case, the spherical image of the surface degenerates to a curve but provides the generators and so the whole information about the surface. Yet, the developable regions of a number of objects, especially the manufactured objects, are parts of planes, cones or cylinders and can be studied in the same manner as the locally convex regions.

We now describe a global approach to obtain the triangulation. Let us consider a given surface that does not consist entirely of parabolic points and that, moreover, contains points of positive as well as negative Gaussian curvature. Since the variation of the Gaussian curvature on the surface is continuous, there must be points on the surface at which the Gaussian curvature vanishes and these points have to form continuous curves separating the regions of positive curvature from the regions of negative curvature. These curves, consisting of parabolic points, are called the parabolic curves of the surface.

The considerations of the last section lead us to separate the initial set of points in different subsets of points belonging to one region where the Gaussian curvature keeps a constant sign added to the points of the parabolic frontier. The different subsets are the equivalence classes for the relation of being neighbour and having the same sign of curvature or zero curvature, provided that the size of the neighborhood is less than the smallest distance between two parabolic curves.

Each subset can be mapped onto the Gaussian sphere by use of the spherical representation and then triangulated by an algorithm looking for the convex hull of the points of that subset. In general the considered region is an open surface limited by one or several polygonal curves. On the contrary, the convex hull is a closed surface separated by the parabolic curves in different open surfaces, one of them being the required triangulation. Thus there remains to eliminate some undesirable edges or equivalently to save the desired edges. This is done by the following procedure:

```
procedure correction;
begin
      C=contour of a triangle T0 with a non parabolic vertex;
      save the edges of T0;
      while (C contains non parabolic points) do
      begin
            save the edges of a triangle T, not yet considered,
            obtaining a non parabolic edge of C;
            update C which becomes the contour of the union of T and the
            previous domain ;
      end;
end;
```

The complete algorithm performing the triangulation of a surface according to our method is roughly described below:

(i) create the graph of the neighbours of the points;
(ii) eliminate the edges between points where the curvature has strictly different signs;
(iii) separate the graph into its different connex subgraphs;
for each subgraph do begin
(iv) map it onto the Gaussian sphere;
(v) compute its convex hull;
(vi) correction;

As previously mentionned, the triangulation performed by the above algorithm is unique whichever way the points are considered, and preserves the Gaussian curvature.

Moreover this algorithm runs in $O(nlogn)$ time as it appears when considering the complexity of the different steps.

(i) Many fast algorithms have been proposed which seek the neighbours of a given point in $O(nlogn)$ time. See for example [5].

(ii) If a point has k neighbours (k independent from n) the number of edges of the graph of the neighbours is $O(n)$ and so the second step can be done in $O(n)$ time.

(iii) This step is a classical problem in graph theory and can also be done in $O(n)$ time [6].

(iv) This step requires $O(n)$ time obviously.

(v) As previously mentionned, the convex hull of a set of points in 3-d space can be computed in $O(nlogn)$ time (prep).

(vi) For the number of triangles is $O(n)$ this last step requires $O(n)$ time.

The number of subgraphs depends only on the surface and not on the number of points, provided that the discretization is fine enough. So the total complexity of this algorithm is $O(nlogn)$ which is optimal.

A triangulation obtained by this technique on a fairly complicated synthetic object is shown in fig.6.

IV) NONPLANAR APPROXIMATIONS :

We present two approaches. A region growing [7] and a region splitting [8] approach.

IV.1) The Oshima and Shirai approach [7] :

The points on the surface of the objects are grouped into surface elements. Each element consists of adjoining 8*8 overlapping squares on the grid defined by the acqisition process. Assuming each element to be a plane, its equation is found by least square approximation. Surface elements at edges yield a poor fit and are ignored in further processing. Next, surface elements are merged into elementary approximately planar regions by iterating the following two steps : (1) search for a region kernel as the most promissing element in terms of the resemblance of its neighbors and their number not belonging to an already found region (2) extend a region around the kernel by merging adjacent elements whose plane equations are similar to that of the region.

At this stage each elementary region corresponds to a plane surface or a part of a curved surface. Regions are therefore classified into plane, curved and undefined based on two parameters : their mean curvature computed as the variance of the orientations of the planar patches in the elementary region and the diameter of the region. The idea is that a plane region has small curvature. Regions with too small a diameter are classified as undefined. At the end of this stage planar regions have been identified.

The next step is an attempt to merge curved and undefined regions into curved global regions. Again we iterate the following two steps : (1) search for a kernel region of a global region as the larger global region already found and (2) merge curved or undefined elementary regions adjacent to the region and smoothly connected to it.

Finally quadratic patches are fitted to the regions by least square approximation and a global description of the scene is constructed in terms of properties of regions (planar, curved, equation of the quadratic patch) and relations between them.

IV.2) The Hebert and Ponce approach [8] :

The preceding approach was bottom-up, we discuss here a top-down approach to segment a set of 3-D point data into a set of planar, cylindrical and conical regions. The key idea is to use the Hough transform (see chapter) on the gaussian sphere to identify these regions.

First, for each point, a normal is estimated by least square approximation on the set of its eight nearest neighbors in a way which is similar to what has been described in Section IV.1. Then these normals are used to identify on the gaussian sphere planes, right cylinders and circular cones. Indeed, the gaussian sphere can be parameterized by two spherical coordinates, u and v. In this space our primitives have simple representations :

.a plane is characterized by a point (u,v) on the sphere.
.a right cylinder has its normals on a circle of radius one and is characterized by the direction (u,v) of its axis.
.a circular cone has its normals on a circle of radius less than 1. The radius of this circle is the cosine of the cone angle.

Therefore the problem of segmenting the data into these primitives is equivalent to finding point and circular clusters of normals on the gaussian sphere. The algorithm iterates the following three steps : (1) detect planes of size larger than some threshold S and remove the corresponding points (2) same for the cylinders and (3) same for the cones. At the end of every iteration S is reduced. At the end of this stage each point has been labeled as plane, cylinder, cone or undefined. Connected components of each set of equally labeled points are then extracted, very small regions eliminated and a global description can be produced. Results obtained by using this technique on the object of Fig.2 are shown in Fig.7.

V) CONCLUSIONS :

We have presented various approaches for going from a point representation of objects in 3-D space to a surface representation. Other techniques exist but much remains to be found in this area.

REFERENCES
1. J.D. Boissonnat and F. Germain, "A New Approach to the Problem of Acquiring Randomly Oriented Workpieces out of a Bin," Proc. 7th Int. Joint Conf. On Artificial Intelligence, pp. 796-802, 1981.
2. B. Bhanu, "Shape Matching and Image Segmentation Using Stochastic Labeling," Ph.D Dissertation, University of Southern California, Los Angeles, Aug. 1981.
3. J.D. Boissonnat, "Representation of objects by triangulating points in 3-D space," Proc. Of ICPR-82, Munich, Oct. 1982.
4. D. Hilbert and S. Cohn Vossen, <u>Geometry and Imagination</u>, Chelsea
5. J.L. Bentley and J.H Friedman, Fast Algorithms for Constructing Minimal Spanning Trees in Coordinate Spaces, IEEE Trans.Comp.,Vol.C-27, No.2,February 1978.
6. A.V. Aho, J.E. Hopcroft and J.D. Ullman, <u>The design and analysis of Computer algorithms</u>, Addison-Wesley, 1974.
7. M. Oshima and Y. Shirai, "A scene description method using three-dimensional information," Pattern Recognition, Vol. 11, pp. 9-17, 1979.
8. M. Hebert and J. Ponce, "A new method for segmenting 3-D scenes into primitives," Proc. Of ICPR-82, Munich, Oct. 1982.

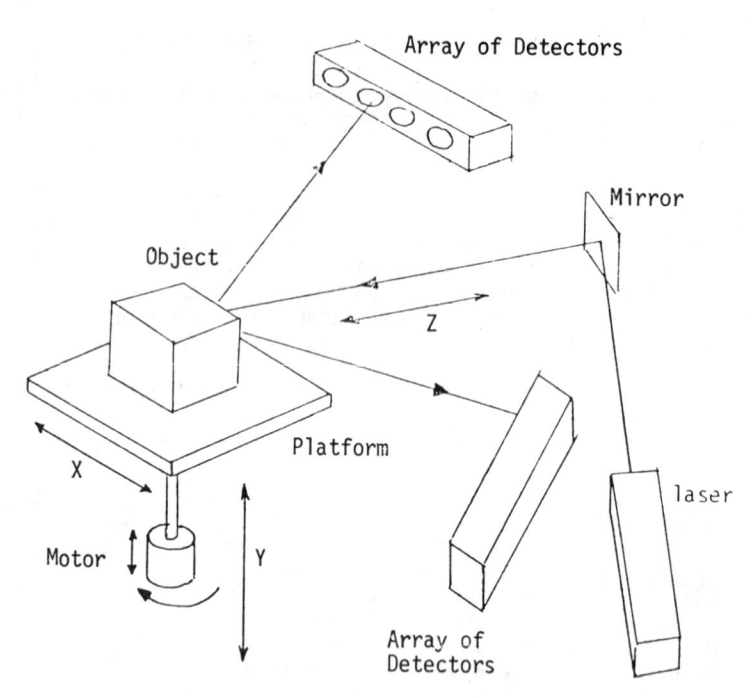

Figure 1. Laser Range Finder System.

Figure 2. Photograph of an automobile part.

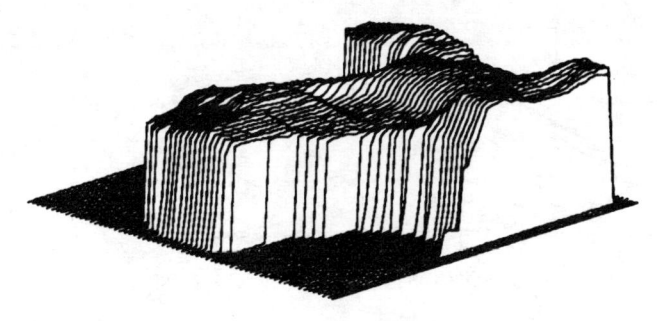

Figure 3. Range data corresponding to Figure 2.

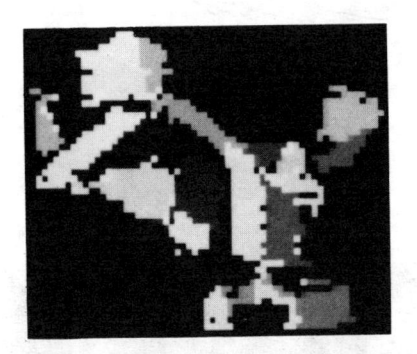

Figure 4. Planar faces found by the 3-point seed algorithm.

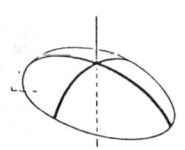

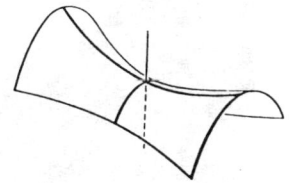

Figure 5. Cup-shaped and saddle-shaped points.

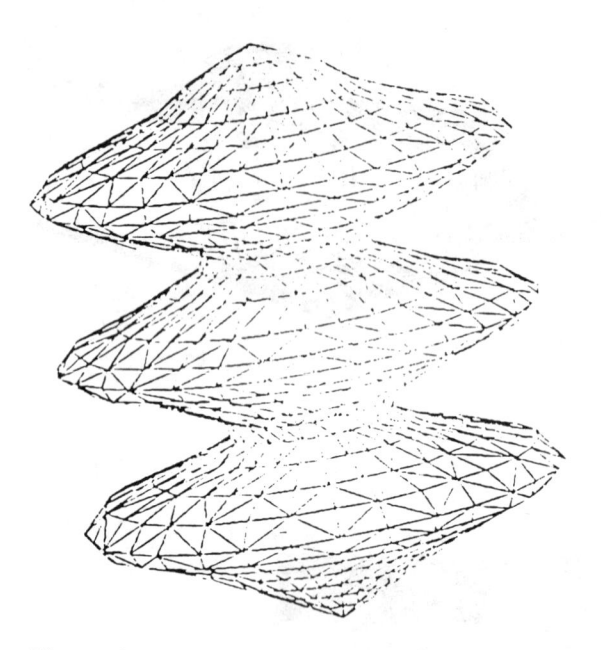

Figure 6. Results of triangulating a complicated object.

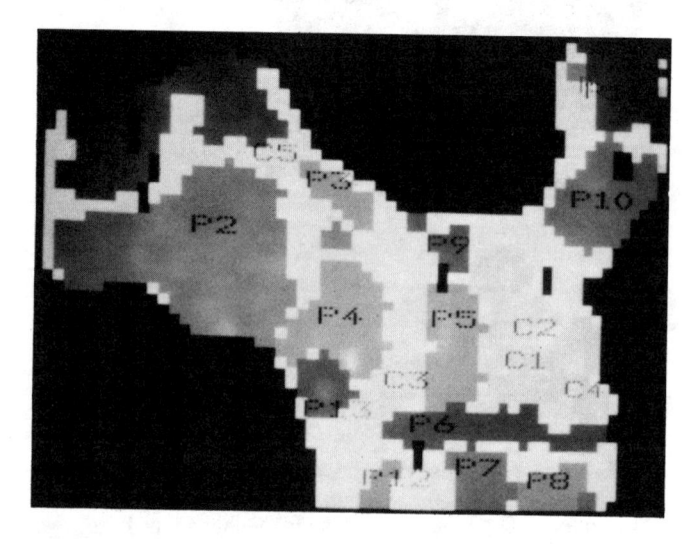

Figure 7. Cones and cylinders found by the Hebert and Ponce approach on the data of Figures 2 and 3.

HIERARCHICAL REPRESENTATION: COMPUTER REPRESENTATIONS
OF DIGITAL IMAGES AND OBJECTS

A. Rosenfeld
Computer Vision Laboratory, Computer Science Center, University of Maryland, College Park, MD 20742

1 *INTRODUCTION*

Algorithms for the processing and analysis of digital images can be designed in various ways, depending on the manner in which the images are represented in the computer. In this chapter we review some basic methods of representing digital images and digital regions, with emphasis on hierarchical methods involving "pyramids" or trees. Extensions of these concepts to three dimensions will also be discussed.

A general review of image representation can be found in Rosenfeld & Kak (1982). References on topics not covered there will be given in later sections. See also Tanimoto & Klinger (1980) for a collection of papers dealing with the uses of (primarily) hierarchical structures in image analysis.

2 *NON-HIERARCHICAL REPRESENTATIONS*

The standard way of representing a digital image is as an array of gray levels, representing the brightnesses of the scene at a discrete set of sample points. Other representations can be defined, which may be more compact if the image is sufficiently simple, i.e., if it consists of only a few regions of constant gray level that have simple shapes. In the following paragraphs we review several of these representations.

An image can be decomposed into connected components of constant gray level, and each of these components is determined by specifying the crack or chain codes of its borders, together with a starting point on each border. (Connected components and border codes were defined in another chapter.) As an example, if there are 2^k gray levels and 100 regions whose average border length is 100, then this representation requires specifying the coordinates of 100 starting points (1800 bits, in a 512 by 512 image) and the associated gray levels (k bits), plus the crack codes of 100 borders of length 100 (20,000 bits), for a total far less than the $k \cdot 2^{18}$ bits needed to define the image as an array. Algorithms for reconstructing regions from the codes of their borders are given in Rosenfeld & Kak (1982); the basic idea is to use the code to "paint" the border pixels with the desired gray level, and then extend this gray level into the interior pixels. Simple examples of border codes were given in an earlier chapter.

Another class of representations is based not on borders, but on blocks of constant gray level. Any row of an image consists of a succession of runs (=maximal sequences of pixels) of constant gray level, and is determined by specifying the lengths and gray levels of the runs. For example, if there are 2^k gray levels in a 512 by 512 image, it requires k+9 bits to specify the gray level and length of each run, so that if there are 20 runs per row we need only 20(k+9) bits to specify the row, far less than the 512 bits needed to specify it in array form. The algorithms to construct run length codes from arrays, and reconstruct the arrays from the codes, are straightforward. A simple example of a run length code for a two-valued image is given in Figure 1.

In the run length representation, the maximal "blocks" are runs, i.e. rectangles of height 1. Another type of maximal block representation, known as the medial axis transformation, uses square blocks of odd side length. For any pixel P let s(P) be the largest such block of constant gray level centered at P. We call s(P) maximal if it is not contained in any other (larger) s(Q). It is easily seen

Figure 1. (a) Simple two-valued image. The run length codes of the rows are (0); (0)2,1,5; (0)5,1,2; (1)7,1; (1)7,1; (1)7,1; (1)6,1; (1)6,1, where the number in parentheses is the value of the initial run on each row, and the following numbers are the run lengths. (b) Centers (underlined in (a)) and radii of the maximal upright squares whose union is the set of 1's in (a).

```
0 0 0 0 0 0 0 0
0 0 1 0 0 0 0 0
0 0 0 0 0 1 0 0
1 1 1 1 1 1 1 0
1 1 1 1 1 1 1 0
1 1 1 1 1 1 1 0
1 1 1 1 1 1 0 0
1 1 1 1 1 1 0 0
```

x	y	r
3	3	2
3	7	0
4	3	2
6	4	1
6	6	0

(a)   (b)

that the image is completely determined if we know the centers, radii, and gray levels of the maximal s(P)'s; in fact, the set of pixels of a given gray level is just the union of the s(P)'s having that gray level. In our standard example of 2^k gray levels and a 512 by 512 image, we need k+18+8=k+26 bits to specify the gray level, center position, and radius of each s(P); if there are only a few hundred maximal s(P)'s, this is far less than the $k \cdot 2^{18}$ bits needed for the array representation. Simple algorithms for finding the maximal s(P)'s are given in Rosenfeld & Kak (1982) for the case of two gray levels; they are based on computing the "chessboard distance" from each pixel to the nearest pixel that does not have the same gray level, since it is not hard to see that s(P) is maximal iff this distance is a local maximum at P. A simple example of a medial axis transformation is given in Figure 1.

Border representations (known as "contour coding") and run length coding are sometimes used as compact representations for image compression, but they (as well as the medial axis transformation) are unlikely to be efficient unless the image has very few gray levels. Their greatest area of potential application is to two-valued ("binary") images. In the binary case, the representations become somewhat simplified. For border representations, we need only specify the borders for 2^k-1 of the gray levels, since after these regions have been "painted", what remains is given the remaining gray level; thus for k=1, we only need to specify one set of borders. In run length coding in the binary case, the runs must alternate between value 1 and value 0, so we need only specify their lengths and the gray level (0 or 1) of the first run; the values of the rest of the runs are then determined. In the medial axis transformation, we need only specify the maximal s(P)'s when P is a 1; their union is then the set of 1's in the image, and the remaining pixels must be the 0's. When we construct a representation in the binary image case, we can regard it as representing a region or image subset, namely the set S of 1's in the image.

Compactness in only one aspect of an image representation; another important aspect is the computational complexity of determining various image properties directly from the given representation. Algorithms for deriving various geometric properties of a set S from the various representations of the corresponding binary image can be found in Rosenfeld & Kak (1982). For example, connected component labeling and counting are immediate from the border representation (particularly if we know which borders are hole borders), since each border belongs to only a single connected region. Area (and other moments) are straightforwardly computed from the border code, and perimeter is immediate, as are curvature, corners, inflections, and convexity; elongatedness, on the other hand, is less easy to determine. From run length code, connected components can be labelled and counted efficiently by detecting run adjacencies from row to row; area and moments are immediate, perimeter is straightforward (using the row to row adjacencies), and curvature, corners, inflections, and convexity are also straightforward, but elongatedness is again relatively difficult. From the medial axis transformation, elongatedness is easier (it corresponds to there being many maximal $s(P)$'s of small sizes), but the other properties are harder because the $s(P)$'s can overlap, and it is hard to determine connectedness, borders, etc. A related issue is that of computing the representations of logical combinations of two-valued images (AND, OR, etc.) from those of the constituent images. This is not difficult for the run length code, but it is much less straightforward for border codes or medial axes.

The representations considered up to now determine the image exactly. If we are satisfied to represent an image only approximately, we can obtain more compact representations even if the number of gray levels is large. For border representations, one possibly is to polygonally approximate the borders, and perhaps to eliminate small regions entirely. Similarly, for the run representations, one can try to eliminate small runs, and for the medial axis representation, one can eliminate small $s(P)$'s. In some cases, we can eliminate all the $s(P)$'s except those for which P lies on a small set of curves; each such curve can then be represented by a chain code together with a "width function" specifying how the radius of $s(P)$ varies along the curve, thus defining a "generalized (=variable-width) ribbon" centered on the curve.

Another possibility for image approximation based on run coding is to piecewise approximate each row of the image, say in the least squares sense, by a sequence of constants to within some error tolerance, and use the sequence of approximating values as the new "runs". On methods of piecewise approximation of waveforms, e.g., of rows of an image, see Pavlidis (1977). Similarly, in the medial axis case, we can redefine $s(P)$ by expanding a square centered at P as long as the standard deviation of gray level does not exceed a threshold (i.e., as long as the gray levels in the square are well approximated, in the least squares sense, by their mean); the image can then be approximated as a superposition of squares of constant gray level, namely the maximal s(P)'s (Ahuja et al., 1978). More generally, we can approximate an image by partitioning it into maximal homogeneous regions, e.g., regions with small standard deviations, using split-and-merge algorithms, as described in another chapter (Pavlidis, 1977); these regions can then be represented in any of the ways described in this section. This last approach involves, at least in part, a hierarchical image representation, as we shall see in the next section.

3 HIERARCHICAL ("QUADTREE") REPRESENTATIONS

In this section we describe a method of exact or approximate region representation based on recursive subdivision into quadrants. It can be thought of as a maximal block representation in which the blocks have sizes and positions that are powers of 2. We shall see that the set of blocks constituting the representation can be compactly defined by specifying a degree-4 tree (a "quadtree"). We assume in this section that the input image is of size 2^n by 2^n.

The exact quadtree representation is constructed as follows: If the entire image has constant gray level, its tree consists of a single (root) node, labelled with that value. If not, we subdivide it into quadrants; we give the root node four sons representing these quadrants; and we repeat the process for each quadrant. This recursive process gives rise to a degree-4 tree in which each leaf represents an image block that has constant gray level, and the image is evidently the union of these blocks. The leaves at level k (where the root is at level 0) represent blocks of size 2^{n-k} by 2^{n-k}, and their positions (x and y coordinates) relative to the edges of the image are multiples of 2^{n-k}. A simple example of a two-valued image and its quadtree representation is shown in Figure 2. Efficient algorithms exist for

Figure 2. (a) Same image as in Figure 1, showing boundaries of quadtree blocks. (b) Quadtree representation; at leaf nodes, solid (hollow) circles represent blocks of 1's (0's).

(a)

```
0 0 0 0 0 0 0 0
0 0 1 0 0 0 0 0
0 0 0 0 0 1 0 0
1 1 1 1 1 1 1 0
1 1 1 1 1 1 1 0
1 1 1 1 1 1 1 0
1 1 1 1 1 1 0 0
1 1 1 1 1 1 0 0
```

constructing the quadtree representation of an image from its array, run length code, or border code representations, and vice versa; see Rosenfeld & Kak (1982) for the references.

Like the other representations defined in Section 2, the quadtree representation is unlikely to be compact if the image has many gray levels; it is more reasonable to use it for two-valued images. It is not generally as compact as the medial axis transformation, since the latter uses square blocks of arbitrary (odd) sizes in arbitrary positions; but this lesser compactness is offset by the fact that computation of some image properties is much easier using quadtrees. In the case of two-valued images, efficient algorithms for connected component labeling and perimeter measurement from quadtrees have been defined (see Rosenfeld & Kak (1982) for the references). Underlying these algorithms is the basic process, given a quadtree leaf representing an image block B, of finding the leaves representing the blocks adjacent to B in the image; it turns out that this can be done by a simple tree traversal scheme. This allows us, in particular, to determine which blocks are on the border (of the set of 1's in the image). Curvature, corners, and inflections can also be derived straightforwardly, once we know how to identify the border in the quadtree representation, and area (and moments) are also straightforward. Elongatedness too should be easier to detect in a quadtree, since the blocks do not overlap; and the quadtrees of logical combinations (AND, OR, etc.) are very easy to determine from those of the original images, using another simple tree traversal process.

We can also define approximate image representations based on quadtrees; we simply modify the tree construction process so that a block is subdivided into quadrants only if it is insufficiently homogeneous (e.g., its standard deviation is above some threshold, implying

Figure 2 (cont'd.)

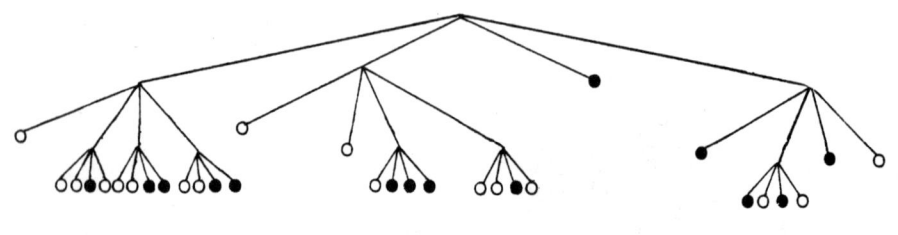

(b)

that it is not a good fit to a constant in the least squares sense), rather than subdividing it as soon as its value is not constant. Note that this is just the splitting process in the standard split-and-merge algorithm for image segmentation. If we allow merging, we can reduce the number of regions in the segmentation by combining adjacent regions if their union is still homogeneous; but by not allowing merging, we retain the advantage that the homogeneous blocks are representable in a structured way by the leaves of a quadtree, which allows straightforward computation of various properties, etc. as described in the preceding paragraph.

4 MULTIRESOLUTION ("PYRAMID") REPRESENTATIONS

For some purposes, it is advantageous to work with an image at more than one resolution. For example, if we want to detect large-scale features such as edges in the image, we can do so very cheaply using a reduced-resolution version of the image, and once an edge has been detected, we can go to the full-resolution image to locate it precisely (Kelly, 1971). Large-scale edge detectors can also be used to identify various types of texture edges that are detectable only when we take averages over large regions (Rosenfeld & Thurston, 1971).

Large-scale feature detectors need not be of all possible sizes; it should suffice to use sizes that are powers of 2, since this will insure that there exist detectors whose sizes are within a factor of $\sqrt{2}$ of any desired object size, and will therefore respond reasonably strongly to it. Moreover, large detectors need not be applied to the image in every position (though this was in fact done in Rosenfeld & Thurston (1971)); even if they are applied in positions that are at a spacing comparable to the detector size, there will still exist detectors that overlap any desired position by (say) at least 50%, and will therefore respond reasonably strongly.

These remarks suggest that a reasonable way of defining feature detectors over a range of sizes is to create an exponentially tapering "pyramid" of reduced-resolution versions of the image, using a reduction by a factor of 2 at each step. Thus if the original image is 2^n by 2^n, the successive levels of the pyramid are 2^{n-1} by 2^{n-1}, 2^{n-2} by 2^{n-2},... . Feature detectors whose sizes are proportional to powers of 2 can thus be applied to the original image, in positions which are multiples of the same powers of 2, by applying local feature detectors of constant size to each level of the pyramid. If several types of feature detectors of the same range of sizes need to be applied to the image, this is evidently an economical way to do it, since the resolution reduction step needs to be performed only once, namely when the pyramid is built. Note that the total number of pixels in a pyramid is only $(1+ \frac{1}{4} + \frac{1}{16} +...) < 1\frac{1}{3}$ times as many as in the original image alone.

In addition to their use for applying large-scale feature detectors, image pyramids have a variety of other uses; see Tanimoto & Klinger (1980) for examples. There has been recent interest (Burt et al., 1981) in methods of image segmentation based on cooperative computation involving "neighboring" pixels at successive levels of a pyramid. The process creates trees, defined by links between the levels, where the set of leaves of each tree represents a homogeneous subpopulation of pixels (or, in a modified version of the process, a compact portion of a homogeneous region); thus the trees define a segmentation of the image. An important feature of this approach is that the trees involve only "vertical" links (=links between levels), so that the time required to construct the trees grows only with the log of the image diameter, even if the regions defined by the trees are very large.

5 REPRESENTATION OF 3D ARRAYS

In this section we briefly discuss how to extend the concepts of this chapter to three-dimensional arrays that represent spatial volumes. On geometrical properties of such arrays see an earlier chapter.

A 3D array can be decomposed into connected components of constant value, and each such component is determined by specifying its border surfaces. However, there is no simple analog of a border code which defines a surface as a simple sequence of moves, so that this type of border representation is not as compact as in the 2D case.

The row-by-row run length code generalizes immediately to 3D, and achieves the same economy as in 2D - namely, the same factor that is achieved on an average row. A better idea in 3D might be to consider the 3D array as a stack of 2D arrays, and apply any of the representation methods defined in this chapter to each of these 2D arrays - i.e., to encode plane by plane rather than row by row.

The medial axis transformation (Morgenthaler, in press) and quadtree (Jackins & Tanimoto, 1980) representations also generalize immediately to 3D. The 3D analog of a quadtree is an octree (i.e., a tree of degree 8), constructed by recursively subdividing the array (which we assume to be 2^n by 2^n by 2^n) into octants. All of the above remarks apply to both exact and approximate representations. The 3D analog of a "generalized ribbon," composed of s(P)'s whose centers lie on a space curve, is known as a "generalized cylinder" or "generalized cone" (Nevatia & Binford, 1977).

The 3D analog of the pyramid representation has not yet been investigated, but it should be noted that a 3D pyramid requires very little more storage space than the original array alone ($1 + \frac{1}{8} + \frac{1}{64} + \ldots < 1\frac{1}{7}$ times as much).

6 CONCLUDING REMARKS

A variety of exact and approximate methods of representing images (or 3D arrays) have been developed. When the input is simple, these representations are not only more compact, but may also permit lower-cost computation of various image properties. Hierarchical representations are of particular interest because they make possible relatively fast operations based on tree traversal. Multiresolution ("pyramid") representations are somewhat less compact than the original image, but they allow efficient implementation of operators having a range of sizes, and they also provide a basis for defining a new class of segmentation processes based on cooperation between adjacent levels of the pyramid.

REFERENCES

N. Ahuja, Davis, L.S., Milgram, D.L., & Rosenfeld, A. (1978).Piecewise approximation of pictures using maximal neighborhoods. IEEE Trans. Computers, 27, 375-379.
Burt, P., Hong, T.H., & Rosenfeld, A. (1981). Segmentation and estimation of image region properties through cooperative hierarchical computation. IEEE Trans. Systems, Man, Cybernetics, 11, 802-809.
Jackins, C.L. & Tanimoto, S.L. (1980). Oct-trees and their use in representing three-dimensional objects. Computer Graphics Image Processing, 14, 249-270.
Kelly, M.D. (1971). Edge detection in pictures by computer using planning. Machine Intelligence, 6, 377-396.
Morgenthaler, D.G. (in press). Three-dimensional simple points: serial erosion, parallel thinning, and skeletonization. Information Control.

Nevatia, R. & Binford, T.O. (1977). Description and recognition of curved objects. Artificial Intelligence, 8, 77-98.
Pavlidis, T. (1977). Structural Pattern Recognition. Berlin: Springer.
Rosenfeld, A. & Kak, A.C. (1982). Digital Picture Processing (second edition), Chapter 11. New York: Academic Press.
Rosenfeld, A. & Thurston, M. (1971). Edge and curve detection for visual scene analysis. IEEE Trans. Computers, 20, 562-569.
Tanimoto, S. & Klinger, A. (1980). Structured Computer Vision. New York: Academic Press.

BASIC NOTIONS IN KNOWLEDGE REPRESENTATION AND CONTROL FOR
COMPUTER VISION

J.C. LATOMBE
Laboratoire IMAG, B.P. 53 X, 38041 GRENOBLE Cedex, FRANCE

A. LUX
Laboratoire IMAG, B.P. 53 X, 38041 GRENOBLE Cedex, FRANCE

Abstract. This essay presents basic methods in knowledge representation and control for computer vision. It includes three main sections on predicate calculus and automated deduction, production systems, plausible reasoning and graph searching, and semantic networks. A final section illustrates the methods introduced in the previous ones on existing vision systems.

A IMAGE UNDERSTANDING AND ARTIFICIAL INTELLIGENCE

Image understanding is the process of constructing a mapping between an image and elements of a domain of discourse, the connection of which with the real-world is known. Such a mapping gives meaning to the image and is called an interpretation of the image. In general, we only want answers to specific questions about the scene represented by the image so that a partial interpretation (i.e. a partial mapping) including these answers is sufficient.

Except when a computer vision system is to be used in a very limited domain (e.g. some industrial vision systems), image understanding is a very complex process. The major difficulty comes from both the amount of data to be processed and the difference of structure between the image -- typically an array of pixels -- and the domain of discourse -- typically classes of objects linked by relations. In order to bridge this gap, it is largely accepted [45] that the mapping should occur at several levels of representation including : lines, regions, surfaces, volumes, objects, spatial relations among objects.

To interpret an image, a system must have a priori knowledge about the types of entities that may occur in the image. Part of this knowledge must be related to the visual appearence of the entities : geometry, size, contrast, color, texture, ... Knowledge may also include pieces of information about the possible functions of objects and about physical laws. Such knowledge is useful to understand and to predict how objects interact in a scene.

A priori knowledge should be organized according to the levels of representation at which mapping is to occur. For instance a contrast line can be described as a sequence of contiguous pixels with high gradient values, a straight segment as a part of a contrast line having zero curvature, a polygonal 2-D surface as a region delimited by connected straight segments, ..., a house as a combination of a parallepiped and a prism.

At the lowest levels, only a few types of entities have to be considered, typically lines and regions. Knowledge about them is domain-independent. It is to be used frequently by the vision system and in a way that is fairly well understood. So, it seems appropriate to encode this knowledge in _procedural_ form, that is as computer programs (these programs can be implemented on a general-purpose computer, or as specific hardware).

At higher representation levels, the situation is very different. A large number of domain-dependent entities have to be considered. Just think of a house scene : classes of objects include grass, tree, drive-way, garage, car, window, door, roof, ... In addition, there are several ways to use knowledge about them. For instance, a piece of model describing a house as a prism supported by a parallepiped may be used to hypothesize the presence of a house in an image, once a prism and a parallepiped have been discovered. It also makes it possible to predict the location of a prism (or a parallepiped) when a parallepiped (a prism) has been recognized. Finally, it can help the system to solve ambiguities of local interpretations. Now, the procedural representation of knowledge is clearly not flexible enough. Indeed, it requires that everything is explicitly anticipated into the code. Furthermore, updating procedural knowledge may require to review a large part of the control structure of the vision system. A more _declarative_ representation of knowledge permitting the system to perform different kinds of inference depending on the current state of the interpretation is recommended.

Until now, most of the efforts devoted to computer vision have concerned the stages of early processing and segmentation [22] [36], for which procedural knowledge is appropriate. Comparatively, there have been only a few attempts to make use of higher-level knowledge. However progress in this direction will be crucial for developping systems capable of understanding complex visual scenes. Therefore, representing and using declarative knowledge become a central issue of computer vision. It requires to answer three questions :

Basic notions in knowledge representation and control

- What language to use for expressing knowledge ?
- How to perform inferences from statements in this language ?
- How to control the flow of inference ?

For more than 20 years, researchers in Artificial Intelligence (AI) have attempted to provide answers to these questions [32] [47]. Important progress has been made through works in various domains including problem solving, robot planning, theorem proving, natural language understanding and expert systems. The goal of this lecture is to introduce some of the principal results that have been obtained so far, as the basis of new developments in the field of computer vision. We will examine :

(1) First-order predicate calculus and automated deduction,
(2) Production systems, plausible reasoning and graph searching,
(3) Semantic networks.

We will present basic concepts with no special commitment to computer vision (other than their potential interest to this domain). A final section will illustrate these concepts on existing vision systems.

B FIRST-ORDER PREDICATE CALCULUS AND AUTOMATED DEDUCTION

1 INTRODUCTION

First-order predicate calculus is a language of symbolic logic in which one can express a wide variety of statements ranging from simple facts to quantified expressions.

Over the last twenty years it has attracted much interest from computer scientists. Important technical results make it possible to implement automated deduction procedures that can prove theorems from axioms stated in this language [28][31].

As a knowledge representation system, predicate calculus (PC) enjoys several attractive features. In particular :
- The syntax is remarkably simple.
- The semantics are completely machine-independent, i.e. PC makes no commitment to the computer processes that will carry out the deduction.
- It permits the user to build a knowledge base incrementally by entering statements independently of each other.
- There exist deduction procedures applicable to PC having nice formal properties like "soundness" and "completeness".

However, PC also has detractors so that today it is not as largely used as one might first imagine (cf. section 5).

Nevertheless, we think that :
(1) PC and deduction procedures based on PC are fundamental concepts in computational logic, which have influenced many knowledge representation paradigms currently used in AI.
(2) Most of the drawbacks that are perceived concerning the use of PC are not fundamental, but are inherent in the state-of-the-art theorem provers.

Thus, this section, although informal and incomplete, should be considered as an introduction to basic notions about knowledge representation.

2 LANGUAGE DEFINITION
2.1 Alphabet

The first-order PC alphabet consists of :
- Pontuation marks :
  ( ) ,
- Logical connectives :

and or not —>
(—> is the implication sign)
- Universal and existential quantifiers :
  $\exists$ $\forall$
- n-place predicate symbols (n$\geq$0)
- n-place function symbols (n$\geq$0)
- variable symbols.

The sets of predicate symbols, function symbols, and variable symbols are any three mutually disjoint sets. 0-place predicate symbols are called propositions. 0-place function symbols are called constants.

In the following we shall reserve the letters u, v, w, x, y, z for variables.

2.2 Syntax

Syntactic constructions in the first-order predicate calculus are terms, atoms, and well-formed formulas.
- A term is a variable, a constant, or any expression of the form $f(t1,...tn)$ where f is a n-place function symbol (n$\geq$1), and $t1,...,tn$ are terms.
- An atom is a proposition or an expression of the form $P(t1,...,tn)$ where P is a n-place predicate symbol (n$\geq$1), and $t1,...,tn$ are terms. $t1,..,tn$ are called the arguments of P in the atom.
- A well-formed formula (wff) is defined as follows :
  . an atom is a wff,
  . if A and B are wffs, then so are not A, A and B, A or B, and A $\rightarrow$ B,
  . if A is a wff and x is a variable, then $(\exists x)A$ and $(\forall x)A$ are wffs,
  . no other constructions are wffs.

In the wffs $(\forall x)A$ and $(\exists x)A$, A is called the scope of the quantifier.

If a variable in a wff is quantified over, then it is said to be a bound variable ; otherwise it is a free variable.

Wffs having all of their variables bound are called sentences. Sentences are the only wffs that we will consider in the sequel.

To avoid ambiguities and in order to reduce the use of extra parentheses in wffs, we will use the following hierarchy among connectives and quantifiers :
  not $\exists$ $\forall$
  and
  or
  —>

Note : The above language is called first-order because its quantifications range only over variables, and not over predicate and function symbols.

Examples of sentences :
- SKY(REGION12)
- (∀x){GREEN(x) → GRASS(x) or (∃y)[TREE(y) and IS-PART-OF(x,y)]}
  (for readability, we use brackets as an alternative to parentheses).
- (∀x){(∃y)BEHIND(x,y) → INVISIBLE(x)}
- TREE(a) and IS-PART-OF(b,a) and GREEN(b) → ABOVE(b,trunk-of(a))

2.3 Semantics

Let us consider a non-empty set D called domain of discourse. For example, D may be the finite set of all the regions extracted from a particular image or the infinite set of all integers.

We note F a set of functions DXDX...XD → D and R a set of relations in DXDX...D.

An interpretation is given to a set S of sentences by assigning :
- an element of D to each constant appearing in S,
- a n-place function of F to each n-place function symbol appearing in S,
- a n-place relation of R to each n-place predicate symbol appearing in S.

Given a variable-free atom A = P(arg1,...,argn), and an interpretation I of A, we say that the value of A in I is TRUE iff the relation assigned to P in the interpretation holds between the elements of D assigned to the arguments arg1,...,argn ; otherwise we say that the value of A in I is FALSE.

Example :

Let us consider the atom GREATERP(a,b) [read "a is greater than b"]. We define an interpretation I1 of this atom by assigning the set IN of positive integers to D, 10 to a, 8 to b, and > to GREATERP. In I1, GREATERP(a,b) has the value TRUE ; we also say that I1 satisfies GREATERP(a,b). If I2 assigns IN to D, 8 to a, 10 to b, and > to GREATERP, then GREATERP(a,b) has the value FALSE in I2 ; I2 falsifies GREATERP(a,b).

Let A and B be two variable-free wffs, and I an interpretation of A and B. The value in I of the wffs made up of A and B are given by the following truth table (T stands for TRUE and F for FALSE) :

A	B	not A	A or B	A and B	A → B
T	T	F	T	T	T
T	F	F	T	F	F
F	T	T	T	F	T
F	F	T	F	F	T

The value of the sentence $(\exists x)A$ in an interpretation I is TRUE iff there exists at least one assignment of an entity of the domain of discourse D to x that gives the value TRUE to A. The value of sentence $(\forall x)A$ in I is TRUE iff all assignments of an entity of D to x give the value TRUE to A. If D is infinite and if quantifiers occur in a sentence, then it is not in general possible to compute the value of the sentence.

Let us consider a set $S = \{S1,...,Sp\}$ of sentences. A sentence X logically follows from S iff every interpretation satisfying S (i.e. satisfying all sentences Si in S) also satisfies X. It is equivalent to say that X logically follows from S iff every interpretation satisfies [S1 and S2 and ... and Sp → X].

Example :
Let consider
S1 = HUMAN(Turing)
S2 = HUMAN(Socrates)
S3 = GREEK(Socrates)
S4 = $(\forall x)\{HUMAN(x) \rightarrow FALLIBLE(x)\}$
and
X = $(\exists y) \{FALLIBLE(y) \text{ and } GREEK(y)\}$

It is easy to see that X logically follows from {S1, S2, S3, S4}.

The goal of an automated deduction precedure is to determine whether some given sentence X follows from a given set of sentences or not.

3 THE RESOLUTION RULE OF INFERENCE

3.1 Notion of rule of inference

Let us consider the problem of showing that a sentence X logically follows from a set S of sentences.

It can be shown that there is no general method to solve this problem. Consequently, first-order predicate calculus is undecidable. However, we will see further that if X actually follows from S, then there exists a general procedure that can prove it within a finite amount of time.

Thus we say that predicate calculus is <u>semi-decidable</u>.

In order to solve the above problem, one can first attempt to use the definition of "logically follows" given in the previous section for showing that every interpretation satistying S also satisfies X. Clearly, this method, called the <u>semantic method</u>, is not applicable but in simple cases since there is an infinity of potential interpretations.

However, it can be shown that it is sufficient to consider only those interpretations which are built over a particular domain of discourse called the <u>Herbrand universe</u>. This universe consists of all variable-free terms that can be constructed from the constant and function symbols occurring in S and X.

Example

Let S = {P(a,b), ($\forall$x)[P(a,x) $\rightarrow$ Q(f(x))]} and X = Q(f(b)). The Herbrand universe for S and X is the infinite set {a, b, f(a), f(b), f(f(a)), f(f(b)), ...}.

Note

If neither S nor X contains a constant, then one should include a single constant symbol in the Herbrand universe.

To apply the semantic method, one now has to consider all the interpretations defined by assigning values TRUE and FALSE to every variable-free atom which can be constructed from the Herbrand universe for S and X and the predicate symbols occuring in X and S. But, whenever there appears a function symbol in S or X, the Herbrand universe is infinite and still an infinity of interpretations remain to be considered.

An alternative method, called the <u>formal method</u>, consists of applying formal rules to infer new sentences from old ones. These rules should be <u>sound</u>, i.e. any sentence derivable from a set of sentences by application of these rules should also logically follow from this set.

A well-known sound rule largely used by humans is modus ponens. From two sentences of the form S1 and S1 $\rightarrow$ S2, modus ponens permits to derive S2.

A set of sound rules is said to be <u>complete</u> if any X logically following from a set S of sentences can be derived from S by applying these rules through a finite chain of derivations.

Most existing automated deduction procedures work according to the formal method and use a unique sound inference rule called <u>resolution</u>

[35]. This rule is complete for refutation proofs (cf. section 4.1).

3.2 Clauses

Resolution applies to a subclass of PC sentences called clauses.

A clause is a disjunction of literal, each literal being either an atom or the negation of an atom ; in addition, all the variables in a clause are universally quantified over the whole formula. Since there is no ambiguity, we will write clauses without universal quantifier.

Examples :
not GREEN(x) or GRASS(x) or TREE(x)
P(x,y) or Q(z,f(x))

It can be shown that any sentence in the first-order predicate calculus can be converted into an equivalent set of implicitly and-connected clauses by an automatic procedure [31]. This transformation consists of successive steps, most of them being straightforward. The only delicate one is the removal of existential quantifiers, which requires the introduction of Skolem functions.

Example :
Let consider the following PC sentence :

$(\forall x)\{GREEN(x) \rightarrow GRASS(x)$ or $(\exists y)[TREE(y)$ and $IS\text{-}PART\text{-}OF(x,y)]\}$.

Converting this sentence into clause form will produce the following intermediate formulas :

- $(\forall x)\{$not GREEN(x) or GRASS(x) or $(\exists y)[TREE(y)$ and $IS\text{-}PART\text{-}OF(x,y)]\}$
  The implication sign has been removed.
- $(\forall x)\{$not GREEN(x) or GRASS(x) or $[TREE(g(x))$ and $IS\text{-}PART\text{-}OF(x,g(x))]\}$
  The existential quantifier has been removed. g(x) is a Skolem function used to replace the "y" that exists. It maps an entity x into another entity representing a "tree" and such that x is "part of" this new entity.
- [not GREEN(x) or GRASS(x) or TREE(g(x))] and [not GREEN(x) or GRASS(x) or IS-PART-OF(x,g(x))].
  The sentence has been put into conjunctive normal form and the universal quantifier has been removed.

From the last formula, we produce the following set of clauses :

not GREEN(x) or GRASS(x) or TREE(g(x))
not GREEN(x) or GRASS(x) or IS-PART-OF(x,g(x))

Clauses may be usefully regarded as a PC sublanguage in which one can directly express logical statements. These statements are usually more clearly understandable than if they were obtained by converting sentences expressed in the full PC language.

Due to the definition of clauses, one should note that :
- two occurrences of the same variable in some clause represent the same entity,
- two occurrences of the same variable in two different clauses may represent different entities,
- it is always possible to rename the variables in a clause, e.g. P(x) or Q(x,y) and P(u) or Q(u,y) are equivalent clauses.

3.3 Instantiation and unification of literals

a) Instantiation of a literal

Instantiation of a literal consists of substituting terms for variables in the literal.

More precisely, let consider a possibly empty set of pairs

$$\theta = \{[V1\ T1],\ [V2\ T2],\ \ldots,\ [Vn\ Tn]\}$$

in which :
- Vi and Ti ($1 \leq i \leq n$) are variables and terms respectively,
- $\forall i \neq j$, Vi $\neq$ Vj.

θ is called a substitution. Each pair [Vi Ti] is called a component of the substitution ; Vi (resp. Ti) is called the variable (resp. the term) of the component.

The instantiation of a literal L by θ is the operation of replacing each occurrence of the variable Vi in L by an occurrence of the term Ti. The resulting literal is noted L.θ and is called the instance of L by θ. Note that, if V2 occurs in T1, then L.{[V1 T1],[V2 T2]} may be different from (L.{[V1 T1]}).{[V2 T2]}.

Example :
Let L = ALONG(x,quay-of(x,y))
and θ = {[x f(a,b)], [y h(a)]}.
Then L.θ = ALONG(f(a,b),quay-of(f(a,b),h(a))).

Let θ and ϕ be two substitutions. The composition of θ and ϕ is the set $\theta' \cup \phi'$, where ϕ' is the subset of all the components of ϕ whose variables are not variables of components of θ ; θ' is the set of all the components [Vi Ti.ϕ] such that [Vi Ti] is a component of θ.

Basic notions in knowledge representation and control

The composition of θ and ϕ is denoted $\theta\phi$.

Example :
Let θ = {[z g(x,y)]}
and ϕ = {[x a], [y b], [u c], [z d]}.
Then $\theta\phi$ = {[z g(a,b)], [x a], [y b], [u c]}.

Note that :
- $(C.\theta).\phi = C.\theta\phi$,
- In general, $\theta\phi \neq \phi\theta$ and $\theta\phi \neq \theta \cup \phi$.

b) Unification

Let E be a set of literals {L1,L2,...,Ln}. A substitution θ unifies E if $L1.\theta = L2.\theta = \ldots = Ln.\theta$.

If there exists a substitution θ that unifies a set E of literals, then E is said to be <u>unifiable</u> and θ is called a <u>unifier</u> of E.

Example :
Let L1 = CONNECTED-BY(x,y,drive-way)
and L2 = CONNECTED-BY(u,road,drive-way)
Then θ = {[x house1], [u house1], [y road]} is a unifier of {L1,L2}.

A unifier θ of a set E of literals is said to be the <u>most general unifier</u> (mgu) of E if for any unifier ϕ of E there exists a substitution ϕ' (possibly empty) such that for all the literals Li in E we have $Li.\theta\phi' = Li.\phi$. One can show that the instance of each literal in E by a mgu is unique except for variants of variable names which are meaningless.

Example :
In the above example, θ is not a mgu of {L1,L2}. A mgu is θ' = {[x u], [y road]}.

The following LISP function UNIFY (L1,L2) produces the mgu of L1 and L2 if they are unifiable ; otherwise it produces "failure". The variables in L1 and L2 should have been renamed so that the same variable does not occur in both literals.

```
(defun UNIFY (el e2)
   (cond ((symbol? el)
          (cond ((eq el e2) nil)
                ((variable? el)
                 (cond ((occurs-in el e2) 'failure)
                       (t (cons (list2 el e2) nil))))
                ((variable? e2) (cons (list2 e2 el) nil))
                (t 'failure)
         ((symbol? e2) (UNIFY e2 el))
         (t (prog (scar scdr)
                  (setq scar (UNIFY (car el) (car e2)))
                  (cond ((eq scar 'failure) (return 'failure))
                        (t (setq scdr (UNIFY (instance (cdr el) scar)
                                             (instance (cdr e2) scar)))
                           (cond ((eq scdr 'failure) (return 'failure))
                                 (t (return (compose-subst scar scdr)]
```

In this function :
- symbol?(e) tests whether expression e is either a predicate symbol, a function symbol, a constant or a variable, or not.
- variable?(e) tests whether expression e is a variable or not.
- occurs-in(v,e) tests whether variable v occurs in expression e or not (indeed in a unifier no component should be such that its variable occurs in its term).
- instance(e,s) computes the instance of e by s.
- compose-subst(s1,s2) computes the substitution s1s2.

The above UNIFY function is rather inefficient and should be considered only as an illustration for unification. There exist algorithms that are less computationally complex.

3.4 Resolution of two clauses

Let us consider two clauses A and B :

A = A1 or A2 or ... or AN
B = B1 or B2 or ... or BP.

The variables in A and B have been renamed so that the same variable does not occur in both clauses.

Suppose that a mgu θ exists for {A1,...,An,not B1,...,not Bp}. Then, the resolution rule applies to A and B. It produces the new clause

[An+1 or ... or AN or Bp+1 or ... or BP].θ, which is called the resolvent of A and B. A and B are called the parents of the resolvent.

One should note that the order to the literals in a clause is meaningless, so that A1, ..., An (resp. B1, ..., Bp) can be any literals among A1, ..., AN (resp. B1, ..., BP).

Example :

If A = [not GREEN(x) or GRASS(x) or TREE(g(x))] and B = [SEA(a) or GREEN(a)], then A and B admit [SEA(a) or GRASS(a) or TREE(g(a))] for resolvent.

It may happen that two clauses have several resolvent. For instance it is the case of [P(x,y) or Q(x)] and [not P(a,z) or not Q(a)]. However the set of all the resolvents of two clauses is finite.

One can easily show that resolution is a sound rule. Indeed, let A = [P or Q] and B = [not P or R]. These two clauses admit C = [Q or R] for resolvent. Any interpretation that satisfies foth A and B must also satisfy either Q or R since P and not P cannot be satisfied simultaneously. Thus, the interpretation satisfies also C.

An important particular case occurs when the resolvent of two clauses A and B is the empty clause, which we write NIL. For example, P(x) and not P(a) resolve to NIL. The empty clause is always the resolvent of two contradictory clauses. Its value is FALSE in any interpretation.

Let CL be a set of clauses. In the following, we will note R(CL) the set of all the resolvents that have their two parents in CL. If CL is finite, then so is R(CL).

4 AUTOMATED DEDUCTION

4.1 Refutation procedures

As we previously said, the goal of an automated deduction procedure is to determine whether a sentence X logically follows from a set of sentences {S1,S2,...,Sn} or not.

This problem is equivalent to showing that S1, S2, ..., Sn and not X form an inconsistent set of sentences, i.e. that there exist no interpretation satisfying these n+1 sentences simultaneously.

Most of the deduction procedures work according to the above paradigm, that is they attempt to show that a set of sentences is inconsistent. They are called refutation procedures.

Let call C1, C2, ..., Cp the clauses obtained by transforming

S1, ..., Sn and <u>not</u> X into clausal form. It can be shown that {S1,..., Sn, <u>not</u> X} is inconsistent iff {C1, ..., Cp} is inconsistent.

A resolution-based refutation procedure applies the resolution rule of inference successively to clauses C1 through Cp, and to those derived from them, until it eventually generates the empty clause NIL, which we noted to be self-inconsistent.

Let CL = {C1, ..., Cp}. We note Rk(CL) = Rk-1(CL)∪R(Rk-1(CL)) (k>0), with R0(CL) = CL. If CL is finite, so is Rk(CL) ∀k.

One can show that if CL is inconsistent, then there exists k such that Rk(CL) contains NIL. This means that the resolution rule is complete for proving that a set of clauses is inconsistent. In addition, since resolution is also a sound rule, if there exists k such that NIL ∈ Rk(CL), the set CL is inconsistent.

The subset of derivations that produces the NIL clause is called a <u>refutation</u>.

Example :

Let consider again the last example of section 2.3. Sentences S1 through S4 can be converted into the following clauses :

C1 = HUMAN(Turing)

C2 = HUMAN(Socrates)

C3 = GREEK(Socrates)

C4 = <u>not</u> HUMAN(x) <u>or</u> FALLIBLE(x)

and <u>not</u> X is converted into :

C5 = <u>not</u> FALLIBLE(y) <u>or</u> <u>not</u> GREEK(y).

Resolution of clauses C4 and C5 produces :

C6 = <u>not</u> HUMAN(x) <u>or</u> <u>not</u> GREEK(x).

Resolution of clauses C2 and C6 produces :

C7 = <u>not</u> GREEK(Socrates).

Resolution of clauses C3 and C7 produces :

C8 = NIL.

If the initial set of clauses is not inconsistent, the refutation procedure may produce new clauses indefinitely. Therefore, a bound should be put on the magnitude of computation in order to avoid perpetual looping. Thus, when a refutation procedure fails to produce the empty clause, there is no guarantee that it would have not been derived if more computation was allowed.

Note :
Resolution-based refutation is not the only way to implement deduction. A discussion of non-resolution theorem proving can be found in [05]. Non-refutation methods are presented in [32].

4.2 Control strategies

In order to derive the empty clause, a refutation procedure usually does not generate R1(CL), R2(CL), ... in turn. It applies a <u>strategy</u> [25] that controls the selection of the successive pairs of clauses to resolve in a more efficient way. Indeed, in the example of the previous section, R2(CL) includes <u>not</u> GREEK(Turing), which is obtained by resolving C1 and C6. This clause then resolves itself with no other clauses. In examples involving a large initial set of clauses, many such unuseful clauses can be generated leading to inefficient refutation.

Control strategies for resolution-based refutation procedures can be classified into refinement strategies and ordering strategies :

- A <u>refinement strategy</u> restricts the pairs of clauses to which resolution can be applied. A typical one is the <u>set-of-support</u> strategy, which requires that one parent of each resolvent is selected among the clauses resulting from <u>not</u> X or from their descendants (cf. the example of section 4.1). The motivation of such a strategy is that it gives a goal-driven behavior to the refutation procedure.

- An <u>ordering strategy</u> orders the pairs of clauses that are candidate parents for resolution. Since the goal is to produce the empty clause, it seems appropriate to attempt to apply the resolution rule to short clauses (i.e. clauses made of a small number of literals) before long ones. For instance, the <u>unit-preference</u> strategy is an ordering strategy in which one tries to select a single-literal clause to be a parent in a resolution. Indeed, every time such a clause is used in resolution, the resolvent is shorter than the other parent.

Of course, these strategies can be combined into more complicated ones. However, although resolution is a complete rule of inference for refutation procedures, not all strategies guarantee that the empty clause can be produced when the initial set of clauses CL is inconsistent. A control strategy is said to be <u>complete</u> if it can produce NIL whenever CL is inconsistent. One can show that the set-of-support strategy is complete.

4.3 Toward logic programming

Let consider a problem stated as a set of wffs H1, H2, ..., Hn and $(\exists x)S(x)$. We interpret H1, ..., Hn as representing the hypotheses of the problem, x as the unknown solution, and S as a condition that should be satisfied by the solution.

This problem can be solved by refuting the set of wffs {H1,.., Hn, not S(x)}. From the refutation (if one can be produced) one can extract a solution by analyzing how the variable x is instantiated in the substitutions which were necessary to resolve the clauses in the refutation.

Example :

The example given at the end of section 2.3 fits exactly this scheme. The problem is to find an individual that is both Greek and fallible. S1 through S4 define the problem conditions. From the substitutions used to apply resolution in the above refutation one can extract Socrates as being a solution.

Stating a problem as a set of sentences is the PC language is called logical programming [26]. A language based on this principle, PROLOG [37], is currently used in the AI field. This language is restricted to a particular class of clauses, called Horn clauses, which have one of the following forms :

- Procedure : [not A1 or ... or not An or B] which is noted [A1 and ... and An → B]
- Goal : [not A1 or ... or not An] which is noted [A1 and ... and An →]
- Assertion : [B] which is noted [→ B]
- Contradiction : NIL.

Resolution between a procedure and a goal derives a new goal and is called backward reasoning. Resolution between a procedure and an assertion is called forward reasoning.

In theory, Horn clauses do not limit the expressive power of full PC. They can be regarded as a well formalized form of "production rules", a class of knowledge representation formalisms to be studied in the next section.

5 CONCLUSION

The major arguments in favor of using PC as the basis of a knowledge representation system have been stated in the introduction :

simple syntax, machine-independent semantics, modularity of the knowledge base, formal properties of the deduction procedures.

PC also presents some disavantages, which may explain why it is not largely used in current AI systems [03] :

- Implementing some common concepts, like equality, is not immediately obvious.

- An action changes one state of the world into another. The problem of specifying which pieces of description should change and which should not is called the "frame problem" [34]. This problem may occur in vision when a sequence of images is to be analyzed. It is difficult to solve when knowledge about possible actions is described as PC sentences.

- Because quantification is restricted to variables, some facts may not be easy to express in the first-order PC. For instance, if x is an element of a class C, which is part of the superclass SC, then it is difficult to state that x should inherit the properties of SC (cf. section D.3).

- Various kinds of reasoning, including plausible reasoning and expectation-driven reasoning, which are both potentially important in computer vision, are not directly addressed by PC.

- Efficient strategies for controlling deduction are frequently non-complete.

However, various tricks make it possible to get round these drawbacks. For example, the "built-in" predicates of PROLOG are a way to implement equality [11], and the "frame predicate" facilitates the solution to the frame problem [26]. In addition, the development of meta-languages to express problem-specific strategies will increase the efficiency of current automated deduction procedure [20].

Logic currently attracts much interest from the AI communauty and its influence on the future knowledge representation systems is likely to increase.

C PRODUCTION SYSTEMS, PLAUSIBLE REASONING, AND GRAPH SEARCHING

1 INTRODUCTION

Production systems (PSs) have deep roots in both Computer Science [33] and Psychology [30]. However, most of their theoretical and application-oriented development took place in the AI field during the last ten years [12] [18]. This development may be partly regarded in the perspective of a reaction against the constraining formalism of predicate calculus, and its subsequent drawbacks.

Knowledge is represented in PSs by separate modules, called production rules. Each rule encodes information about what action it is relevant to perform under specific conditions ; thus it is a "condition-action" pair.

"Good" application-oriented PSs offer the following attractive features :
- Unlike some predicate calculus clauses, each production rule is directly intelligible by humans and it is intended to be similar to the kind of statements that are frequently used by humans to explain how they solve problems.
- A primitive step of reasoning of a PS-based system consists of making use of one rule. Thus, such a system can explain its behavior step by step in terms that are natural to humans.
- The language in which rules are expressed is not unique. Therefore, it can be tailored to a particular application domain. In fact, unlike predicate calculus, PSs form a class of different knowledge representation systems, which are all based on the notion of "condition-action" pairs.
- The behavior of a PS-based system can be improved through interactive experimentation by updating the set of available production rules. Within certain limits, it is not necessary to understand how the inference engine of the PS operates to perform such modifications.

These features explain why PSs are currently so popular in AI, in particular in the field of the so-called "expert systems".

However, the price paid for them is high :
- The semantics of the rules is usually not completely independent from the programs that work on them.
- In general one cannot prove formal properties of a PS like predicate calculus soundness and completeness.
- Since the language in which production rules are stated is not unique,

some work must be redone at almost each new implementation.

2 BASIC STRUCTURE OF A PRODUCTION SYSTEM

A production system describes a computation by the means of three basic components : a data base, a set of production rules, and an interpreter.
- The data base is a set of facts relevant to a given problem (context, current solution). Its structure is dependent on the problem domain.
- Each production rule has the general form "When this condition holds then this action is relevant" and is usually written

$$LHS \rightarrow RHS$$

where LHS (left-hand side) defines a condition on the data base, and RHS (right-hand side) specifies an action on the data base. At any moment, the condition of the LHS of a rule may be satisfied, or not, by the data base. If it is satisfied, the rule is said to be applicable, and the action given by the rule's RHS should or could be performed. The precise meaning of "the condition in LHS is satisfied by the data-base" and "the action in the RHS is performed on the data base" can vary from one application of production systems to another. Note however that RHS cannot explicitly call an other rule.
- The interpreter makes use of the production rules to iteratively modify the initial data base until some goal condition becomes true in the data base. In first approximation, the overall structure of an interpreter can be described by the following non-deterministic procedure :

```
PROCEDURE SP1;
    DB := initial data base;
    WHILE NOT GOAL(DB) DO
        BEGIN R := CHOICE (RULES-APPLICABLE-TO(DB));
            DB := PERFORM (RHS(R),DB)
        END;
```

The choice by the interpreter among several applicable rules is a question of control strategy. It also depends on the PS application.

Using production rules from left to right, as it has been presented above, is called forward reasoning. It is also possible to define a rule as being applicable when its RHS represents a subgoal that we want to reach. Then applying the rule consists of making true the condition in its LHS ; it may require the generation of new subgoals. This type of reasoning is called backward (or goal-driven, or top-down) reasoning. Both forward and backward types of reasoning may be combined in the same interpreter.

3 EXAMPLES OF PRODUCTION RULES

3.1 A first example

In this example, production rules are used to encode the laws of a well-known game, the 8-puzzle (see Nilsson [31]).

The puzzle consists of 8 tiles that can be moved vertically and horizontally on a 3 by 3 square board. In each board state, 8 positions are occupied by the tiles ; the ninth is free and permits motion. The game starts from some initial board state, for instance

```
2 5 8
1 . 3
4 7 6
```

and the goal is to establish a goal configuration, for example :

```
1 2 3
4 5 6
7 8 .
```

Hereafter we present three ways to define production rules for this game. For all of them, we consider the data base of the production system as being a board state description.

a) A production rule is used to encode one possible move, e.g. :

```
2 . 8       2 5 8
1 5 3   →   1 . 3
4 7 6       4 7 6
```

Using this kind of rules, the applicability test is trivial. One just has to test the identity between the LHS of the rule and the current data base The RHS defines the action to be performed by specifying where to move the free position.

However, the number of rules, even for this simple game, is enormous (about one million).

b) The rules in a) "look all the same". We can reduce their numbe by introducing variables, e.g. :

```
a . b       a d b
c d e   →   c . e
f g h       f g h
```

The number of rules becomes very manageable (24), but the applicability test and the interpretation of the RHS are more complicated. The applicabi lity test matches the LHS against the current state and furnishes a binding list (a substitution) needed to execute the RHS. In the above example, the binding list is

((a 2)(b 8)(c 1)(d 5)(e 3)(f 4)(g 7)(h 6)).

c) The most elegant way to describe the game laws uses just 4 rules :

$$a \;.\; ==> .\; a$$
$$.\; a ==> a \;.$$

$$\begin{matrix} a & . \\ & ==> \\ . & a \end{matrix}$$

$$\begin{matrix} . & a \\ & ==> \\ a & . \end{matrix}$$

The LHS gives an incomplete description which does not apply to all the board states. For example, the first rule does not apply when the empty tile is in the rightmost column.

3.2 Other examples

- Shortliffe [42] use production rules to encode knowledge in the field of bacterial infections. An example rule is :

    IF 1) the site of the culture is blood, and

       2) the gram stain of the organism is gramneg, and

       3) the morphology of the organism is rod, and

       4) the patient is a compromised host

    THEN

       there is suggesting evidence (.6) that the identity of the organism is pseudomonas-aeruginosa.

- Davis [13] uses production rules in a stock investment advisor. One of his rule is internally represented as

    ($AND (SAME OBJCT RISK-LEVEL HIGH-RISK)

       (SAME OBJCT TIMESCALE SHORT)

       (SAME OBJCTINV-AREA GAMBLING))

    (CONCLUDE OBJCT STOCK-NAME BALLY .6)

and is to be read

    IF 1) the risk level of the investment should be high risk, and

       2) the time-scale of the investment is short-term, and

       3) the investment area is gambling stocks,

THEN
> Bally is a likely (.6) choice for the investment.

- Tsuji and Nakano [44] use rules to define the interpretation of lines in cine-angiograms as artery branches. One such rule is

    IF a candidate line
    1) branches at 0 % - 20 % of the left arterial descending coronary artery (LAD), and
    2) branches from LAD with a 20-90 degrees angle, and
    3) is shorter than the left circumflex artery, and
    4) is shorter than LAD,
    THEN
    > there is strong (.9) evidence that the candidate line is a diagonal branch.

- Descotte [14] designed a rule-based system for planning how to machine mechanical parts. An example of rule that he uses is

    (<= (SURFACE-QUALITY &X) 3.2)
    (>= (EXTRA-THICKNESS &Y) 3.0)
    (SUPPORTED-BY (ROUGHING-CUT &Y) &X)
    ==>
    (8 (> (FINISHING-CUT &Y) (ROUGHING-CUT &Y)))

which is to be read

    IF 1) the surface quality of an entity &X is lower (better) than 3.2, and
    2) the extra-thickness of an entity &Y is greater than 3.0, and
    3) the part is supported by &X during the roughing cut of &Y,
    THEN
    > one is advised (8) to perform the finishing cut of &X after the roughing cut of &Y.

4 TEST FOR RULE APPLICABILITY

The test for rule applicability can differ very much from one PS to the other. At an extreme, it may involve a complex deduction procedure to determine whether the conditions in the LHS of a rule logically follow from the content of the current data base. But, more frequently, it is based on a simpler pattern-matching operation. The LHS of a rule is said to match the current data base if there is a structural correspondance between the LHS and a subset of the data base. Clearly, this loose definition

Basic notions in knowledge representation and control   347

leaves room for important variations.

The result of a pattern-matching operation is either TRUE if the LHS matches the data base (then the rule is applicable) ; otherwise, it is FALSE. If it is TRUE a side effect of the operation is to establish one or several binding lists that define the values of the variables for executing the action specified by the LHS (cf. section 3.1). If several binding lists are furnished, then it means that several instantiations of the rule are applicable.

It frequently occurs that both the LHS and the data base are made of LISP-style expressions. Then the test for rule applicability consists of matching each expression of the LHS with an expression in the data base. In this case, a typical pattern-matching operation is unification (cf. section B.3.3). Atoms in the LHS are either variables or non-variables. Variables can be distinguished from non-variables by a particular prefix, for instance '?'. They can be bound to any element of a list.

Example
(ALONG ?X (QUAY-OF ?Y))
matches (can be unified with)
(ALONG B1 (QUAY-OF T2))
and the match furnishes the binding list (substitution)
((?X B1) (?Y T2)).

One may attempt to generalize unification by distinguishing between simple variables (prefixed by '?') and segment variables (prefixed by '::') [48]. By definition simple variables can be bound to any element of a list ; segment variables can be bound to any sequence of consecutive elements of a list.

Example
(CONNECTED-BY ::X ?Y)
matches
(CONNECTED-BY HOUSE ROAD DRIVE-WAY)
and the match produces the binding list
((::X (HOUSE ROAD) (?Y DRIVE-WAY)).

The introduction of segment variables has far-reaching consequences :

- First, variables must occur at most in one of the two expressions being matched. Indeed, what would it mean to match (CONNECTED-BY ::X DRIVE-WAY) and (CONNECTED-BY HOUSE ::Y)? So this type of pattern matching is not actually a generalization of unification.
- Second, a match between two expressions may not be unique as soon as an expression contains more than one segment variable on the same level. For example, (A ::X ::Y) matches (A B C D) in four different ways with the following binding lists :

>((::X NIL) (::Y (B C D)))
>((::X (B)) (::Y (C D)))
>((::X (B C)) (::Y (D)))
>((::X (B C D)) (::Y NIL))

The pattern-matching operation may also be extended by permitting the user to attach boolean functions to variables [48]. These functions act as additional filters during matching and make it possible to enforce conditions on the match which are not structural. For instance, they may impose that the value of a variable is an even number, or that it belongs to a particular set.

>Example :
>(A (restrict ?X even-numberp))

does not match

>(A 3)

because ?X must be bound to an even number.

A sophisticated form of pattern matching has been integrated in the interpreter of some AI languages [06], [39].

5 PLAUSIBLE REASONING WITH PRODUCTION SYSTEMS

Production rules are particularly well suited to encode knowledge for plausible reasoning. Then the rules take the following form : "From pieces of evidence E1 and E2 and ... and En, one can hypothesize H with some degree of confidence p" or more symbolically :

>E1, E2, ..., En ==> H
>(p)

Many AI systems, in particular expert consultant systems [15], make use of rules of this form (cf. the first two examples of section 3.2). We believe that they should also be very useful to model uncertain knowledge in vision ; in section 3.2 we showed one such rule intended to help a vision system interpret cine-angiograms. It is likely that similar rules can have applications to the interpretation of aerial images [17], outdoor scenes, and industrial scenes.

A plausible reasoning procedure makes use of these rules to combine forward and backward inferences :
- Forward use of rules permits the procedure to hypothesize new facts from pieces of observed evidence to apply other rules.

Usually pieces of observed evidence have a degree of confidence associated with them. Thus, it is necessary to choose a mathematical model for propagating degrees of confidence through inferences. Models based on probability or fuzzy sets are possible, but they quickly lead to very complex computations. Therefore, more or less empirical formulas are used [27], e.g. : if $E1$ through En have degrees of confidence $p1$ through pn, then if the rule
$$E1, \ldots, En ==> H$$
$$(p)$$
is applied, the degree of confidence of H may be computed as
$$p \times \min\{pi\}.$$
- Backward use of rules permits the procedure to focus its attention on interesting hypotheses. For example, the top-level hypothesis H with the highest degree of confidence can be selected as the current focus of attention. Then all the rules which bear on it are examined. The piece of evidence in their LHSs whose observation would cause the greater change in the degree of confidence of H, may be selected as the new focus of attention. The process is repeated until the selected piece of evidence can be directly observed.

Forward and backward inferences can be combined in many ways. For example, some pieces of evidence may be first observed using a special initialization procedure. Then, forward inferences permit the system to derive successive hypotheses. When all possible inferences have been performed, and if no conclusive (that is with a sufficiently high degree of confidence) top-level hypothesis has been derived, backward inferences can serve to suggest new observations, which in turn cause new forward inferences.

6 GRAPH SEARCHING WHITH PRODUCTION SYSTEMS

6.1 State-space representation

Let consider a set STATES of data structures called states. One state, S0, is known explicitly and is called the initial state.

Let now consider a set RULES of production rules. Each RHS's rule specifies a function mapping a subset of STATES into STATES. The applicability condition of the function is defined in the LHS.

The initial state S0 and the set RULES implicitly define a graph, the state-space graph, the nodes of which are states. In this graph, state s2 is the successor of state s1 if there exists r∈RULES such that :
- the LHS of r is satisfied by s1,
- the RHS of r maps s1 into s2.

Given S0 and RULES, one defines a search problem by specifying a subset GOALS of STATES. Elements of GOALS are called goal states.

If the specification of GOALS is an explicit description, i.e. if the goal states are knwon, a solution to such a problem is a path linking the initial state to a goal state in the state-space. The 8-puzzle example introduced in section 3.1 corresponds to exactly this case.

If GOALS is specified by a condition, then solving the problem may only consist of making a goal state explicit. For example, a scheme to express a vision task as a search problem is the following :
- each state is a set of hypotheses about an image,
- the initial state consists of a priori knowledge about the image content,
- each rule derives new hypotheses from existing ones and from the image data,
- a state containing contradictory hypotheses has no successor,
- each hypothesis in a state is justified by other hypotheses and/or image data (these justifications are extracted from the LHSs of the applied rules) and a goal state is one containing strongly justified hypotheses asserting facts of interest with respect to the particular task that is considered.

6.2 A search procedure

Let call G the state-space graph implicitly defined by the initial state S0 and the set RULES. Searching G consists of making explicit a portion of G until it includes a goal state. It is performed by the production system's interpreter, which is now called a search procedure. At each step of the execution of this procedure, the portion of G that has been explicited so far is called the search graph.

Different search procedures can be implemented. The choice of one of them depends on the answers to several questions, in particular :
- Is the solution a path from the initial state to a goal state, or an explicited goal state ?
- Are the successors of the state to be generated all at once, or one at a time ?
- Is it worthwile to test for state identities (this test may be computationally expensive) ? Or do we prefer to avoid this test (but then there may be several identical states in the search graph) ?

We present below a simple search procedure (SEARCH) assuming that :
(1) we are interested in making a goal state explicit,
(2) it is not worthwile to test for state identities,
(3) the successors of a state are generated all at once.

Assumption (2) implies that the search graph is a tree. Assumptions (1), (2) and (3) permit us to save only the tip states of this tree, whose successors have not been computed yet. These states are recorded into a list named "tipstates", which is initialized to (S0) by the main call to the procedure.

```
(defun SEARCH (tipstates)
   (prog (s)
      loop (cond ((null tipstates) (return 'fail))
                 (t (setq s (CHOICE tipstates))
                    (setq tipstates (remove s tipstates))
                    (cond ((GOAL s) (return s))
                          (t (setq tipstates (append tipstates (EXPAND s))))
                    (go loop)]
```

This procedure makes use of three domain-dependent functions :
- GOAL, which tests whether a state is a goal state or not ;
- EXPAND, which generates the successors of a state s by performing the actions specified by the RHSs of the rules that are applicable to s (it may happen that a node has no successor) ;
- CHOICE, which selects the tip state to be expanded next.

Clearly, the efficiency of SEARCH depends very much on the CHOICE function. This function must order the tip states according to their promise of being on a path to a goal state. Ordering can be based on the computation of an heuristic merit function (cf. [31]). Such a function may take very different forms. In a well-formalized application (e.g. the 8-puzzle), it may be a mathematical function computing some kind of metric difference between the current state and a goal state. In less structured domains (vision is typically one of them), it may consist of production rules. Such rules, which are different from those used by the EXPAND function to compute the successors of a state, are a kind of <u>meta-rules</u> (cf. [13]). An example of such a rule may be :

" IF the goal is to locate a docked ship in an aerial photograph, THEN a state containing a strongly justified hypothesis about the location of docks should be considered with high priority".

Another type of search, called <u>automatic-backtracking</u>, consists of generating one successor of a node at a time, and of following each path until success or failure. When a failure occurs, the search procedure goes back to the most recent node where an alternative path is still unexplored. A number of programming languages have been proposed in AI whose interpreters can perform graph searching in this way [06].

6.3 Comparison with automated deduction and plausible reasoning

It may be interesting to briefly compare graph searching with automated deduction of section E.4 and plausible reasoning of section 5. To that purpose let introduce first the following definition [32] :

A production system defining a state-space graph is said to be <u>commutative</u> iff :
(1) $\forall r \in$ RULES, $\forall s \in$ STATES, if r is applicable to s, then it is also applicable to any successor of s in the state-space graph G.
(2) $\forall r1$ and $r2 \in$ RULES, $\forall s \in$ STATES, if r1 and r2 are both applicable to s, then the state obtained by executing the actions A1 and A2 specified by the LHSs of the two rules is the same whether A1 is executed before A2, or A2 before A1.

Automated deduction and plausible reasoning can be regarded as graph searching with commutative production systems. Then, the commutativity property permits us to save only the current state, which contains all the facts derived so far. Each inference adds new facts to the current state without restricting possibilities for future derivations. In many cases, the difficulty is that the set of facts grows very large.

In non-commutative graph searching, each path in the search tree can be regarded as a commitment to certain hypotheses, which form the context of the tip node on the path. Such a commitment is useful to limit both the size of the description of the states and the number of their successors. But, of course, there may be no solution in a particular context, so that the search procedure may waste some time moving from one context to an other (that is from one path of the search tree to an other).

Sometimes a good compromise is to mix different reasoning schemes. For example, a system may carry deduction on a state before generating its successors.

7 CONCLUSION

A major argument in favor of PSs is that for many applications production rules are a natural way to express knowledge in a form directly intelligible to humans. In addition, since the interpreter of a PS applies one rule at a time, it can explain its behavior step by step in terms of the rules content. Thus, one can improve the behavior of a PS-based system through successive experimentation by simply updating the set of production rules. It does not require a detailed understanding of the interpreter.

Compared to algorithmic programming (e.g. PASCAL programming), a PS permits the user to separate problem knowledge that is expressed in the form of independent production rules and control that is the responsability of the interpreter. This subdivision, which is crucial for easily extending the performance of a system, is very different from the modularity provided by algorithmic procedures. Indeed, in order to add a procedure to a program one has to insert procedure calls in the right places ; and to remove a procedure it may require to review a large part of the program logic.

Predicate calculus also permits the user to input knowledge in modular form. In fact, clauses may be regarded as a particular case of production rules ; it is particularly obvious with Horn clauses (cf. section B.4.3). However, in general, predicate calculus clauses do not permit us to express knowledge in a form as intelligible as production rules. In counterpart, predicate calculus and automated deduction procedures like refutation procedures form a coherently defined framework ; the behavior of a predicate calculus system can be characterized by a number of formal properties. In general, one cannot prove that PSs enjoy similar properties.

D SEMANTIC NETWORKS

1 INTRODUCTION

Both predicate calculus and production systems encapsulate knowledge into separate modules. Interactions among these modules occur when they are used to solve problems. For instance, in a particular example, two PC clauses, which were never related to each other before, may interact through resolution in order to produce a new clause that will resolve with another one, etc.

Semantic netwoks (SNs) proceed from a very different perspective. They represent knowledge as a highly interconnected structure described as a labelled graph. Nodes may represent such entities as physical objects, concepts, situations, processes, classes, ... and links describe binary relations among them.

Examples :

- The following network expresses knowledge about dock scenes in a computer vision system [02]. It encodes that dock ships are on lines parallel to the intersection of coastlines with dock areas at a small distance.

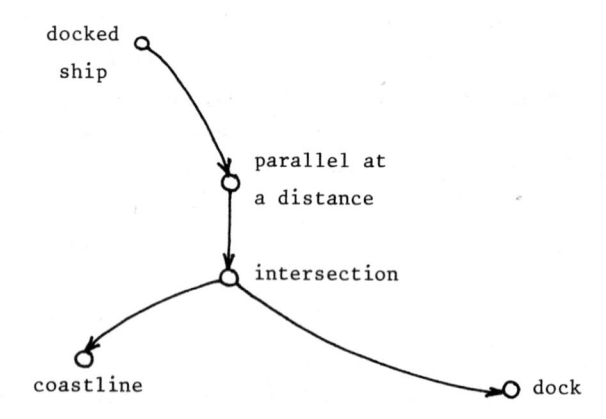

- The following figure shows another SN that has been used by Winston [46] to represent knowledge about block scenes.

```
              house
               ○
               ↑
              is-a
               |
               ○
   part-of  ╱     ╲  part-of
          ╱         ╲
         ○           ○
         |           |
        is-a        is-a
         |           |
         ↓           ↓
         ○           ○
       wedge       brick
```

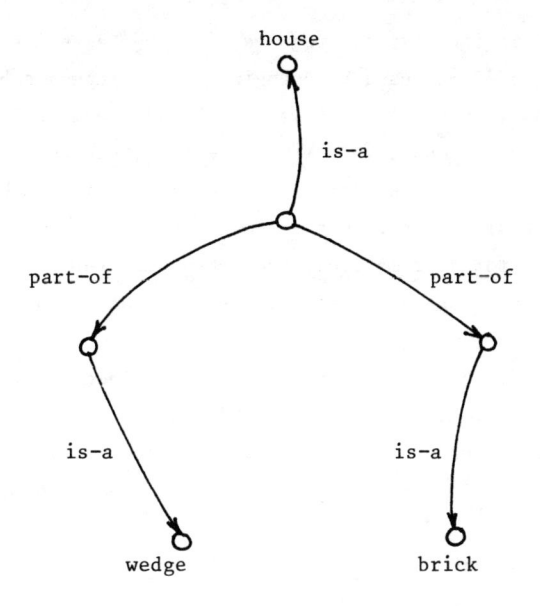

The problem with semantic networks is that they form a very broad class of knowledge representation systems, much broader than production systems. In fact, the graphic notation is about all they actually have in common. Their semantics is very dependent on the programs that access them, and there are great variations among these programs.

Nevertheless, SNs are popular in the AI communauty and have been used in a number of systems including some vision systems. Indeed, SNs are attractive in several ways :
- It is often appealing to provide a system with a global view of what it knows. Predicate calculus and production systems fail to that purpose, whereas the structure of a SN does not put restrictions on the association of entities (including substructures) that are conceptually relevant. Even more, it encourages building such associations. In addition, it can be shown that SN can be used to simulate the full expressive power of the first-order predicate calculus [24] and to implement production rules [16].
- The favorite AI language, LISP, provides facilities for handling data bases, which are appropriate to implement SN. One can encode a node as a LISP atom and links starting from this node as properties of the atom ; the values of these properties are atoms encoding other nodes of the network. Furthermore, LISP facilities permit the user to easily extend the basic graph formalism of SNs. For example, one may attach other pro-

perties to node-atoms like types, programs, ... The semantics may also be tailored to the problems at hand by changing of the access subprograms.

Thus, a SN is potentially a very rich and flexible formalism to implement knowledge in a computer. It may be well-suited when one has to deal with more or less unexplored domains, like vision, in which it would be difficult to constrain oneself to the limitations of representation systems having strict formal semantics. Even when such representation systems are used, SNs may be usefully regarded as a method to index knowledge in the computer memory.

Anyone who actually tries to describe knowledge in the graph notation of SNs rapidly face difficulties [49]. This section is a short introduction to some representation issues when using SNs. Its purpose is only to give the flavor of some of the immediate problems one may encounter.

2 DISTINCTION BETWEEN INSTANCES AND CLASSES

Consider the following SN :

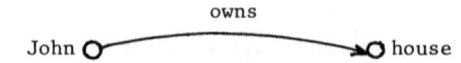

```
                    owns
        John O───────────────►O house
```

A rather natural interpretation of that net is that "John owns a house". Except if we use a very specialized access program including a priori knowledge about "John", "owns" and "house", these are only symbols ; their meaning is only that "John" and "house" are element of a domain of discourse and that "owns" is a binary relation between them.

However, the representation may become somewhat unclear if we consider a little more complex SN :

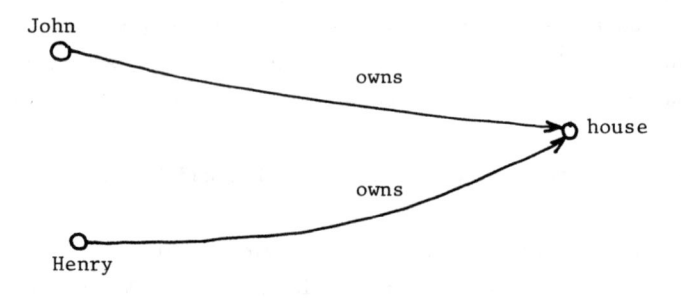

```
        John
         O
          \         owns
           \──────────────►O house
           ┌──────────────►
          O         owns
        Henry
```

Does it mean that John and Henry own the same house or two different houses ? If it means that they own two different houses, then how one can encode that John's house has a red roof ?

These simple examples illustrate a first difficulty that can be easily solved by distinguishing between two types of nodes at least : <u>instance</u> (or <u>token</u>) nodes and <u>class</u> (or <u>type</u>) nodes. Each instance node should be related to a class node by a link that may be labelled by "is-a". Then, we obtain the following network :

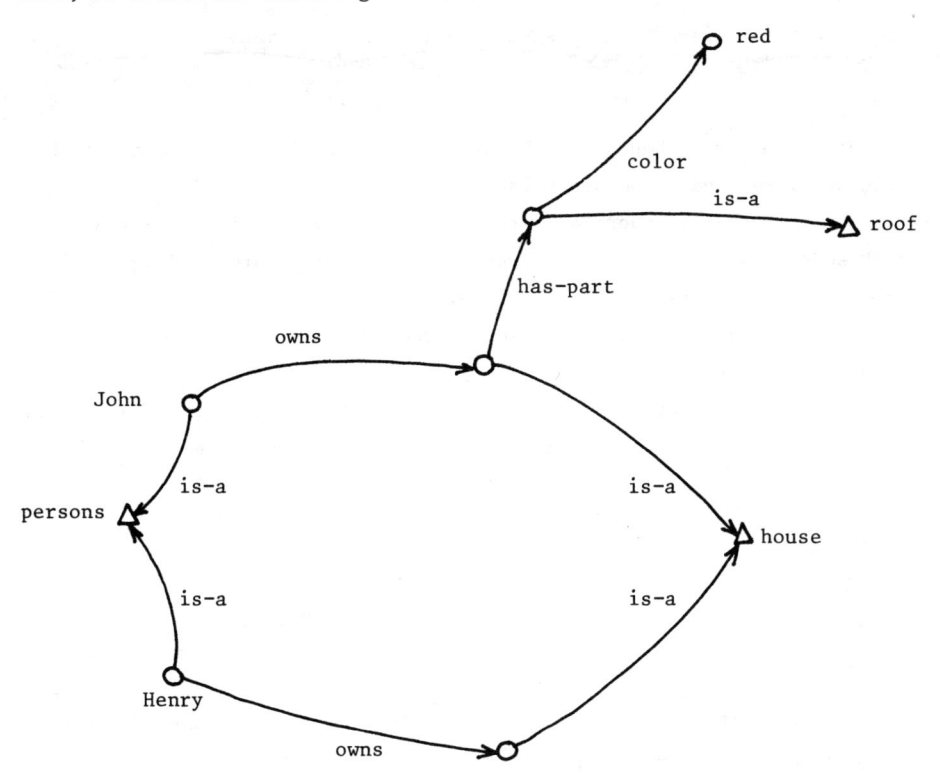

[Instance nodes and class nodes are noded O and Δ respectively].

3 PROPERTY INHERITANCE

One can attach pieces of information to a class node, for example :

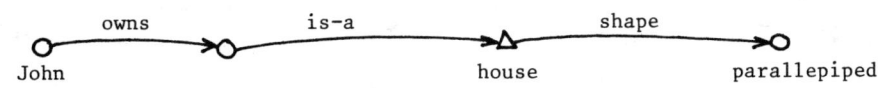

From such a net we can infer that any house instance has got a parallepipedic shape. This kind of reasoning is called <u>property inheritance</u> [04] [32]. For the program accessing the network, this means that it will have to follow "is-a" link assessing that any assertion about a class is also an assertion about its instances. Thus, some kind of special semantics is given by the access program to "is-a" links, class nodes and instance nodes.

This concept can be enlarged to a hierarchy of classes. For example, from :

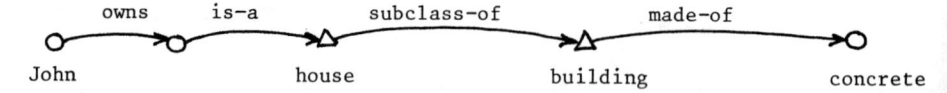

```
         owns        is-a      subclass-of           made-of
    John           house              building              concrete
```

the access program should infer that John's house is made of concrete by tracing "is-a" and "subclass" links.

Property inheritance reasoning is a typical illustration of how a SN subset can carry important semantics, its structure leading to simple computations. The same kind of reasoning is not straightforward to implement when using such representation systems as first-order predicate calculus or production systems.

However, too simple property inheritance reasoning may rapidly lead to some difficulties. For instance, one might interpret

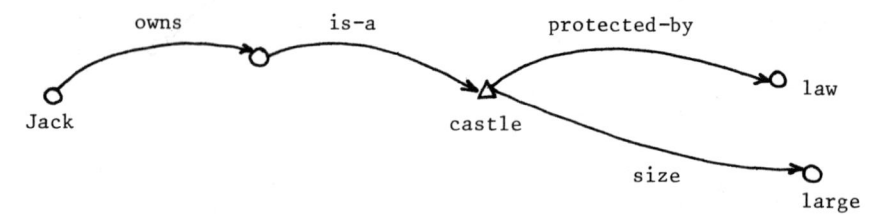

```
         owns              is-a            protected-by
                                                              law
    Jack              castle
                                       size
                                                              large
```

a saying that the castle of Jack is protected by law and that it has a large size. But may be we only want to say that castles in general, not any castle, are protected by law, so that Jack's castle may not be protected by law. This kind of problem can be solved by introducing a new type of node, called the prototype node. The properties that an instance node should inherit from a class are now attached to the prototype nodes attainable from the corresponding class nodes by "prototype" links.

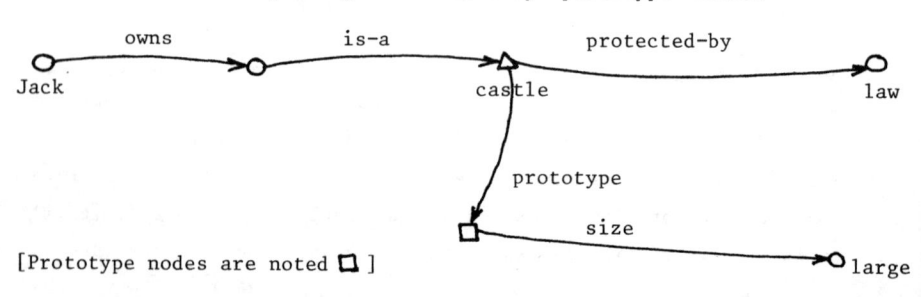

```
         owns              is-a            protected-by
    Jack              castle                              law
                                  prototype
                                          size
                                                              large
```

[Prototype nodes are noted ▫]

Another problem occurs when an instance node is connected to two classes nodes.

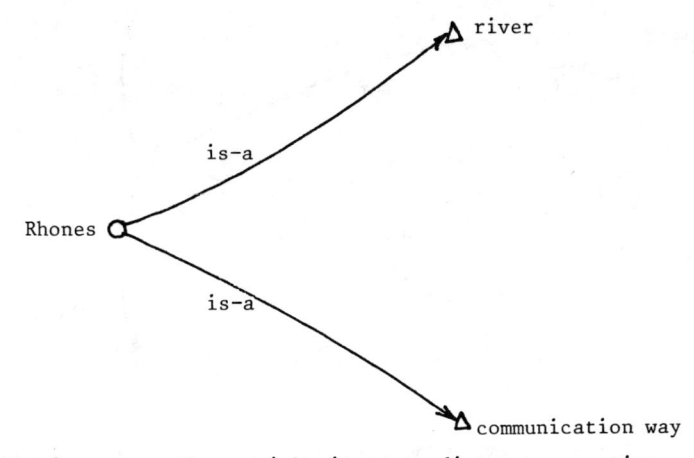

```
                              △ river
                             ↗
                       is-a
            Rhones ○
                       is-a
                             ↘
                              △ communication way
```

Then the instance node may inherit contradictory properties.

However, the relative ease with which properties can be inherited by some nodes from other nodes is one of the advantages of using SNs.

4 SITUATIONS

Links represent binary relationships between entities. Let call the substructure made of two nodes connected by a link a <u>situation</u>. The two nodes are the situation arguments.

In many cases it is necessary to represent situations with more than two arguments. Often, it is also useful to distinguish between a situation instance and a situation class.

Thus, during the development of a SN-type representation formalism, it frequently occurs that a situation that was initially represented by means of a single link becomes a node.

<u>Example</u> :

The following net

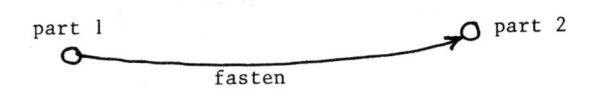

```
        part 1                    ○ part 2
          ○————————————————→
                  fasten
```

may become

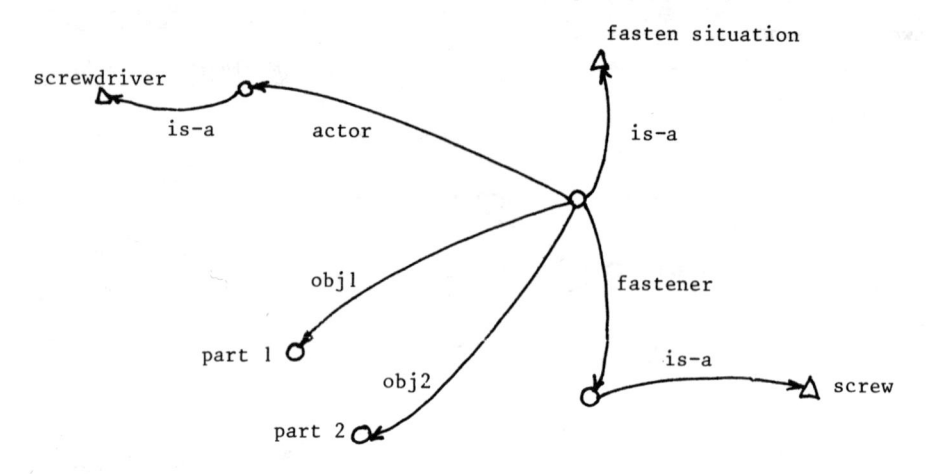

[Figure: semantic network showing "fasten situation" with is-a links to screwdriver (actor), screw (fastener), and obj1/obj2 linking to part 1 and part 2]

In computer vision, situations may typically be used to structure knowledge about scenes.

5 NET ENCLOSURES

Constructing a SN formalism by adding new types of nodes and links with labels that are meaningful to the programs can become difficult to handle. Then, a powerful construct, named <u>enclosure</u>, becomes useful. Its main purpose is to delimit a subset of interest in a net. It helps very much to structure complex knowledge.

Enclosures can be used in a variety of ways. For instance, to express disjunction and conjunction of facts, quantification of variables, implication rules, production rules, hypothetical worlds, ...

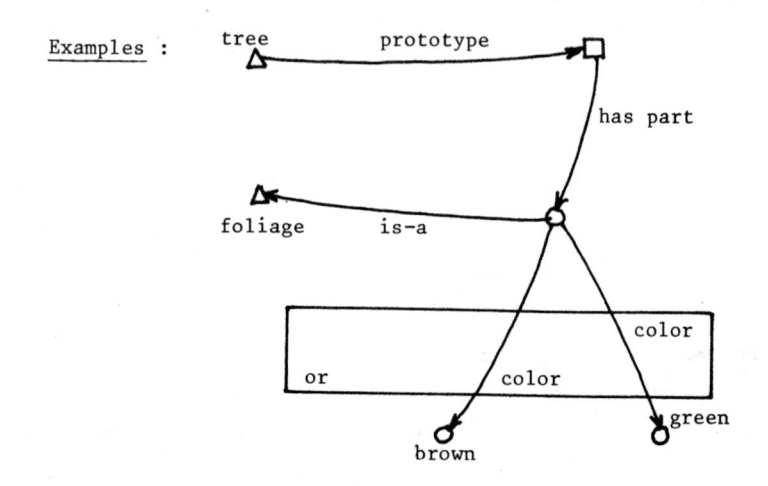

Examples :

[Figure: enclosure example showing "tree" with prototype link to a node with "has part" to another node. The second node has "is-a" link to "foliage", and within an "or" enclosure, two "color" links go to "brown" and "green".]

"the foliage of a tree is either green or brown". [The "or" symbol labelling the enclusure should be specially known to the access program.]

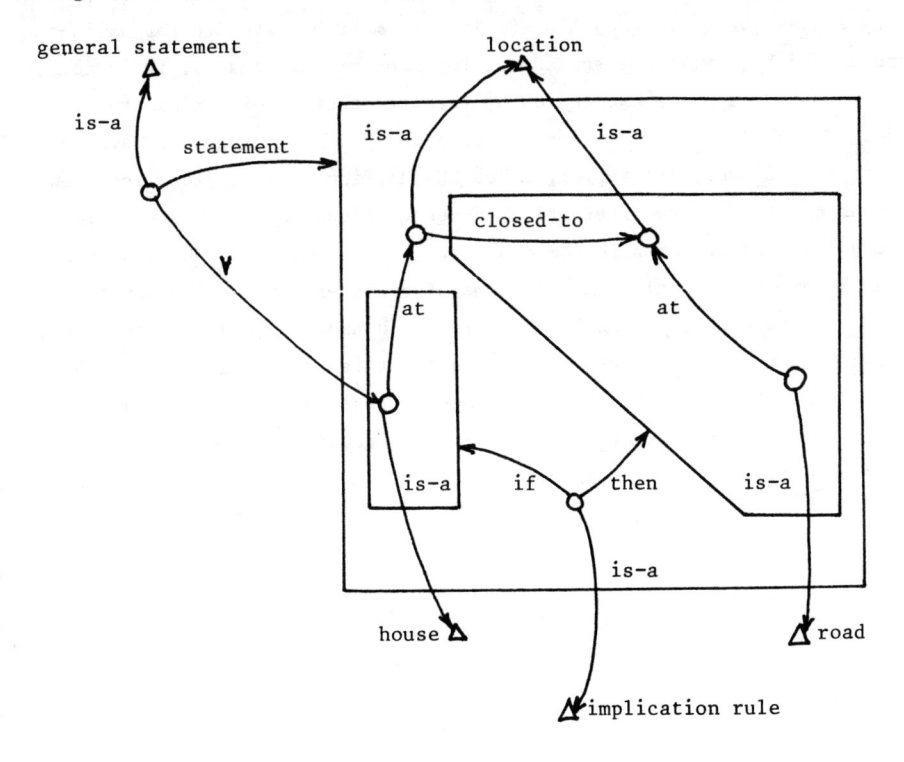

[Figure: semantic network diagram with nodes "general statement", "location", "statement", "closed-to", "at", "at", "is-a" relations, "if", "then", "house", "road", "implication rule"]

"Close to every house there is a road"
[here the access program should have a priori knowledge about nodes labelled "general statement" and "implication"].

In vision, enclosures can be useful to define "frames" [29] encoding expectations about objects in specific situations.

An implementation of enclosure named <u>partitioning</u> has been studied in detail by Hendrix [24], who suggested many other applications.

6 CONCLUSION

The most attractive feature of SNs is their ability to associate conceptually relevant entities into a graph structure. The basic tenet of this kind of representation is that the graph structure can carry important semantics, so that certain operations (typically property inheritance) can be performed by simple procedures. With other knowledge representation systems, similar operations would require special inference rules and numerous derivations.

However, the representation of knowledge by nodes and links lacks expressive power and this simple idea must be extended for having some applicability. As we saw above, for non-trivial domains, it is necessary to enrich the representation by the introduction of various types of nodes and labels having a priori meaning to the access programs. This enrichment of the semantics rapidly makes SNs difficult to handle : representations are ambiguous, alternative representations look equally valid, access programs are complicated to update, ... In addition, SNs implementations become very costly both in terms of computer time and memory space.

Nevertheless, much recent research has been based on the elabotation of semantic networks. In particular, this is the case of KRL [07]. It is likely that this type of knowledge representation will be very important in the future. But undoubtly it will require a more formal basis and more unifying concepts than there exist now.

E A SURVEY OF SOME EXISTING SYSTEMS

Image understanding has been a major objective of AI since the mid sixties, and vision research holds an important place in AI, as can be seen, in particular, from the proceedings of the IJCAI conferences. However, much of this work has to do with low-level vision and is not directly related to the general concepts developped in this lecture.

This section presents a number of systems which do apply general AI principles to vision tasks, thus illustrating the claim that AI is relevant to vision. These examples do by no means exhaust the subject. Indeed, the application of AI concepts to vision still is very much a research problem, where just some general guidelines have emerged.

1 ARCHITECTURE OF A GENERAL IMAGE UNDERSTANDING SYSTEM

The VISIONS [23] system is proposed as a general ambitious system for the interpretation of outdoor scenes. As shown by figure E.1, the system is composed of a segmentation subsystem, capable of extracting features from an image, and an interpretation subsystem. The latter subsystem interprets features in terms of world knowledge and proposes new features to be looked for by the segmentation process.

The interpretation subsystem of VISIONS draws from AI the following characteristics :
- Declarative knowledge is stored in a semantic network of class nodes. Class elements are created during the interpretation of an image. The final result of the analysis is a net of token (instance) nodes.

As shown by figure E.2, the nets used by VISIONS contain knowledge at different levels of representation. Nodes at the highest level, called "schemas", correspond to typical situations involving a number of physical objects. A schema indicates default values that one can expect for various parameters in the situation. This notion is often called "frame" in the AI literature [29]. If facilitates a kind of "expectation-driven" reasoning, looking for things that are expected based on the context one thinks one is in [04].
- There is a clean separation between (cf. figure E.1) :
  :: declarative knowledge encoded in the semantic network,
  :: procedural knowledge about how to use the elements of the network to construct hypotheses about an image,

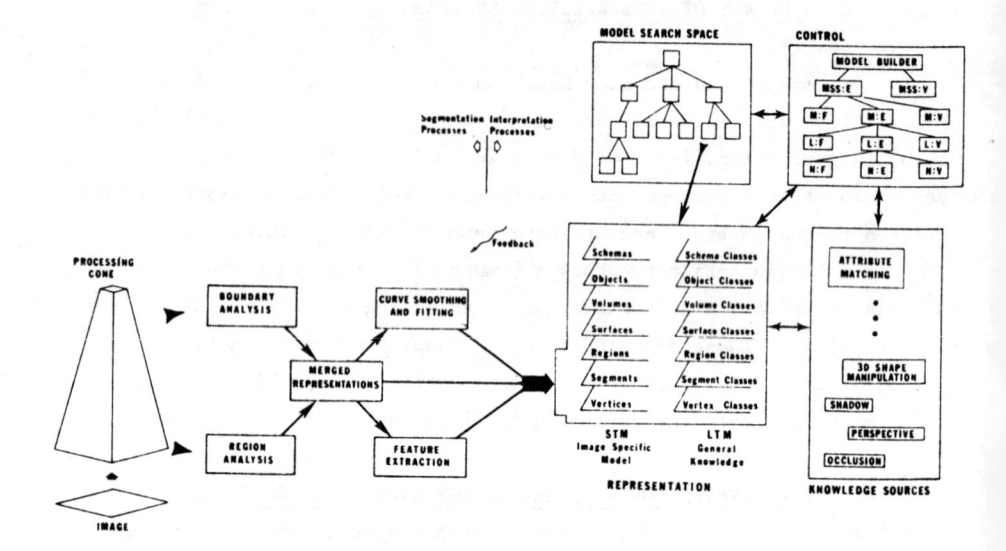

Figure E.1 [23]

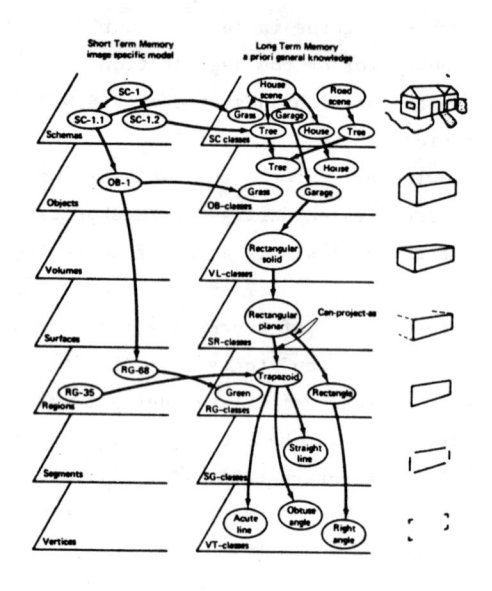

Figure E.2 [23]

:: control knowledge about how to proceed in the search space formed by
different hypotheses.

2 KNOWLEDGE DIRECTED IMAGE ANALYSIS

Another general-purpose image analysis system has been experimented on several different applications at the University of Rochester : aerial photographs and radiographic chest images [02] [03] [41].

The system contains a semantic network as its central data structure :
- The net directly represents geometrical constraints on possible object locations (it is called a "location net"). During image analysis, these constraints are evaluated to narrow down the part of the image where an object is searched for. An example of location net was given is section D.1 (first example).
- Mapping procedures are attached to the nodes of the net representing object classes. They are specialized procedures for finding an object in an image, given certain preconditions. Different mapping procedures can be attached to a single node, in order to describe the object class in various contexts. When several mapping procedures can be activated, the image analysis control program makes a choice based on the expected reliability and cost associated with each procedure.

This kind of knowledge directed image analysis using procedural knowledge is related with the notion of "active knowledge" proposed by Freuder [19]. Knowledge is active in the sence that it constructs hypotheses from image features, and does search to confirm an hypothesis.

3 GEOMETRICAL REASONING IN VISION

A major problem for a vision system using object models is the following : according to the viewpoint, an object may appear in very different ways in an image. The ACRONYM system [09] [10] tackles this problem with the introduction of 3-D object models, using generalized cones. Figure E.3 shows examples of generalized cones, and figure E.4 gives three models of electric motors composed of generalized cones. Since generalized cones are not directly visible in an image, the system has to decide how observable features are related to object models. It is then capable to compute the 3-D position of the object. ACRONYM has an important geometric reasoning capability, which manipulates sets of geometric constraints derived from the models and the image.

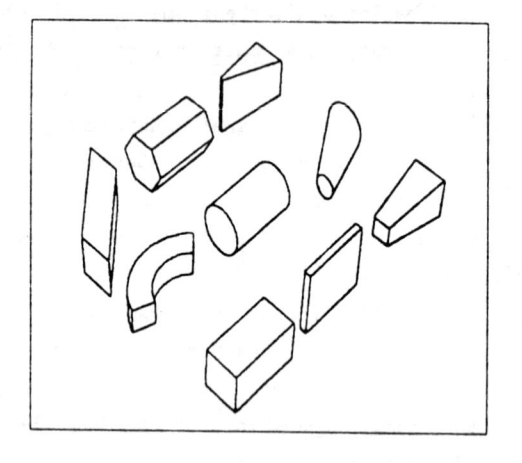

Figure E.3 [10]

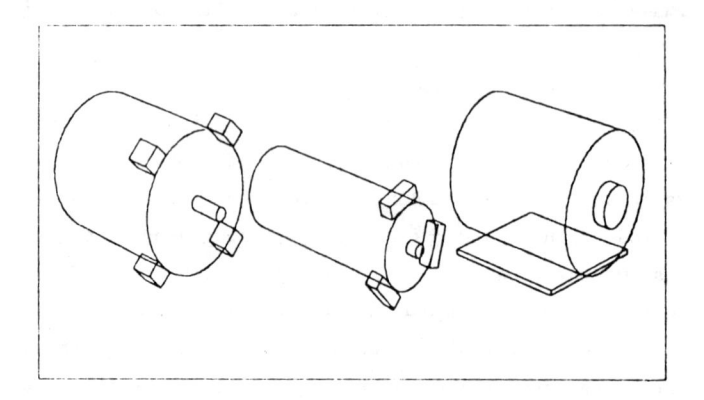

Figure E.4 [10]

4 SEARCH

Matching image features with object models (scene knowledge) leads to the generation of a space of hypotheses. Any vision system containing general knowledge does some kind of graph search.

The ARGOS system [38] explores this aspect more specifically, considering image analysis exclusively as a search problem. It uses a two-pass non-backtracking search technique. During the first pass, a graph containing the most plausible hypotheses is constructed ; the second pass, proceeding in reverse order, selects one solution.

ARGOS interprets view of a city (Pittsburgh) and can determine the angle of view around the city within 40 degrees of error.

5 PREDICTION

Once an hypothesis about a scene is established, knowledge can be applied to predict new image features, which have to be verified if the hypothesis is correct. Prediction provides a feedback from interpretation to segmentation (cf. figure E.1).

When much a priori knowledge is available about a scene, this feedback can significantly cut down the low level processing time by proposing relevant features to look for [40]. Location networks (cf. section 2) are an example of this kind of feedback.

In the PVV system [43], which is used to interpret 2-D industrial scenes, the interpretation program completely guides the segmentation process : segmentation (contrast lines extraction) is carried out only when it is known to bring something to the interpretation process. As a result, the segmentation process analyzes as little of the image as is necessary for the given image-understanding task.

The optimal use of a priori knowledge to globally cut down image analysis cost has been studied by Garvey [21]. He uses plan generation techniques [32] to derive an image analysis strategy. When required to locate an object, his system computes the distinguishing features of the objects, combining them into a strategy represented by a planning graph. The graph contains optimal subgoals for achieving the goal of locating the object. An execution routine selects the best subgoal, executes it, rates its effect, and selects the next best goal, continuing with the process until either the object is located, or no more options remain.

An important case of highly constrained scenes is found in automatic assembly. Bolles [08] has constructed an interactive system to enable non-specialists to generale programs solving "verification vision" tasks. Based on a priori knowledge about a scene he uses Bayesian techniques to construct an optimal combination of operators to find object features, and then uses a least squares fit to locate the object.

6 USE OF PRODUCTION SYSTEMS IN VISION

Tsuji and Nakano [44] describe an image understanding system using a production system for the interpretation of cine-angiograms (radiographic images of coronary arteries). The task is the identification and classification of arteries in a line image. This task, currently executed by medical experts, is difficult for several reasons :
- image contrasts are low and noise is high,
- one often has to combine information from several views in order to obtain a satisfying interpretation,
- a small change in the viewpoint may result in drastic changes in the image,
- images are time varying.

The program does not use a complete geometrical model of the coronary system. Production rules are used to encode pieces of such a model. The interpreter analyzes line segments found by a first image-processing stage (segmentation). It uses production rules of different types :
- Identification rules establish an interpretation for a line. They are affected with a plausibility factor. An example of such a rule has been given is section C.3.2.
- Judgement rules are used to compare the interpretations of two different views, and to choose between contradictory interpretations.
- Control rules specify a subset of the previous rules that can be activated at a given moment. This is a kind of meta-rules representating the order in which one should search for different parts of the arterial system.

Production rules have also been used to encode knowledge in a vision system that interprets outdoors scenes [01].

BIBLIOGRAPHY

CVS = Computer Vision Systems, Edited by A.R. HANSON and E.M. RISEMAN, Academic Press, 1978.
PCV = The Psychology of Computer Vision, Edited by P.H. WINSTON, McGraw-Hill, 1975.
IJCAI = International Joint Conference on Artificial Intelligence.
AIJ = Artificial Intelligence Journal.

[01] Bajcsy, R., A.K. Joshi : "A partially ordered world model and natural outdoor scenes", in CVS.
[02] Ballard, D.H., C.M. BROWN, J.A. FELDMAN : "An approach to knowledge-directed image analysis", in CVS.
[03] Ballard, D.H. and C.M. BROWN : "Computer vision". Technical report, Computer Science Department, University of Rochester, 1980.
[04] Barr, A. and E.A. Feigenbaum : "The handbook of Artificial Intelligence". Volume 1, William Kaufman Inc., 1981.
[05] Bledsoe, W.W. : "Non-resolution theorem proving". AIJ, 1977.
[06] Bobrow, D.G. and B. Raphael : "New programming languages for Artificial Intelligence research". ACM Computing Surveys, Vol. 6, 1974.
[07] Bobrow, D.G. and R. Winograd : "An overview of KRL, a knowledge representation language". Cognitive Science, 1977.
[08] Bolles, R.C. : "Verification vision for programmable assembly". 5th IJCAI, 1977.
[09] Brooks, R.A., R. Greiner, T.O. Binford : "The ACRONYM model-based vision system". 6th IJCAI, 1979.
[10] Brooks, R.A. : "Symbolic reasoning among 3-D models and 2-D images". AIJ, August 1981.
[11] Clocksin, W.F. and C.S. Mellish : "Programming in PROLOG". Springer-Verlag, 1981.
[12] Davis R., J.J. King : "An overview of production systems". In Machine Intelligence 8, edited by E. Elcock and D. Michie, Ellis Horwood, Chichester, England, 1977.
[13] Davis R. : "Meta-level knowledge : overview and applications". 5th IJCAI, 1977.
[14] Descotte Y., J.C. Latombe : "GARI : a problem solver that plans how to machine mechanical parts". 7th IJCAI, 1981.
[15] Duda R.O., P.E. Hart and N.J. Nilsson : "Subjective bayesian methods for rule-based inference systems". National Computer Conference, 1976.
[16] Duda R.O. et al. : "Semantic network representation in rule-based inference systems". In Pattern-directed inference systems, edited by D. Waterman and F. Hayes-Roth, Academic Press, 1978.
[17] Duda R.O. and T.D. Garvey : "A study of knowledge-based systems for photo interpretation". Artificial Intelligence Center, SRI International, June 1980.
[18] Feigenbaum E.A. : "The art of Artificial Intelligence : Themes and case studies of knowledge engineering". 5th IJCAI, 1977.
[19] Freuder E.C. : "A computer system for visual recognition using active knowledge". 5th IJCAI, 1977.
[20] Gallaire H. and C. Lasserre : "Controlling knowledge deduction in a declarative approach". 6th IJCAI, 1979.
[21] Garvey T.D. : "Perceptual strategies for purposive vision". Technical Note 117, Artificial Intelligence Center, SRI International,

September 1976.
[22] Hall, E.L. : "Computer image processing and recognition". Academic Press, 1979.
[23] Hanson, A.R., E.M. Riseman : "VISIONS : a computer system for interpreting scenes". In CVS.
[24] Hendrix, G.G. : "Expanding the utility of semantic networks through partitioning". 4th IJCAI, 1975.
[25] Kowalski, R. : "Search strategies for theorem-proving". In Machine Intelligence 5, edited by B. Meltzer and D. Michie, Edinburg University Press, 1970.
[26] Kowalski, R. : "Logic for problem solving". North Holland, 1979;
[27] Laurière, J.L. : "Représentation et utilisation des connaissances. -- Première partie : les systèmes experts". Technique et Science Informatiques, n° 1, 1982 (in French).
[28] Loveland D.W. : "Automated theorem proving : a logical base". North Holland, 1978.
[29] Minsky, M. : "A framework for representing knowledge". In PCV.
[30] Newell, A. : "Production systems : models of control structure". In Visual information processing, edited by W. Chase, Academic Press, 1973.
[31] Nilsson, N.J. : "Problem-solving methods in Artificial Intelligence". Mc Graw-Hill, 1971.
[32] Nilsson, N.J. : "Principles of Artificial Intelligence". Tioga, Palo Alto, 1980.
[33] Post, E. : "Formal reductions of the general combinatorial problem". American Journal of Mathematics, 1943.
[34] Raphael, B. : "The frame problem in problem-solving system". In Artificial Intelligence and Heuristic Programming, edited by N.V. Findler and B. Meltzer, American Elsevier, 1971.
[35] Robinson, J.A. : "A machine-oriented logic based on the resolution principle". Journal of ACM, 1965.
[36] Rosenfeld, A. and A. Kak : "Digital picture processing". Academic Press, 1976.
[37] Roussel, P. : "PROLOG : manuel de référence et d'utilisation". Groupe d'Intelligence Artificielle, Université d'Aix-Marseille, Luminy, 1975 (in French)
[38] Rubin S. : "The ARGOS image understanding system". Doctoral Dissertation, department of Computer Science, Carnegie-Mellon University, November 1978.
[39] Sacerdoti, E.D. et al. : "QLISP -- A language for the interactive development of complex systems". AFIPS National conference, 1976.
[40] Shirai, Y. : "Analyzing intensity arrays using knowledge about scenes". In PCV.
[41] Selfridge, P.G. and K.R. Sloan : "Reasoning about images : using meta-knowledge in aerial image understanding". 7th IJCAI, Vancouver, 1981.
[42] Shortliffe, E.H. : "A computer-based medical consultations : MYCIN". North Holland, 1976.
[43] Souvignier, V., "Prédiction et vérification en vision". Rapport de DEA, Laboratoire IMAG, Grenoble, June 1981 (in French).
[44] Tsuji, S. and H. Nakano : "Knowledge-based identification of artery branch in cine-angiograms". 7th IJCAI, 1981.
[45] Tenenbaum, J.M., H.G. Barrow, R.C. Bolles : "Prospects for industrial vision". In Computer vision and sensor-based robots, Edited by G.G. Dodd and L. Rossol, Plenum Press, 1979.

[46] Winston, P.H. : "Learning structural descriptions from examples".
    In PCV.
[47] Winston, P.H. : "Artificial Intelligence". Addison-Wesley, 1981.
[48] Winston, P.H., B.K.P. Horn : "LISP". Addison-Wesley, 1981.
[49] Woods, W. : "What's in a link : foundation for semantic networks".
    In Representation and Understanding, Edited by D.G. Bobrow and
    A. Collins, Academic Press, 1975.

RELAXATION: PIXEL-BASED METHODS

A. Rosenfeld
Computer Vision Laboratory, Computer Science Center, University of Maryland, College Park, MD 20742

1 INTRODUCTION

"Relaxation" (Rosenfeld et al., 1976) is a method of using contextual information as an aid in classifying a set of interdependent objects, by allowing interactions among the possible classifications of related objects. If the objects are pixels, for example, improved classification results can be obtained using interactions among neighboring pixels. This approach can be applied to pixel classification into subpopulations, as used for image segmentation, as well as to local feature detection by recognition of special types of pixels and also to image matching (these topics are all treated in separate chapters).

Section 2 describes several types of relaxation algorithms, some of them based on probabilistic arguments or on an optimization approach. Section 3 illustrates the application of such algorithms to various pixel classification tasks, including thresholding, edge detection, and matching. Relaxation can also be applied to region classification (using neighboring or related regions as context) and to relational structure matching; these topics will be discussed further in later chapters.

An overview of relaxation methods and their uses in image analysis can be found in Davis & Rosenfeld (1981). References on specific algorithms and applications will be given in Sections 2 and 3.

2 RELAXATION ALGORITHMS

Let $A_1,\ldots,A_n$ be a set of objects (e.g., pixels), and let $C_1,\ldots,C_m$ be a set of classes. In the relaxation approach to classification, we begin with a set of estimates of the "probabilities" $p_{ij} \geq 0$ that object i belongs to class j, $1 \leq i \leq n$, $1 \leq j \leq m$. (We assume here that the classes are mutually exclusive and exhaustive, so that $\sum_{j=1}^{m} p_{ij}=1$ for each i; in this sense we can think of the p's as probabilities. Other interpretations are also possible, as we shall see below.) We also begin with a set of measures of the "compatibilities" c_{ijhk} between the pairs of events $(A_i \in C_j, A_h \in C_k)$, $1 \leq i,h \leq n$, $1 \leq j,k \leq m$. The relaxation process then iteratively adjusts each p_{ij}, based on the values of the other (relevant) p_{hk}'s and the compatibilities, so as to obtain estimates of the class membership probabilities that are more mutually consistent.

The following is one way of designing such an adjustment process (Rosenfeld et al., 1976). Suppose that $-1 \leq c_{ijhk} \leq 1$, where -1 indicates strong incompatibility, 0 indicates irrelevance, and 1 indicates compatibility. The effect of p_{hk} on p_{ij} should intuitively be as follows: If p_{hk} is high and c_{ijhk} is close to 1, p_{ij} should be increased; if p_{hk} is high and c_{ijhk} is close to -1, p_{ij} should be decreased; if p_{hk} is low, or c_{ijhk} is close to 0, p_{ij} should be relatively unaffected. We can achieve these effects by using a quantity proportional to $c_{ijhk} p_{hk}$ as an increment (or decrement) to p_{ij}. The total effect of all the p's on p_{ij} is then proportional to $q_{ij} \equiv \sum_{h,k} c_{ijhk} p_{hk}$, and we can achieve this effect by multiplying p_{ij} by $1+q_{ij}$. The adjusted p's no longer sum to 1, but we can renormalize them by dividing by their sum. The adjustment process can then be iterated.

The c's used in this algorithm can be defined on the basis of a model for the given classification task, as we will illustrate in Section 3. A general way of defining them is as follows: Let p_{ijhk} be (an estimate of) the probability that $A_i \in C_j$ and $A_h \in C_k$. Then we have $\frac{p_{ijhk}}{p_{ij} p_{hk}} > 1$ if these two events support one another; =1 if the events are independent; and <1 if they contradict one another. Thus $\log \left(\frac{p_{ijhk}}{p_{ij} p_{hk}}\right)$, which is known as the mutual information between the two events, is positive if they are compatible, zero if independent, and negative if incompatible; we can thus normalize these logs, so they take on values in $[-1,1]$, and use them as c's (Peleg & Rosenfeld, 1978).

A closely related relaxation algorithm can be derived using a Bayesian argument (Peleg, 1980); it leads to the adjusted p's being obtained from the current p's by an expression of the form

$$\frac{p_{ij} \sum_k c_{ijhk}}{\sum_{j,k} c_{ijhk} p_{hk}}$$

In this expression the c's are nonnegative, and in fact turn out to be just the probability ratios introduced in the preceding paragraph. Thus in this algorithm the increments are all nonnegative, but when we renormalize the p's to sum to 1, the ones that were incremented only by small amounts will decrease, so that qualitatively the result is the same as if we used compatibilties in $[-1,1]$. The results obtained using this method are generally quite similar to those obtained using the algorithm described in the preceding paragraphs.

In some cases it is not appropriate to assume that the classes are mutually exclusive and exhaustive; in these situations one should use a somewhat different type of relaxation algorithm. As an example, consider a pattern matching task in which the A's are pixels and the C's denote local matches to pieces of a given pattern, i.e., $A_i \in C_j$ means that subpattern C_j is present in location A_i. Here the initial p_{ij}'s would depend on the degrees of match to the C_j's at each pixel A_i, and the c_{ijhk}'s would depend on the relative positions in which the matches to the C_j's should ideally occur - e.g., if C_k should be in a given position relative to C_j, then p_{hk} should support p_{ij} to the extent that A_h and A_i are in (approximately) that relative position. Note that here, however, it makes no sense to add the contributions of the p_{hk}'s for all A_h, since only one of them can represent the desired C_k; it is more reasonable to take their maximum over h. It is more reasonable to add them for all C_k (the more C_k's are present in good relative positions, the more support $A_i \in C_j$ gets); but there are other possibilities, e.g., taking their minimum over k, which amounts to saying that we may not want to support the hypothesis $A_i \in C_j$ if any of the necessary C_k's is missing. It is also not necessary to normalize the adjusted p_{ij}'s so that they sum to 1, since the classes are not exhaustive - A_i may not be any of the C_j's. In fact, in this case the p_{ij}'s might be regarded as fuzzy set memberships rather than as probabilities. On the use of relaxation algorithms of this (max, min) type for pattern matching and for relational structure matching, see Ranade & Rosenfeld (1980), Kitchen & Rosenfeld (1979), and Kitchen (1980).

When a relaxation algorithm is iterated, we expect that the classifications should become less ambiguous; if we regard the p's as probabilities, this means that one of them should tend toward 1 and the others toward 0, while if we regard them as fuzzy set memberships, it means that at most one of them should remain high and the others tend toward 0. At the same time, we expect the values of the p's to stabilize - in other words, the process should converge. Some empirical studies of the behavior of relaxation processes are presented in Fekete et al. (1981).

Based on the remarks in the preceding paragraph, we can define a combined cost function, e.g., involving both the entropy of the p's and their rate of change, and use it to evaluate the performance of a relaxation process. (See, however, Peleg & Rosenfeld (1981) for some remarks on the limitations of such cost functions.) In fact, one can regard the goal of relaxation as the minimization of such a cost function. Thus, given a cost function, one can use standard minimum-seeking methods, e.g., steepest descent, as relaxation algorithms. This approach to the design of relaxation processes will be described in a later chapter.

3 APPLICATIONS

The simplest pixel classification task to which relaxation methods can be applied is thresholding (Rosenfeld & Smith, 1931). Here there are (e.g.) two classes, "light" and "dark", and at pairs of neighboring pixels light is compatible with light and dark with dark. We can define the initial light and dark probabilities of a pixel as normalized versions of its gray level, i.e., proportional to its distances from the ends of the grayscale; and we can use mutual information estimates of c's. Figure 1 shows typical results obtained in this way; for display purposes we have reconverted the light probabilities to brightnesses. Note that after a few iterations, most of the pixels are nearly certain of being light or of being dark, and the object (an infrared image of a tank) has been extracted.

Relaxation: Pixel-based methods

A somewhat more complex example is that of edge or curve detection (Zucker et al., 1977; Schachter et al., 1977). One approach to this task is to use a set of classes representing edges (or curves) in various orientations, and another class representing "no edge". The initial probabilities of the edge classes at a given pixel can be taken as proportional to the relative magnitudes of the responses of a set of edge masks at that pixel; and the initial "no edge" probability can be defined in terms of the degree to which the largest of these responses

Figure 1. Pixel classification into "light" and "dark" using relaxation: original and seven iterations, with their histograms.

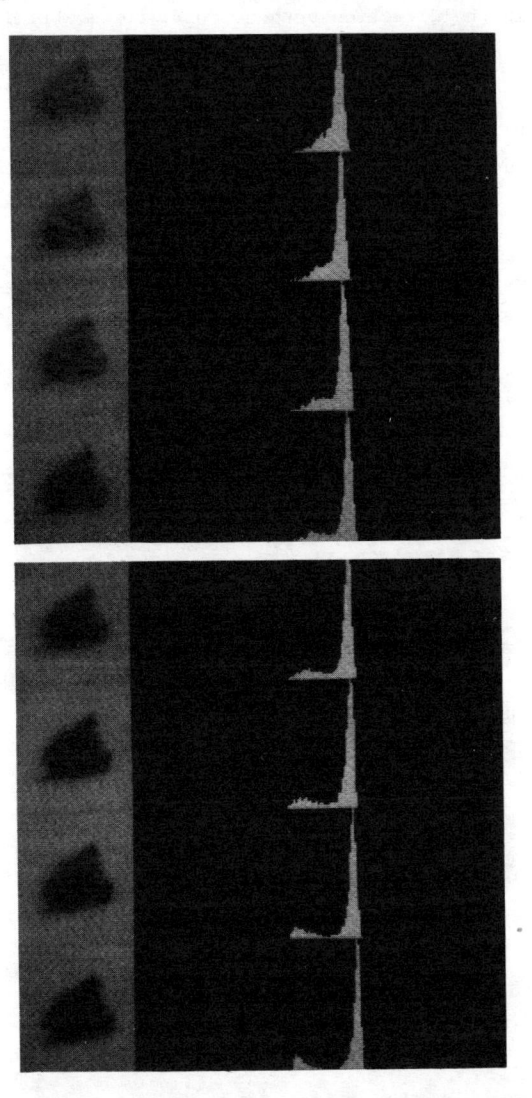

falls short of the maximum possible edge response. The c's in this case depend in part on orientation; two edge classes at neighboring pixels are compatible to the extent that they smoothly continue one another. (For example, if the direction from A_i to A_h is θ, and the classes correspond to edge slopes θ_j and θ_k, we might use $c_{ijhk} = \cos(\theta-\theta_j)\cos(\theta-\theta_k)$; note that this is +1 when θ_j and θ_k are both collinear with θ, drops to 0 as either of them becomes perpendicular to θ, and drops to -1 if one of them is collinear with θ and the other is anticollinear, representing the fact that these edges have the same orientation but opposite senses.) Similarly, an edge at θ_j is compatible with "no edge" at a neighbor in the direction perpendicular to θ_j, but incompatible with "no edge" in direction θ_j; and so on. Alternatively, c's derived from mutual information estimates can be used (Peleg & Rosenfeld, 1978). Figure 2 shows some results using this approach; here the highest edge probability at each pixel has been displayed as a gray level (the higher the darker), but the pixel has been left blank if its no-edge probability is higher than any of its edge probabilities. One can also use a simplified process in which there is only one edge probability (initially taken to be proportional to the gradient magnitude), and an orientation parameter (initially determined by the gradient direction) is associated with it; here again, edge reinforces edge to the extent

Figure 2. Edge detection using relaxation: (a) image; (b) gradient magnitudes; (c) pixels whose highest probability is an edge probability, at iterations 0,1,2, and 3.

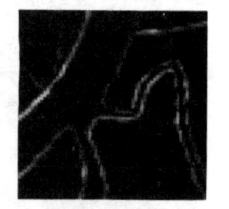

(a) (b)

Relaxation: Pixel-based methods

that their orientations smoothly continue one another, and we can also adjust the orientations (toward collinearity) at each iteration.

As a final example, we consider a simple point pattern matching task (Ranade & Rosenfeld, 1980), where "$A_i \in C_j$" means 'point i of the given pattern is point j of the reference pattern'. Here, the support given to $A_i \in C_j$ by $A_h \in C_k$ depends on the agreement between the position of A_h relative to A_i and that of C_k relative to C_j; e.g., we can define the support as $\frac{P_{hk}}{1+\delta^2}$, where δ is the difference between these relative positions. Initially, all the p's are 1's, since the points are indistinguishable; and we use a relaxation algorithm of the (max,min) type, as described in Section 2. Figure 3 shows an example of this method; here the displays are essentially cross-correlations, where for each relative displacement of the two point patterns that takes one point into another, we have displayed the sum of the p_{ij}'s (rescaled) for which A_i and C_j have that given displacement. We see that after a few iterations, there is a cluster of high values near the correct displacement (which happens to be (0,0)), and there are no high values elsewhere.

4 CONCLUDING REMARKS

Relaxation is a useful technique for obtaining consistent classifications of interrelated objects. It has many applications in image segmentation, feature detection, and matching, as well as in many other areas.

Figure 2, cont'd.

(c)

Relaxation algorithms involve a substantial amount of computation, but since this consists primarily of iterated local operations, such algorithms are very well suited for implementation on special-purpose hardware.

Figure 3. Point pattern matching by relaxation. (a) Two point patterns. (b) Initial values of displacements. (c-f) Values at iterations 1-4.

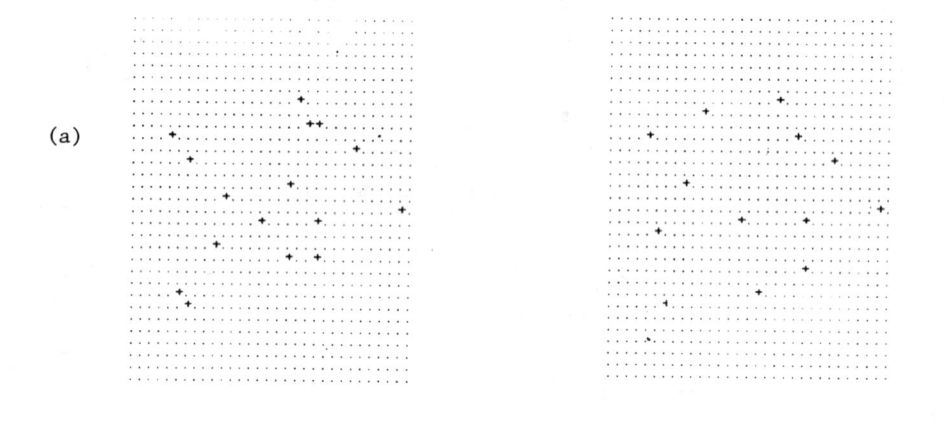

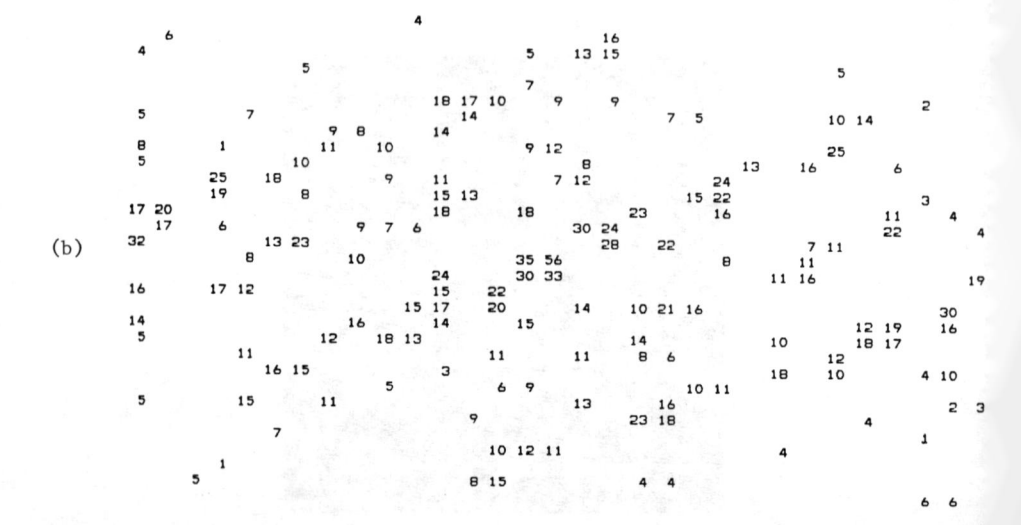

Relaxation: Pixel-based methods

Figure 3, cont'd.

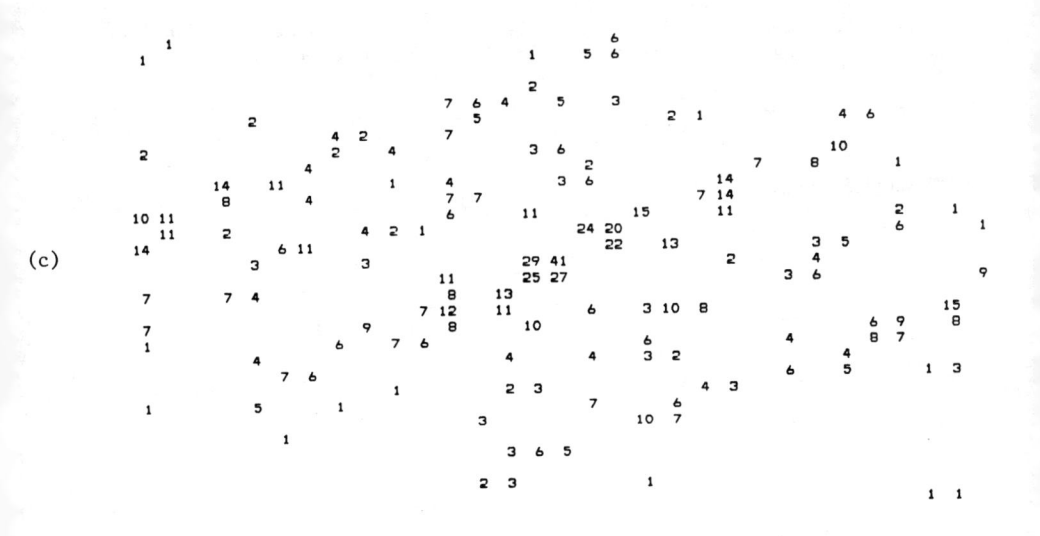

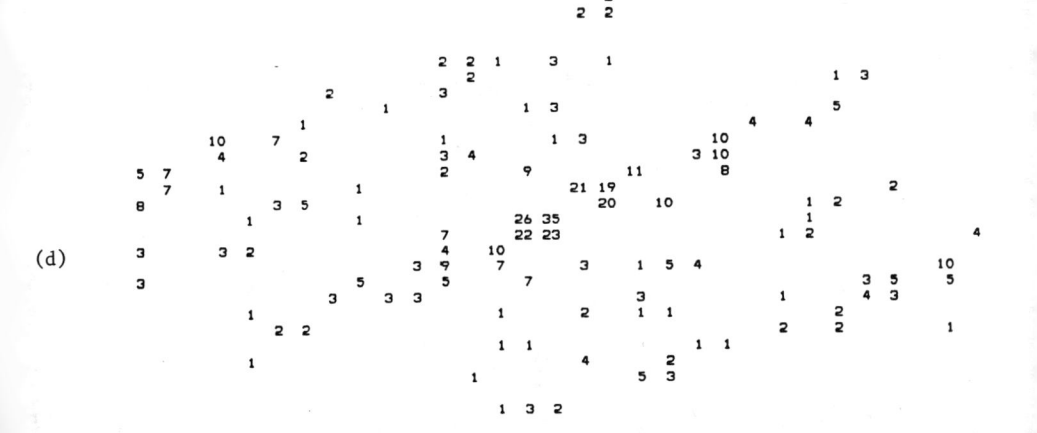

REFERENCES

Davis, L.S., & Rosenfeld, A. (1981). Cooperating processes for low-level vision: a survey. Artificial Intelligence, 17, 245-263.
Fekete, G., Eklundh, J.O., & Rosenfeld, A. (1981). Relaxation: evaluation and applications. IEEE Trans. Pattern Analysis Machine Intelligence, 3, 459-469.
Kitchen, L., (1980). Relaxation applied to matching quantitative relational structures. IEEE Trans. Systems, Man, Cybernetics, 10, 96-101.
Kitchen, L., & Rosenfeld, A. (1979). Discrete relaxation for matching relational structures. IEEE Trans. Systems, Man, Cynernetics, 9, 869-874.
Peleg, S., (1980). A new probabilistic relaxation scheme. IEEE Trans. Pattern Analysis Machine Intelligence, 2, 362-369.

Figure 3, cont'd.

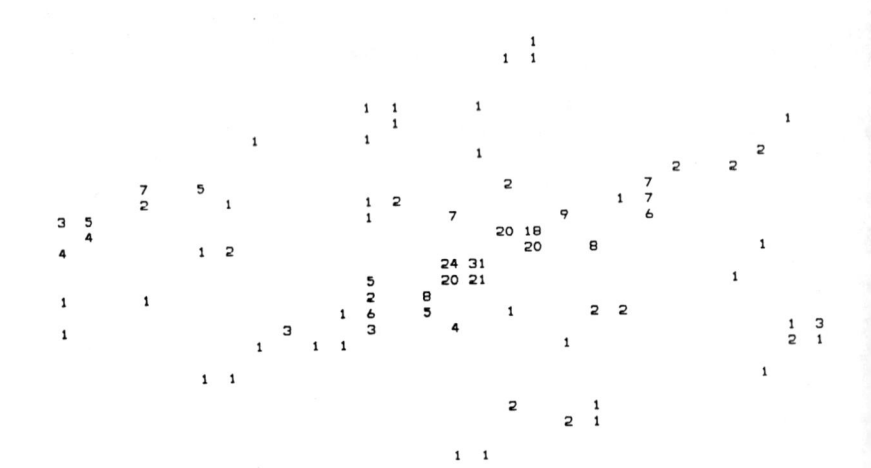

Peleg, S., & Rosenfeld, A., (1978). Determining compatibility coefficients for curve enhancement relaxation processes. IEEE Trans. Systems, Man, Cybernetics, 8, 548-555.
Peleg, S., & Rosenfeld, A., (1981). A note on the evaluation of probabilistic labelings. IEEE Trans. Systems, Man, Cybernetics, 11, 176-179.
Ranade, S., & Rosenfeld, A., (1980). Point pattern matching by relaxation. Pattern Recognition, 12, 269-275.
Rosenfeld, A., Hummel, R.A., & Zucker, S.W., (1976). Scene labeling by relaxation operations. IEEE Trans. Systems, Man, Cybernetics, 6, 420-433.
Rosenfeld, A., & Smith, R.C., (1981). Thresholding using relaxation. IEEE Trans. Pattern Analysis Machine Intelligence, 3, 598-606.
Schachter, B.J., Lev, A., Zucker, S.W., & Rosenfeld, A., (1977). An application of relaxation methods to edge reinforcement. IEEE Trans. Systems, Man, Cybernetics, 7, 813-816.
Zucker, S.W., Hummel, R.A., & Rosenfeld, A. (1977). An application of relaxation labeling to line and curve enhancement. IEEE Trans. Computers, 26, 394-403, 922-929.

RELATIONAL STRUCTURE MATCHING AND RELAXATION LABELING

O.D. FAUGERAS
INRIA
Domaine de Voluceau - ROCQUENCOURT
B.P. 105 - 78153 - LE CHESNAY CEDEX - FRANCE

1. - INTRODUCTION

As we saw in another chapter relaxation is a useful technique for many low-level vision tasks. In this chapter, we introduce a new algorithm for relaxation labeling, describe some of its computational properties and give two concrete examples of its applications. Our relaxation technique is based on the optimization of a global criterion defined on a set of objects. More details can be found in [1-2].

An overview of relaxation methods and their uses can be found in [3]. An approach similar to the one presented here can be found in [4].

The two applications we discuss then are examples of the use of such a continuous relaxation procedure in symbolic matching tasks. The first application is the Semantic Description of aerial images [5] and the second is the recognition of partially occluded objects [6-7].

2. - OPTIMIZATION APPROACH TO RELAXATION LABELING

Let $A_1, \ldots, A_n$ be a set of objects (not necessarily pixels) and $C_1, \ldots, C_m$ a set of classes or names. We associate with every object A_i an m-dimensional vector $\vec{p}_i = [p_{i1}, \ldots, p_{im}]^T$. The numbers p_{ij}, $j = 1, \ldots, m$ are positive, add up to one and can be considered as the probabilities of object A_i to belong to the various classes C_j. They are computed initially from measurements performed on the data and then iteratively updated to reduce ambiguity and improve consistency, a notion we introduce next.

We also have a "world model" as a set of measures of the "compatibilities" c_{ijhk} between the pairs of events $\{A_i \in C_j, A_h \in C_k\}$, $1 \leq i$, $h \leq n$ and $1 \leq j, k \leq m$. These numbers take their values between -1 and +1 or between 0 and 1. Here 0 indicates complete incompatibility and 1 complete compatibility. These numbers also define implicitly a notion of neighborhood on the set of objects. The neighborhood V_i of object A_i is the set of related objects A_k for which the c_{ijhk}'s are defined. This allows us to associate with every object A_i another vector $\vec{q}_i = [q_{i1}, \ldots, q_{im}]^T$ called a compatibility vector whose components are given by [1-2-3]:

$$q_{ij} = \sum_{A_h \text{ in } V_i} \sum_{k=1}^{m} c_{ijhk} p_{hk}$$

Clearly, q_{ij} is high if for most objects A_h in V_i the label(s) C_k compatible with C_j in terms of the c_{ijhk} have high probability. Intuitively the vector $\vec{q}_i$ is a summary of the "opinions" of the neighbors of object A_i as to its labeling. If we call the set of vectors $\vec{p}_i$, $i = 1, \ldots, n$ a labeling of the set of objects A_i, then the compatibility of this labeling can be defined loosely as the agreement between the pairs of vectors $(\vec{p}_i, \vec{q}_i)$ and formally [2-3-4] as:

(1) $$\sum_{i=1}^{n} \vec{p}_i \cdot \vec{q}_i$$

This criterion has the nice property of being maximum when $\vec{p}_i = \vec{q}_i$ = unit vector, $i = 1, \ldots, n$. The first equality indicates maximum compatibility and the second minimum ambiguity.

The labeling problem can therefore be formally defined as that of maximizing criterion (1) [1]. Due to the particular structure of (1) as a sum of usually loosely coupled terms this maximization can be performed efficiently either by steepest ascent techniques [1] or by decomposition and decentralization techniques [2,7]. In both cases, the algorithm starts with an initial labeling $\vec{p}_i^{(0)}$, $i = 1, \ldots, n$ of the set of objects and iteratively adjusts the vector $\vec{p}_i$ to converge to a local maximum of criterion (1).

This is achieved in the steepest ascent technique by defining a sequence $\vec{p}_i^{(\ell)}$ of labelings as:

$$\vec{p}_i^{(\ell+1)} = \vec{p}_i^{(\ell)} + \rho_\ell \, \text{Proj}_\ell \{\vec{g}_i^{(\ell)}\}$$

where ρ_ℓ is a positive step size, the vector $\vec{g}_i^{(\ell)}$ is the value of the partial derivative of criterion (1) with respect to the vector $\vec{p}_i$ computed for the ℓth labeling, and Proj is a projection operator that ensures that $p_i^{(\ell+1)}$ is still a probability vector. The computation of $\vec{g}_i^{(\ell)}$ and Proj_ℓ can be done in parallel for all objects A_i and the optimization procedure could be implemented on fairly simple specialized parallel hardware.

For the decomposition technique this can be achieved by defining a sequence $\vec{p}_i^{(\ell)}$ of labelings as :

$$\vec{p}_i^{(\ell+1)} = \underset{\vec{p}_i}{\text{Max}} \, [\vec{p}_i \cdot (\vec{q}_i^{(\ell)} + \vec{v}_i^{(\ell)}) - \frac{g}{2} \, ||\vec{p}_i - \vec{p}_i^{(\ell)}||^2]$$

$\vec{v}_i^{(\ell)}$ is computed from the partial derivatives of criterion (1) and g is a positive constant. Here the global problem has been completely decomposed into n subproblems which can be solved in parallel on fairly simple hardware.

The key advantage of this approach is that since we are maximizing a criterion that explicitely takes into account the consistency (ambiguity) of the labeling these quantities are increasing (decreasing) on the average at each criterion.

3. - APPLICATION TO THE DESCRIPTION OF AERIAL IMAGES

We have shown in other chapters various techniques for image segmentation that take an array of pixels and produce a symbolic description represented as a labeled graph or semantic network. To proceed any further toward an interpretation of the image we must assume that we have access to another body of knowledge containing a priori information about the expected content of the images of interest. This information is also represented as a semantic network. The process of obtaining a semantic description of a given image can then be viewed as finding the solution of a graph matching problem : either match the image onto the model or match the model onto the image. Barrow et al [8] have applied a relaxation procedure to a much simpler problem in symbolic matching than the one described here.

The matching system to be described uses a feature-based, symbolic, description of an idealized scene (the model) and of the image of a portion of this scene [9]. The image description is derived automatically and the model is developed by the user through an interactive procedure. The basic objects used for the image description are the segments of the image generated both by a general region-based image segmentation procedure (see chapters on segmentation and [10]) and by a linear feature extraction procedure (see chapter on edges and [11]). The regions are derived by locating connected areas which are uniform with respect to some feature in the input image. Typically the images that are used have a total of 100-200 segments. The symbolic description is completed by computing various features of the segments and relations between them.

The features used for the symbolic description are those which can be reliably computed from the available data (the input image and the region or line description). They include properties such as average color and texture (currently only simple texture measures), size, position, two-dimensional orientation, and simple shape measures. Also included are various relations between image segments such as adjacency, nearby, and relative positions (above, below, etc.). With all these relations, a segment may easily be related to as many as 100 other segments. This description is not intended to be used for reconstruction of the original image ; it is meant to capture the important, observable, information contained in the image.

The model description is identical to that used for the image feature-based descriptions of basic region-like and line-like elements including relations between them. Additionally, the basic elements in the model are grouped into more complex objects, associated with generic descriptions, and referred to by actual names. The feature values in the model will not correspond exactly to image values, but are approximations of the likely values, such as those derived from map-like data, so that one model of a scene can be used with many similar images of the same scene. The relations between elements, which are included in the model, are the "important" ones, that is, if a relation appears in the model description then it is expected to occur in the image description, but if no relation occurs in the model then nothing may be said about its appearance in the image (negative relations could be used, such as must not be adjacent, etc.). Similarly, only the important objects are described in the model ; thus it is not a complete description of the entire scene. Generally, the model description is smaller than the image description containing 20-30 basic elements.

The nodes of the model are the objects A_i of section 2 and the names are the nodes of the image. Initial likelyhoods $\vec{p}_i^{(o)}$ and compatibilities c_{ijhk}, $1 \leq i, h \leq n$ and $1 \leq j, k \leq m$ are computed using the same basic matching technique which combines differences in all available features and relations to produce a single rating of the quality of the match. Briefly we combine differences in all feature values which are weighted to account for the different ranges of values. The rating for match between a model objet (A_i) and image segment (C_j) is given by :

$$(2) \qquad R(A_i, C_j) = \sum_{k=1}^{p} |V_{ik} - V_{jk}| w_k$$

where p is the number of features being considered, $V_{ik}(V_{jk})$ is the kth feature value of the model (image) element and w_k is the weighting for the given feature. This produces a positive number from 0, for good matches up to 1000 or more, for bad matches. These values are transformed to the range [0, 1] by :

$$f(A_i, C_j) = \frac{1}{1 + R(A_i, C_j)}$$

and we define $p_{ij}^{(o)}$ as proportional to $f(A_i, C_j)$ for $j = 1, \ldots, m$.

The second operation is the compatibility measure, which computes the effect of making an assignment of one element on the assignment for another element. The interaction between objects and their assignments is through the relations between them, so that the compatibility measure is based on these relations.

Figure 1 illustrate how the relations between objects are used in the matching procedure. The compatibility between model elements i and h for a particular assignment of each is based on the number of relations (neighbors, below, etc.) between i and h, which also occur between the potential assignments j and k. That is, if i and h are neighbors, then for the assignments (j and k) to add to the compatibility measure, they must also be neighbors ; if not, the compatibility contribution will be low (i.e. near 0). The results from relations are combined as in (2), with the feature values being the number of relations which occur between the two model objects or the two image segments.

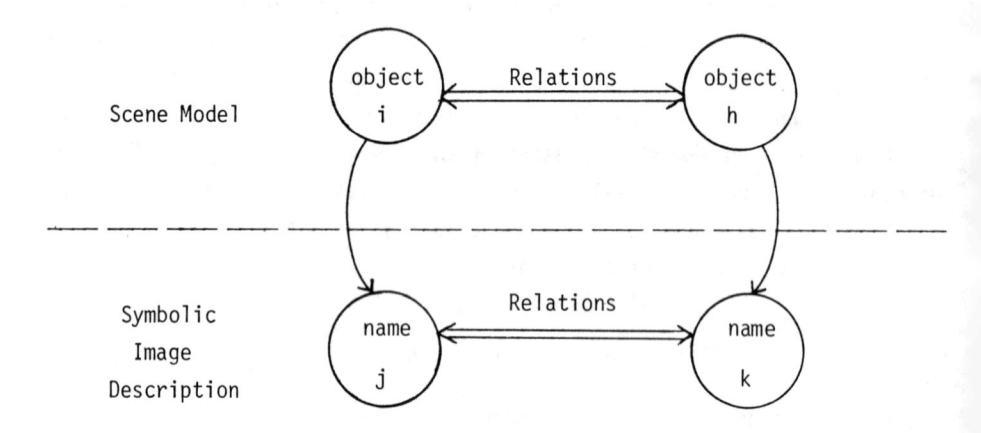

Figure 1.

The relaxation matching system has been applied to several different scenes. Results are presented from a high altitude aerial image with a few major regions (a city and rural areas) and a number of linear features (a major highway, river channels, and roads along the channels) ; see figure 2. The complete results are shown in figure 3. All the major elements are correctly matched.

4. - APPLICATION TO THE RECOGNITION OF PARTIALLY OCCLUDED 2-D OBJECTS

Recognizing partially visible objects is becoming a crucial problem in many applications requiring image-understanding capabilities such as Robotics, Biomedical Analysis, Industrial Inspection. Part of this problem is to identify planar shapes as approximate matches of larger shapes, the so-called inexact shape matching problem. For other references on this subject see [12-18].

The shapes to be considered are compact, connected and bounded by simple closed curves. These curves are approximated by polygons. Let $C = \{C_1, \ldots, C_{m-1}\}$ and $A = \{A_1, \ldots, A_n\}$ be the polygonal lines corresponding to an object and a template, respectively. The problem then is to identify part of the polygonal line A (template or model) within the polygonal line C (object or observation). Therefore, we are trying to label each segment A_i, $1 \leq i \leq n$ as being either one of the segments C_j, $1 \leq j \leq m-1$ or as not belonging to C in case of occlusion (this is the mth name or class). For this problem the relaxation labeling formalism can be applied to the segments A_i, $1 \leq i \leq n$ (objects) and the segments C_j, $1 \leq j \leq m$ (classes).

There are various ways of obtaining polygonal approximations of discrete curves. A good review can be found in [19]. This polygonal approximation can be viewed as a segmentation of the curves into primitives which are chosen in this example to be line segments but could possibly be more general curves such as conics or splines.

We must now specify the subjects of related primitives attached to each model primitive. They can be either some privileged segments which are characteristic of the shape to be recognized corresponding for example to linear parts, regions of high curvature, teeth, notches, etc. or if such segments are not present simply segments adjacent to the primitive at hand. Just as in the case of the example described in section 3 this results in a graph which has been so far constructed manually. Usually it appears than an average of 2 or 3 related segments per primitive is sufficient to provide both acceptable results and a reasonable computation time.

The compatibilities c_{ijhk} are computed in the following manner. A_i and A_h are two related model segments and C_j and C_k two arbitrary object segments which we assume first to be distinct from the "missing" segment C_m. First we compute the rating for the match between the similitude s_1 that takes segment A_i onto segment C_j and the similitude s_2 that takes segment A_k onto segment C_k. This can be done for example by applying s_2 to A_i and comparing $s_2(A_i)$ with C_j. The rating $R(A_i, C_j, A_h, C_k)$ is proportional to the sum of the distances between corresponding ends of $s_2(A_i)$ and C_j and c_{ijhk} is defined as :

$$c_{ijhk} = \frac{1}{1 + R(A_i, C_j, A_h, C_k)}$$

When C_j or C_k is the missing segments c_{ijhk} is given a small value (close to 0).

The initial probabilities $p_{ij}^{(o)}$ are taken proportional to :

$$c_{i-1, j-1, i, j} + c_{i+1, j+1, i, j}$$

in the case where the related segments are adjacent and using a slightly more complicated technique described in [7] in the case of privileged segments.

Figure 4 shows the outlines of two planar tools, which are models 1 and 2. Figure 5 shows the outlines of object 3, where models 1 and 2 partially occlude each other. Both models have undergone a scaling by a factor of 1.4, and have undergone planar rotations of 121 and 263 degrees respectively.

Figures 6 and 7 show the polygonal approximations of the models and object contours. The dissimilarities between the object and models segmentations mainly comes from the lack of robustness of the segmentation algorithm. Nevertheless, experiments showed that segments 1, 5, 9 and 13 of model 1 and segments 5, 6, 8, 9, 17 and 18 of model could be selected as privileged segments.

Results of the labeling are shown in tables 1 and 2.

5. - CONCLUSIONS

This chapter has introduced another way of thinking about relaxation labeling and demonstrated on two examples that this formalism can be efficiently used to solve complex relational structure matching tasks.

The implementation os such algorithms on parallel hardware would improve further their performances.

REFERENCES

[1]   O.D. FAUGERAS and M. BERTHOD, "Scene labeling : an Optimization approach", IEEE Trans. Pattern Anal. Mach. Intell., vol. PAMI-3, pp. 412-424, July 1981.

[2]   O.D. FAUGERAS, "Decomposition and decentralisation techniques in relaxation labeling", CGIP.16, pp. 341-355, 1981.

[3]   O.D. FAUGERAS, "Optimization techniques in image analysis", Proc. 4th Int. Conf. on Analysis and Optimization of Systems, pp. 790-823, Lecture Notes in Control and Information Sciences, Springer Verlag, 1980.

[4]   R. HUMMEL and S.W. ZUCKER, "On the foundation of relaxation labeling", IEEE Trans. Pattern Anal. Mach. Intell., to appear.

[5]   O.D. FAUGERAS and K. PRICE, "Semantic description of aerial images using stochastic labeling", IEEE Trans. Pattern Anal. Mach. Intell., vol. PAMI-3, pp. 633-642, November 1981.

[6]   B. BHANU, "Shape matching and segmentation using stochastic labeling", USCIPI Report #

[7]   N. AYACHE and O.D. FAUGERAS, "Recognition of partially visible planar shapes", Pattern Recognition Letters, 1982.

[8] H.G. BARROW and J.M. TENENBAUM, "MSYS : a system for reasoning about scenes", AIC-SRI Int., Menlo Park, CA, Tech. Note 108, 1975.

[9] K. PRICE and R. REDDY, "Matching segments of images", IEEE Trans. Pattern Anal. Mach. Intell., vol. PAMI-1, pp. 110-116, Jan. 1979.

[10] R. OHLANDER, K. PRICE and R. REDDY, "Picture segmentation using a recursive region splitting method", Comput. Graphics Image Processing, Vol. 8, pp. 313-333, 1978.

[11] R. NEVATIA and K.R. BABU, "Linear feature extraction and description", Comput. Graphics Image Processing, vol. 13, pp. 257-269, 1980.

[12] L. DAVIS, "Shape matching using relaxation techniques", IEEE Trans. Pattern Anal. and Mach. Intel., Vol. 1, N° 1, January 79, pp. 60-72.

[13] W. PERKINS, "A model-based vision system for industrial parts", Trans. on Computers, Vol. c_27, N° 2, February 78, pp. 126-143.

[14] W. PERKINS, "Model-based vision system for scenes containing multiple parts" fifth Int. Joint Conf. On A.I, 1977, pp. 678-684.

[15] L. SHAPIRO, R. HARALICK, "Structural descriptions and inexact matching", IEEE Trans. Pattern Anal. and Mach. Intel., Vol. 3, N° 3, September 1981, pp. 504-519.

[16] O. FAUGERAS and B. BHANU, "Reconnaissance des formes planes par une méthode hiérarchique d'étiquetage probabiliste", 3ème congrès de reconnaissance des formes et d'I.A., Nancy, Sept. 1981.

[17] O. FAUGERAS and B. BHANU, "Recognition of occluded tow dimensional objects" the Second Sandinavian Conf. on Image Analysis, Helsinki, Finland, June 15-17, 1981, pp. 72-77.

[18] T. PAVLIDIS, " A review of algorithms for shape analysis", Comput. Graphics Image Processing, Vol. 7, 1978, pp.243-258.

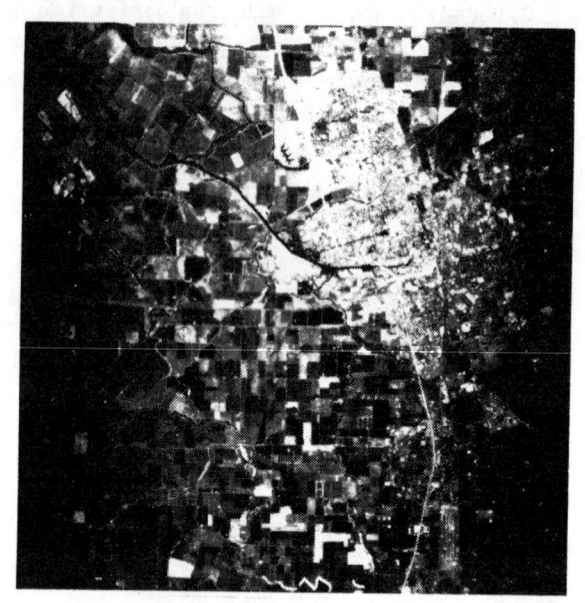

Fig. 2. Original high altitude image of Stockton, CA, area.

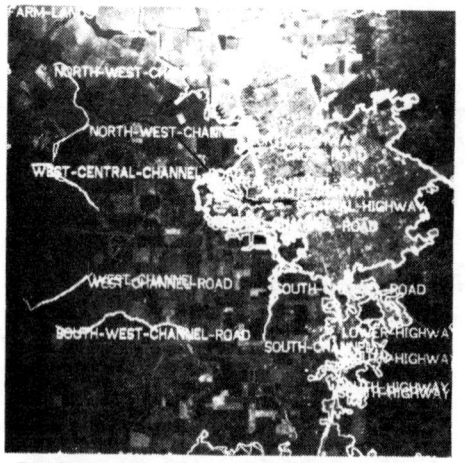

Fig. 3. Final assignments for Stockton image.

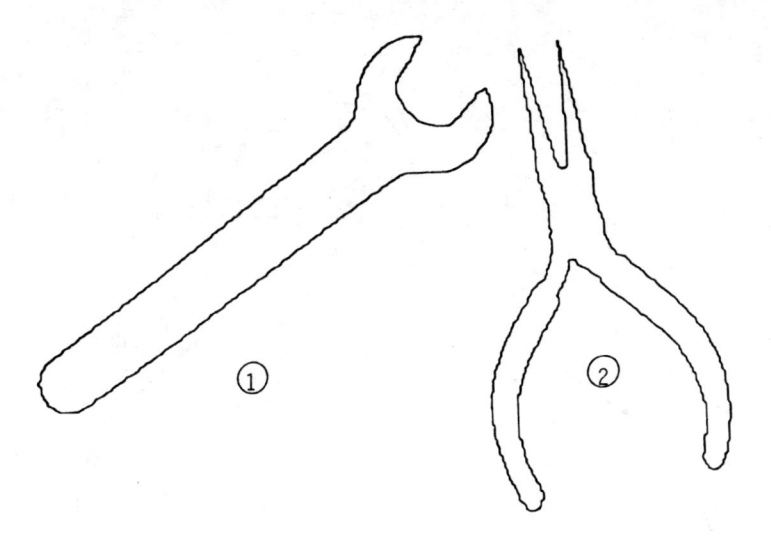

Figure 4. Outlines of Models 1 and 2.

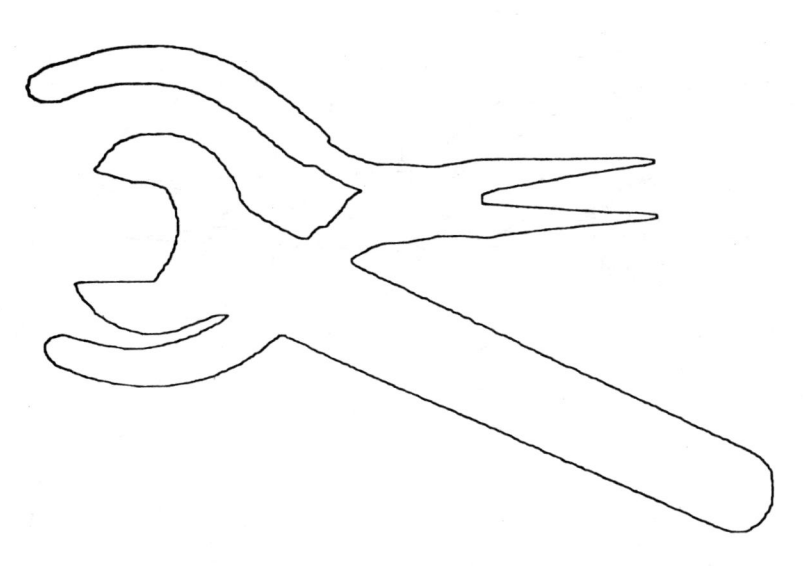

Figure 5. Object 3 Outlines.

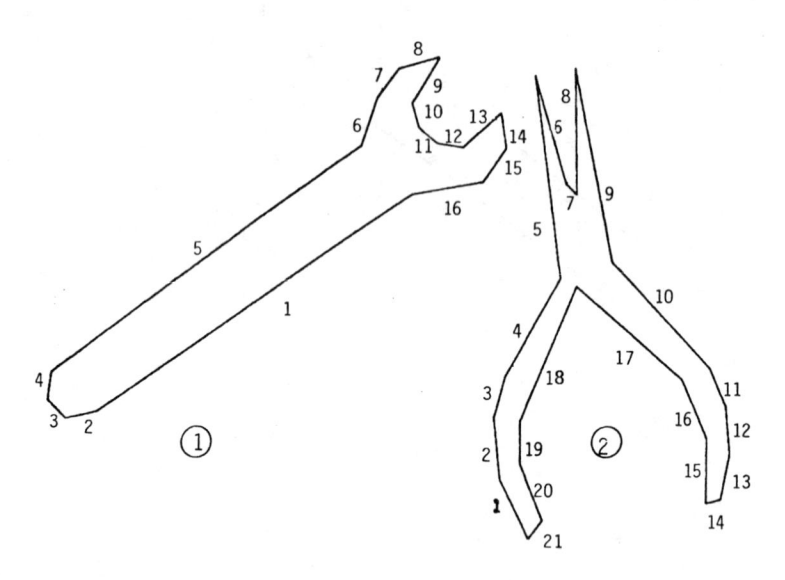

Figure 6. Segmentation of Models 1 and 2 (16 seg. and 21 seg.).

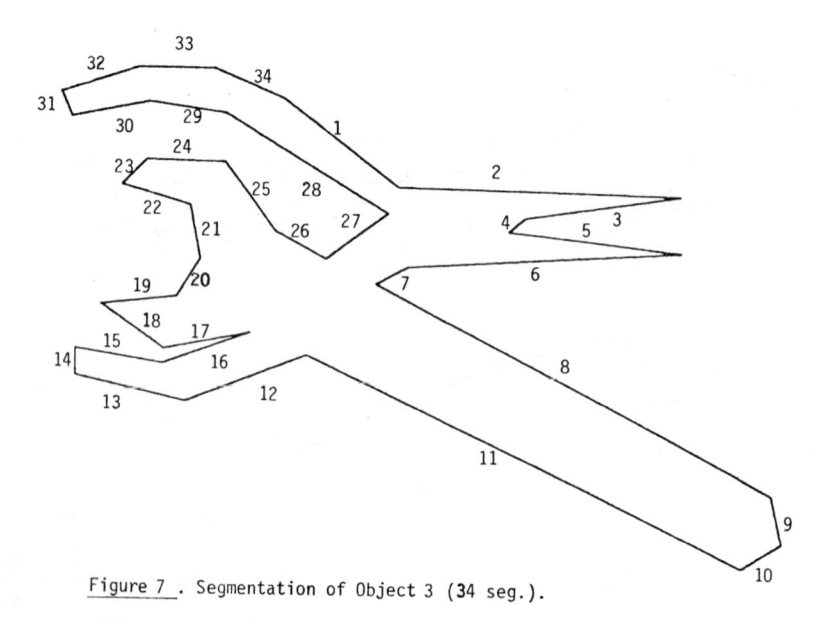

Figure 7. Segmentation of Object 3 (34 seg.).

Table 1. Labels of segments of model 2 at different iterations of the maximization of criterion (1), and final validation of the solution.

Segments M_i	Labels at different iterations 0	Labels at different iterations 2	Final validation *
1	NIL (1.0)	NIL (1.0)	NIL
2	9 (1.0)	9 (1.0)	9 (4)
3	9 (1.0)	9 (1.0)	9 (42)
4	10 (0.9)	10 (1.0)	10 (42)
5	11 (1.0)	11 (1.0)	11 (6)
6	17 (1.1)	17 (1.0)	17 (42)
7	23 (0.75)	23 (0.78)	NIL
8	18 (0.94)	18 (1.0)	18 (91)
9	22 (0.90)	19 (1.0)	19 (96)
10	20 (0.94)	20 (1.0)	20 (95)
11	20 (0.50)	21 (.59)	21 (57)
12	21 (1.0)	21 (1.0)	21 (72)
13	22 (1.0)	22 (1.0)	22 (70)
14	23 (0.87)	23 (1.0)	23 (81)
15	24 (1.0)	24 (1.0)	24 (56)
16	25 (1.0)	25 (1.0)	25 (55)
Value of criterion (1)	4.1	8.0	—

Table 2. Final validation of the labels of Model 3 segments after 1 iteration.

Segments M_i	Final labeling
1	32 (87)
2	33 (92)
3	34 (93)
4	1 (96)
5	2 (99)
6	3 (99)
7	4 (100)
8	5 (99)
9	6 (98)
10	NIL
11	12 (1)
12	13 (1)
13	13 (33)
14	14 (80)
15	15 (77)
16	16 (73)
17	NIL
18	28 (91)
19	29 (87)
20	30 (84)
21	31 (80)

ALGORITHMS AND HARDWARE

B. Kruse
IMTEC, Image Technology AB, Box 5047, S-58005
Linköping, Sweden

ALGORITHMS AND HARDWARE

Introduction

Picture processing is still a very young and immature field of research. Traditional signal processing techniques have proved insufficient in many respects. A deep understanding of the processing involved in natural vision systems is still lacking. Therefore it is not surprising that the picture processing field is still moving fast. In the light of two decades of development in both theory and practise the early approaches to the problems may seem somewhat naive. This fact, however, will not discourage us, from first describing some early approaches to the problems especially since similar ideas still emerge.

Most computational problems in picture processing are very resource demanding in terms of information storage and computation speed. Therefore the inherently sequential general purpose computer fall short of capacity in non-trivial real-time processing. In natural vision systems the low speed of the electro-chemical processes has been compensated by extreme parallelism. This has inspired the design of large arrays of identical processing elements interconnected to each other in a regular pattern. The strive for increased speed and efficiency has led to a wide variety of approaches to the basic problems.

The history of picture processors

Early approaches to picture processing architectures assume that the pictorial information can be given a simple binary foreground/background representation. The digitization into binary form is often supposed to be done already in the sensor. This may be possible in for example simple optical character recognition applications but certainly not in more complex tasks. Later designs do include provisions for more efficient arithmetic operation and one of the machines is in fact designed for greylevel image representation.

The first attempt to consider parallel or array computer architectures is Unger (1958) who designed what he calls a "Spatial Computer". In the following we will see that his basic ideas reappear in many later designs.

Unger proposed a distributed architecture in which a large set of identical modules are interconnected to form a regular orthogonal pattern. The computing capacity of any module is restricted to logical operations on the contents of its 1-bit accumulator, those of its four neighboring modules, and a set of three 1-bit general purpose registers. The processing element is computationally complete so that any boolean function can be computed. The main restriction on the processing power is the small amount of internal storage available. A common master controller issues identical commands to all modules. Thus individual operation of the modules is impossible.

Unger realized that the synchronous way in which the modules operate may restrict the speed in some cases. Segmentation of an image into its connected parts for example calls for spatial signal propagation within the array structure in order to label all modules belonging to the same connected segment. He introduced signal paths, the link circuits, that allow asynchronous signal propagation between all modules (flash through) which is inherently much faster than clocked operation.

At the time when this design was proposed (1958) the technology was insufficient for such massive investment in hardware, and therefore the machine was never realized. However, Unger's work raised important questions and his work has influenced most of his successors in the field.

The need for efficient processing of bubble-chamber pictures motivated a group at the University of Illinois to actually build a picture processor Illiac III, McCormick (1963). The heart of Illiac III is called the Pattern Articulation Unit (PAU), a two-dimensional array of identical modules called stalactites. Compared to Unger's design the PAU uses the same basic approach to array instruction and neighborhood accessing. In addition to the orthogonal interconnection pattern the modules can communicate hexagonally in the array structure. Also, the PAU includes a counting mechanism besides the logical processing unit. The elaborate design which also provided for efficient loading of the array was unfortunately not completed to an extent that permitted actual use.

In the late sixties M. Golay proposed a set of hexagonal parallel pattern transformations on which the Golay processor is based, Golay (1969). While the previous described processors could perform any neighborhood operation the Golay processor operates isotropically. Golay discovered that only 14 classes of hexagonal neighborhoods exist if equivalence under rotation is enforced. This simplifies the construction but introduces other problems. Segmentation of connected components, for example, cannot be performed on the complete array without subdividing it into distinct subsets of non-adjacent modules. The operation is applied to each subset in sequence.

The Golay processor introduces a new approach in that it doesn't employ physical parallelism. Instead a single processing module is used covering the array by sequential application to all neighborhoods. Conceptually, however, the operation is still parallel. The complexity, in terms of number of components, is drastically reduced although some buffer storage has to be added. The prize is of course a significant reduction of the processing speed. Due to the sequential application of a single module fast propagation in other paths than those coinciding with the scanning sequence is impossible.

The Golay processor has been successfully applied to pictorial pattern recognition problems, most significantly in white blood cell analysis.

Locally computable properties comprise one aspect of pattern classification that was more or less neglected in the earlier discussed designs. S. Gray proposed a set of locally computable properties that form the basis of the Binary Image

Processor (BIP), Gray (1971).

Gray described how certain measurements that are valuable for examination of connectivity relations may be very easily computed serially. The interesting fact behind the properties is that although they are locally countable they carry global information about the topology of the objects in a picture. Other more obvious properties are for example area and perimeter length. Another interesting fact is that these properties are not easily computed in parallel.

The realization of BIP depends in many ways on a sequential implementation. This is the first processor that explicitely incorporates counters for sequential accumulation of measurements over the entire array. The correlation of two images, for example, is recorded in eight adjacent locations. BIP also includes facilities for thresholding a graylevel image with possibilities for shading correction. The BIP has been successfully applied to advanced optical character recognition problems such as multifont recognition and recognition of handwritten characters.

The large scale integration possibilities renewed interest in physically parallel machines. M. Duff designed the Clip 3, Duff et al (1974), as a rather small pilot machine for a large truly parallel machine, Clip 4, Duff (1976).

The structure of the computing element consists of the neighborhood interconnection network in which an arbitrary neighborhood configuration can be selected within a three by three binary neighborhood, the boolean processor, an internal memory of 16 bits per module and two 1-bit registers. The modules are interconnected in an array of 12 x 16 elements. The common control unit issues commands that select the desired neighborhood, boolean function and operands identically for all modules. Using the 16 binary image planes of the memory graylevel operations such as thresholding and convolution can be programmed. Clip 3 does also allow for fast information propagation through the array. The design is however more functionally sophisticated than the earlier machines in that neighborhood and boolean conditions may be programmed to influence the propagation.

The only true graylevel processor among the early machines is PICAP I, Kruse (1973) and Kruse (1976). The originally completely parallel concept got a sequential implementation and was soon equipped with sequential operations in addition to the parallel ones. The instruction set includes operations for two-dimensional convolution of graylevel images (2-D filters), template matching operations on neighborhood relations, iterative operations, recursive operations and linear operations on two or more images. In addition to the image-into-image transformations, information is gathered about the graylevel distribution in the form of a histogram, the occurrencies of a programmable set of neighborhoods are recorded and positional information is extracted. The execution of an instruction presents a vector of such information to the controlling host computer upon which decisions about the continued processing can be made.

The PICAP I processor has been successfully used in a variety of pattern recognition problems such as malaria parasite detection, fingerprint analysis, moving white bloodcell analysis, printed circuitboard inspection and many others.

Parallel-sequential

The introductory remarks about picture processing as an immature field of research are especially true for the hardware in which the picture processing algorithms are to run. Clearly, the hardware choice influences the algorithms. This becomes evident for example in the choice between parallel or sequential algorithms. In the following we will describe two basic functions of every pictorial pattern recognition system, functions belonging to completely different categories. To be able to go into sufficient detail we will use a very simple machine model that can be implemented either as a physically parallel array of identical modules or with minor modifications as a physically sequential machine using one module only. We will use these machines to illustrate how different the algorithms become in the two architectures. We will also give coarse estimations of the processing speed.

The functions that we will use are
1) Isolation of a connected component in a binary image
2) Computation of the area of a connected component in a binary image.

Component extraction is a type of operation that preferably is performed in the spatial domain (the image itself) while area computation belongs to the category of feature extraction functions from the spatial domain to the set of integers.

The parallel machine model

We will describe the machine models in terms of automata and sequential network nomenclature. To emphasize the issue we will not describe every detail that would be required to make the machines realistic but rather comment on the important differences. Thus, details about control, timing and array border behaviour for example will be neglected.

The array, fig. 1, is built from identical modules shown in fig. 2. The module consists of a look-up-table (LUT) unit that can be loaded with any boolean function of the inputs.

Figure 1   Parallel Processor Array

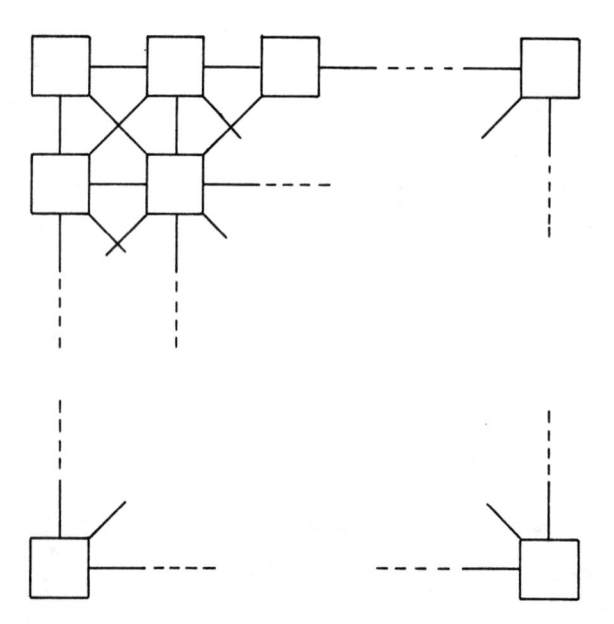

The function is tabulated for every combination of the input variables. The output of the module can be chosen to be either the LUT output or the output from one of the internal 1-bit registers Q and R depending on the setting of the control input which is common to all modules of the array. Thus

$$u = \begin{cases} Q & \text{if } p = 0 \\ F(Q,R,x_1,\ldots,x_8) & \text{if } p = 1 \end{cases}$$

The input variables x_i, $i = 1,8$ are the outputs u from the eight nearest neighbors, fig. 3, and R is an auxilliary 1-bit register. Either the Q- or the R-register can be set from the function-table output. For example

$$Q(t+1) = F(Q(t), R(t), x_1(t),\ldots,x_8(t))$$

This is the synchronous mode of operation. For asynchronous signal propagation p = 1 and the Q- and R-registers

Figure 2   Parallel machine model

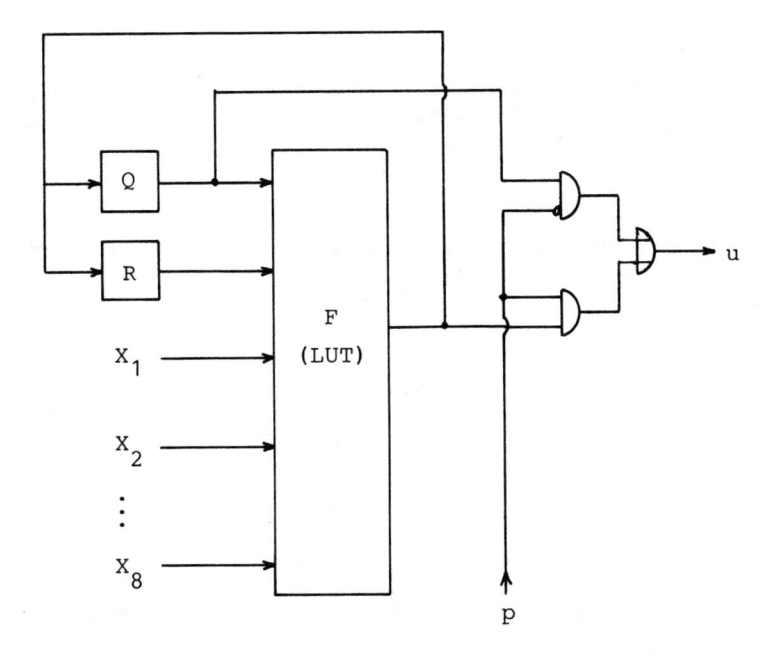

B. Kruse   408

Figure 4 Extraction of uppermost leftmost component

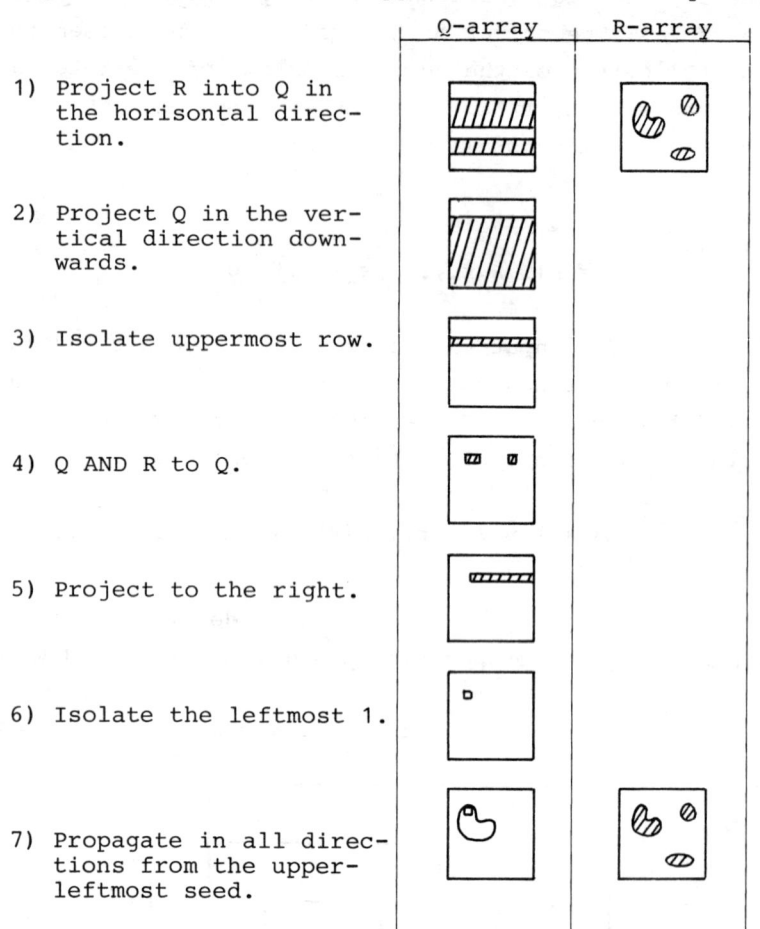

	Q-array	R-array
1) Project R into Q in the horisontal direction.		
2) Project Q in the vertical direction downwards.		
3) Isolate uppermost row.		
4) Q AND R to Q.		
5) Project to the right.		
6) Isolate the leftmost 1.		
7) Propagate in all directions from the upper-leftmost seed.		

Figure 3 Processor array interconnections

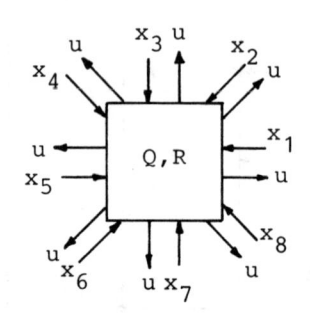

Algorithms and hardware

can be set when the network has stabilized. Note that careless programming can cause the entire array to oscillate without stabilizing in this mode. The function-table is reloaded for every new operation that is to be performed. For operations that are iterated the stopping criteria is that a stable condition is reached. This type of operation can also cause infinite iteration in an oscillating array if caution is not used. We will now show how the first problem can be programmed in the parallel machine model. A similar procedure has been used on the Clip 4 machine.

The procedure extracts the uppermost leftmost connected component of the R-register array. To describe the operations we only show the deviations from a regular table. For example, the entries that cause the state of a Q- or R-register to be set or reset are listed while the other entries that cause the state to remain the same are implicit. The procedure starts by projecting all elements of the R-array in the horisontal direction. The operations are shown in table I.

Table I

		F	Q	R	x_1	x_2	x_3	x_4	x_5	x_6	x_7	x_8
1	Q: p = 1	1	-	1	-	-	-	-	-	-	-	-
		1	-	-	1	-	-	-	-	-	-	-
		1	-	-	-	-	-	1	-	-	-	-
2	Q: p = 1	1	1	-	-	-	-	-	-	-	-	-
		1	-	-	-	-	1	-	-	-	-	-
3	Q: p = 0	0	1	-	-	-	1	-	-	-	-	-
4	Q: p = 0	0	1	0	-	-	-	-	-	-	-	-
5	Q: p = 1	1	-	-	-	-	-	-	1	-	-	-
		1	1	-	-	-	-	-	-	-	-	-
6	Q: p = 0	0	1	-	-	-	-	-	1	-	-	-
7	Q: p = 1	1	1	1	-	-	-	-	-	-	-	-
		1	-	1	1	-	-	-	-	-	-	-
		1	-	1	-	1	-	-	-	-	-	-
		1	-	1	-	-	1	-	-	-	-	-
		⋮										
		1	-	1	-	-	-	-	-	-	-	1

The first operation asserts that the look-up-table equals 1 whenever R or x_1 or x_5 equals 1. Since p = 1 the table output is distributed to the surrounding modules. Therefore, every row of the R-array that includes a 1 causes the corresponding Q-array row to be all 1's. It is not difficult to see that the following operations 2 to 7 causes the desired upper-leftmost component to be isolated in the Q-array. Fig. 4 shows schematically the operation of the program. We leave as an exercise to show that several of the steps in the program can be combined into one instruction.

The second problem, to compute the area of a component, is completely different since it amounts to producing a scalar value. We will assume that, external to the array, there exists an accumulator which can be incremented in various ways according to the outcome of the array operations. The idea of the program is to convert the component to a certain form using projections in order to be able to compute the area in a regular way.

Table II

		F	Q	R	x_1	x_2	x_3	x_4	x_5	x_6	x_7	x_8
1	Q: p = 0 iterated	0 1	1 0	-	0 -	- -	- -	- -	- 1	- -	- -	- -
2	Q: p = 0 iterated	0 1	1 0	-	- -	- -	- 1	- -	- -	- -	0 -	- -
3	Q: p = 0 iterate and count until sense = 1	0 1	- -	-	0 1	- -	- -	- -	- -	- -	- -	- -
4	Q: p = 0 iterate and accumu- late base until sense = 0	0 1	- -	-	- -	- -	0 1	- -	- -	- -	- -	- -

Algorithms and hardware

The first two steps of the program in table II projects the component first onto the right border, then down to the bottom border without altering the number of elements (fig. 5). This ensures that the number of elements in an arbitrary row is always less or equal to the number of elements in a row below. The following steps 3 and 4 are part of a program loop in which the area is computed as a "line integral" following the border of the surface.

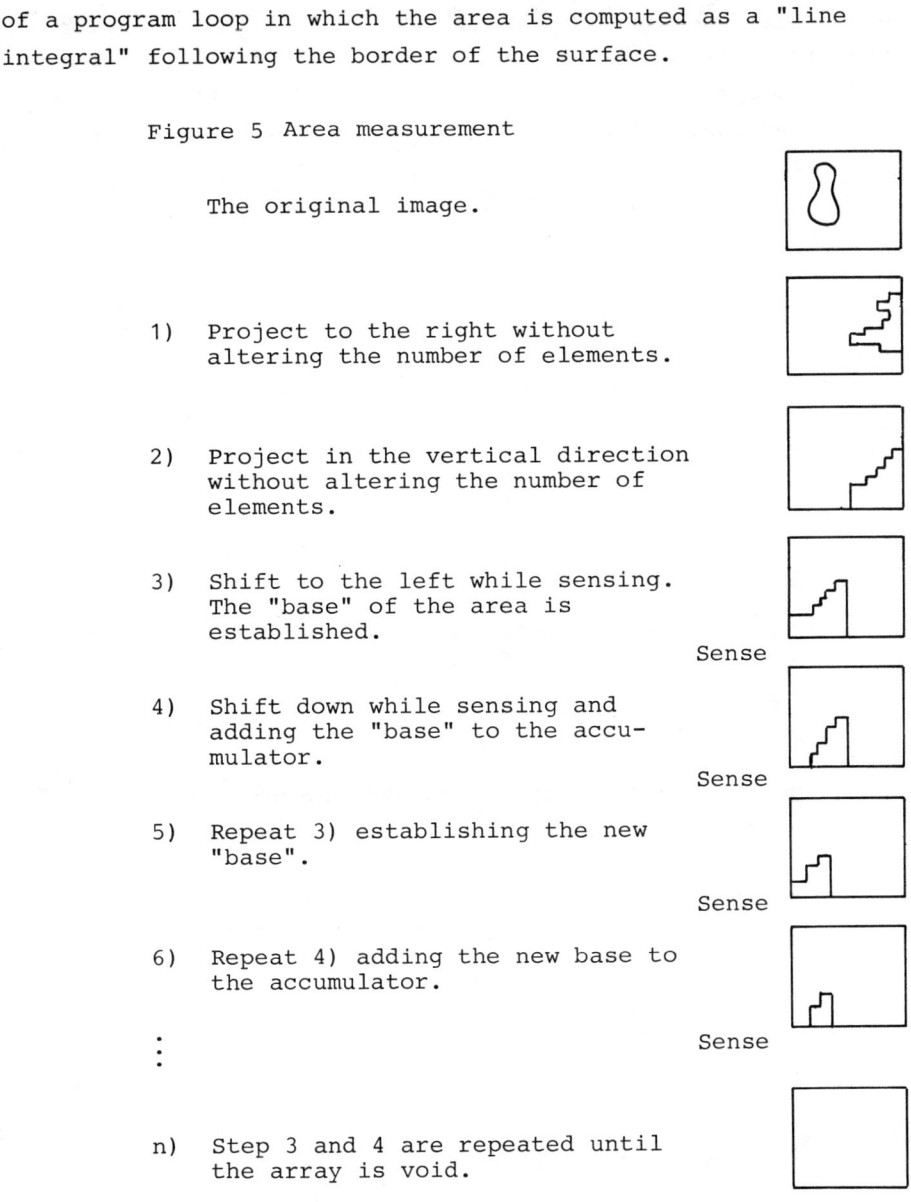

Figure 5 Area measurement

The original image.

1) Project to the right without altering the number of elements.

2) Project in the vertical direction without altering the number of elements.

3) Shift to the left while sensing. The "base" of the area is established.
   Sense

4) Shift down while sensing and adding the "base" to the accumulator.
   Sense

5) Repeat 3) establishing the new "base".
   Sense

6) Repeat 4) adding the new base to the accumulator.
   Sense

⋮

n) Step 3 and 4 are repeated until the array is void.

The sequential machine model

The physically parallel implementation requires heavy investment in hardware, which can be avoided at the prize of lower performance by a sequential construction that only requires some additional buffer storage. In the parallel array the images are distributed among the modules in the Q- and R-registers. Since there is only one module in the sequential implementation the images are stored in memory outside the module (fig. 6, fig. 7). Another difference is that the asynchronous propagation cannot be realized. It has to be

Figure 6 The sequential machine model

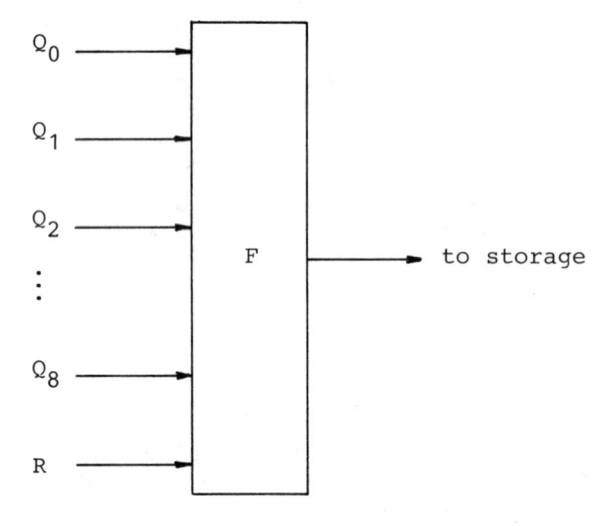

Figure 7 Memory of the sequential model

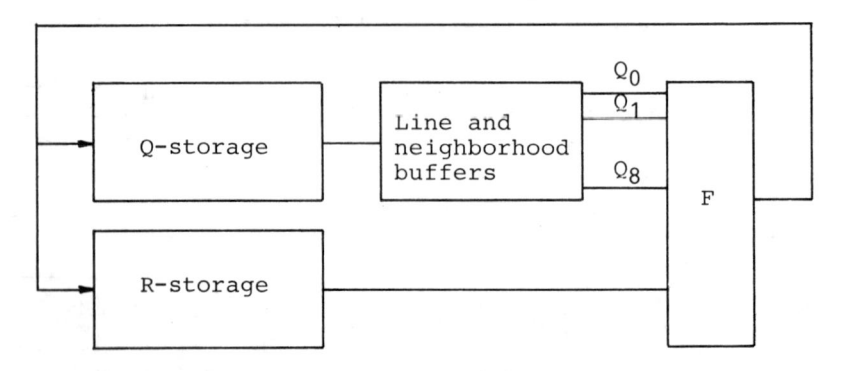

Algorithms and hardware

performed synchronously one step at a time. The operation is now described by

$$Q(t+1) = F(Q_0(t), Q_1(t), \ldots, Q_8(t), R)$$

where the subscripts refer to the position of the elements in the neighborhood (fig. 8).

Figure 8

Q_4	Q_3	Q_2
Q_5	Q_0	Q_1
Q_6	Q_7	Q_8

For the first task of extracting the uppermost leftmost connected component we will assume that a tagging function exists which is not shown in the figures. The operation is such that the first and only the first occurrence of a 1 in the sequence of elements from the R-storage will set the corresponding element in the Q sequence. The program is shown in table III. The iteration of the second instruction causes the "seed", implanted in the Q-image by the tagging operation, to grow within the bounds of the component in the R-image.

The second task to compute the area is even more simple. Again, we assume additional functions not shown in the figures. Since the image is scanned sequentially all that has to be done is to increment a counter for every 1-occurrence in the Q-image at virtually no additional cost.

Table III

		F	Q_0	Q_1	Q_2	Q_3	Q_4	Q_5	Q_6	Q_7	Q_8	R
1	Q: Tagging	1	-	-	-	-	-	-	-	-	-	1
2	Q: iteration	1	-	1	-	-	-	-	-	-	-	1
		1	-	-	1	-	-	-	-	-	-	1
		⋮										
		1	-	-	-	-	-	-	-	-	1	1

From the above descriptions it should be evident that the "programming style" is completely different at least for the chosen task examples.

Execution-time

Justification for the higher cost of the physically parallel machine comes from its superior processing capacity. Without going into too much detail we will give the worst case execution-time for the two given tasks in both machines. A similar evaluation for a parallel machine and a general purpose computer on a larger set of tasks can be found in Cordella et al. (1976).

In the estimations, the array will be assumed to be square and $N \times N$ elements large. We also assume similar technology with a clockperiod of T in synchronous operation and an intermodule propagation time of αT where α is a constant in the range $(0.2, 1.0)$. We denote the estimates t_{p1}, t_{p2}, t_{s1} and t_{s2} for the parallel respectively the sequential solution to task 1 and 2.

The basic speed improvement for operations that are perfectly matched to the architecture should naturally be N^2. This is true for simple neighborhood operations such as thinning and shrinking. In the parallel machine it takes one instruction only and hence the time-consumption $t_p = T$. The sequential approach requires the whole array to be scanned and hence $t_s = N^2 \cdot T$ and the ratio $t_s/t_p = N^2$. The basic speed improvement will not always be possible to reach, however. For the first task it is not hard to see that $t_{p1} = 3N \cdot \alpha T$ (the instructions of table I can be reduced to three propagations). The corresponding figure t_{s1} equals $(1+N) \cdot N^2 \cdot T$, where the second factor is due to the scanning of the entire image for every instruction. The ratio t_{s1}/t_{p1} for large N's equals $N^2/3\alpha$. For the second problem it is clear that $t_{s2} = N^2 \cdot T$ since the image has to be scanned once only. The corresponding figure t_{p2} equals $(4N-3) \cdot T$. (It is easily shown that the two first operations of table II always requires N-1 iterations.) In this case the ratio $t_{s2}/t_{p2} \sim N/4$ for large N's.

Algorithms and hardware

An interesting fact is that by introducing recursive operation and both forward and backward scanning direction, the estimate t_{s1} equals $K \cdot N^2 \cdot T$ where K is a small integer depending on the topology of the connected component. The modification yields a ratio of $t_{s1}/t_{p1} = K \cdot N/3\alpha$ as compared to $N^2/3\alpha$.

The conclusions we can draw from the figures above is that the speed improvement of parallel over sequential implementation is not necessarily proportional to the number of processors. Such an expectation is obviously far too optimistic. However, provided the input/output-complex has an efficient design the speed improvement may be substantial.

HARDWARE STRUCTURES FOR PARALLEL PICTURE PROCESSING

B. Kruse
IMTEC, Image Technology AB, Box 5047, S-58005
Linköping, Sweden

HARDWARE STRUCTURES FOR PARALLEL PICTURE PROCESSING

The picture processing field is not alone in requiring supercomputing resources. There are many already established scientific and engineering problem areas that are very demanding. Meteorological forecasting, mechanical engineering design, designrule checking in integrated circuit manufacturing, many-body problems in astronomy and control optimization, just to mention a few examples. These areas belong to a class of problems that are intractable by the ordinary von Neuman computer. Not because of the problems in principle but simply because of their limited speed. - It may be interesting to be able to make a 24-hour wheather-forecast in 30 hours for a researcher, but certainly not for the public. - Although the throughput of the general purpose computer (GPC) has been increasing eversince the first machines became available the demands seem insatiable.

The basic motive for research in new computer architectures is efficiency in terms of capacity versus economy. We will not discuss the entire spectrum of possibilities but rather concentrate on the high-capacity end where parallelism in some form or other seems to be necessary due to the physical speed limitations of available components.

Levels of parallelism

In general there are several levels in a system where parallelism may be introduced. Flanders (1979) discusses four levels

- Complete jobs
- Task within a job
- Basic operations within a task
- Sub-operations

In general, the possibilities for exploitation of parallelism become larger at a low level than at a high level. Normally different jobs have little in common which means that parallelism at this level requires a number of concurrently operating systems. At the task level, in one job, the information may be the same but usually the task instruction streams are different. Thus parallelism at this level takes the form of multiple-instruction multiple-data operation. Within a task there are usually many different ways to make use of parallelism. The large number of operations to be performed usually means that a substantial fraction of them are independent enough to allow concurrent execution. The lowest level, the sub-operation level, is not clearly defined but large numbers of bit-serial processors is surely one example of parallelism at this level.

Limiting the scope to picture processing problems one can be more specific. In Danielsson (1981) four dimensions of parallelism are discussed that to some extent correspond to the four levels already mentioned.

- Sequence of operations
- Image dataspace
- Neighborhood dataspace
- Pixel bits

The first dimension clearly corresponds to the "Task within a job" level. The restriction to a sequence of operations is unnecessary and somewhat misleading since independent operations that can be executed concurrently are frequent in picture processing. At the level of complete jobs there is no correspondance. The two following dimensions both belong to the level "Basic operations within a task" while the last belongs to the "Sub-operations" level.

When the problem domain is narrowed the computing problems become more and more alike. This usually means that one can identify a large number of processes in which parallelism may be useful. The basic levels of parallelism as discussed in Danielsson (1981) presupposes a high degree of regularity in the problem. Properties of this kind are often present in picture processing problems. However, there are problems in which neither the image dataspace nor the neighborhoodspace are regular. Resampling and geometric rectification of images, for example, involve position- and data-dependent neighborhood processing.

Architectures

A data-flow architecture is best described as a reconfigurable set of processors that can be interconnected according to the problem in question (see fig. 9). As data flows through the network intermediate results are formed that in their turn are processed and passed along downstream. The algorithm or program manifests itself in the interconnection pattern of the processors. The interconnection pattern and the operations performed by the individual processing units cannot be altered during operation which usually means that conditional execution of alternatives is prohibited.

The processing units are usually thought of as being quite simple, only capable of doing lowlevel arithmetic. The different datastreams must not necessarily originate from

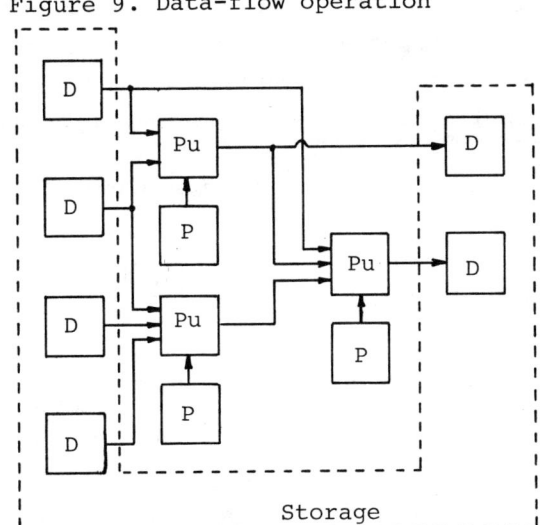

Figure 9. Data-flow operation

separate physical sources. They may come from common storage, provided there is a fast enough information channel to supply the data.

The data-flow architecture is conceptually very attractive since the program is mapped directly onto the network of processors. In practise difficulties arise in the "switchboard", where the network is set up, since the number of processors that can be randomly interconnected is very restricted. In a given application area, however, where only minor modifications of the program is necessary, this technique may prove to be optimal.

Pipe-lining is a technique that can be used where there is only one data-stream. It is essentially a restricted data-flow architecture with only one processing unit at each level (see fig. 10). The program has the form of a sequence of fixed operations. Contrary to the case of a general data-flow machine there is no restriction on the number of processors in a pipeline. Their interconnection pattern is fixed and does not have to be changed. A program is loaded by issuing different commands to each processor so that the appropriate operations are performed in sequence on the data.

The advantage of a pipeline structure is simplicity. The basic speed is constant and independent of the program in question. The delay from input to output is a constant that only depends on the number of processors. Furthermore, common input and output devices such as TV-cameras and monitors can easily interface to a pipeline since the data-flow is in common.

Figure 10. Multiple-instruction single-data stream operation (pipelining)

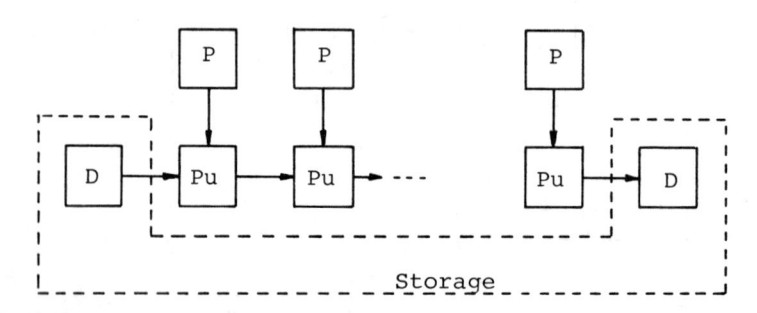

If a program has fewer steps than the pipeline has processors the "idle" processors copy the data they receive to the output. Should there not be enough processors available for the complete program, the pipelining is ruined and the information has to be recycled with the rest of the program loaded. To be able to do that, the output flow has to be delayed in a buffer until it can be reinserted in the input stream and be processed. Since most picture processing operations require access to a neighborhood the processors themselves are equipped with buffer storage for one or more rows of data. Given that the program can be written as a fixed sequence of operations the pipeline is optimal for many practical problems.

The architecture that in a way is complementary to the pipeline is the single-instruction multiple-data (SIMD) structure (see fig. 11). A common instruction-stream is fed to a set of processors that operate identically on separate data. The parallel array processors are of this type. Sometimes they are referred to as cellular logic machines (Preston (1979)). Contrary to the data-flow approaches, where the program is mapped in hardware, the array processors constitute a mapping of the array data. Although by no means necessary, the mapping usually allocates one processing unit to each data item in the array. In Danielsson (1981) another structure is proposed in which a number of processors (less than the number of data items) are evenly distributed over the array in a regular

Figure 11. Single-instruction multiple-data stream operations (SIMD).

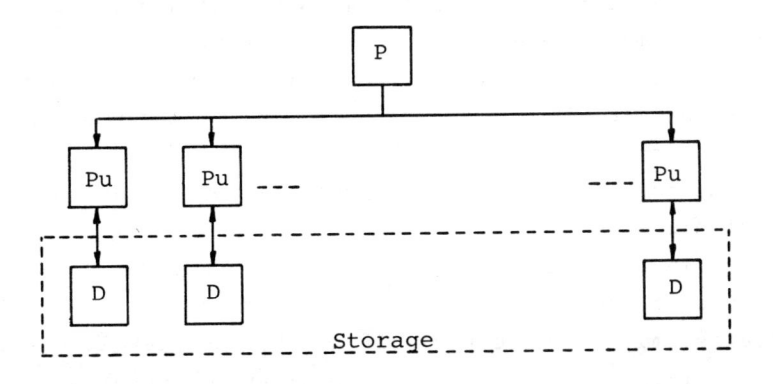

fashion. The advantage of less overhead becomes apparent when large neighborhood operations are considered.

The potential speed improvement in this type of architecture is much greater than in the data-flow structures, simply because the data items usually are far more numerous than the operations in a program. However, the large amount of data to be input and output from the array adds to the overhead. In order to keep the speed improvement high, the input/output complex of the array computer need special consideration. Since most sensors are delivering data sequentially an efficient conversion to parallel format is usually necessary.

The control structure of an array processor is very simple since there is only one instruction-stream. This does not mean, however, that conditional execution necessarily is prohibited. The instruction stream may include the computation of logical data-dependent conditions which can be used to inhibit application of an operation. Array computers usually incorporate such functions in the hardware.

In neighborhood operations the processing unit have to communicate with each other in order to access data in a surround. Therefore they are usually interconnected in a regular way. Orthogonal and hexagonal interconnection patterns between the immediate neighbors in the array are used in practise. These communication paths offer another form of parallelism in which each processing element has direct and simultaneous access to its neighbors output data. The possibility of this form of parallelism is not unique to array computers but may be used in almost any architecture.

The necessarily very regular mapping of the data onto the hardware may cause problems that we touched upon in the introductory remarks. If the data items are structured in a way that does not match the hardware layout, they have to be reorganized.

In problems with little or no structure, parallel approaches of the above types are not applicable. However, parallelism may still be possible at the task level in the form of multiple-instruction multiple-data computer structures (see fig. 12). At this level complete tasks are executed in parallel with moderate or little communication between the processors. Often, the only communication that exists is passing the output data to another task. To avoid transportation of the data, a large common memory is usually used as intermediator of information. If there is a large number of interconnected tasks going on simultaneously a multiport memory is required on some kind of bussystem.

A discussion in general on the effectiveness of the different architectures can be found in Flynn (1972).

Figure 12. Multiple-instruction multiple-data operation (MIMD)

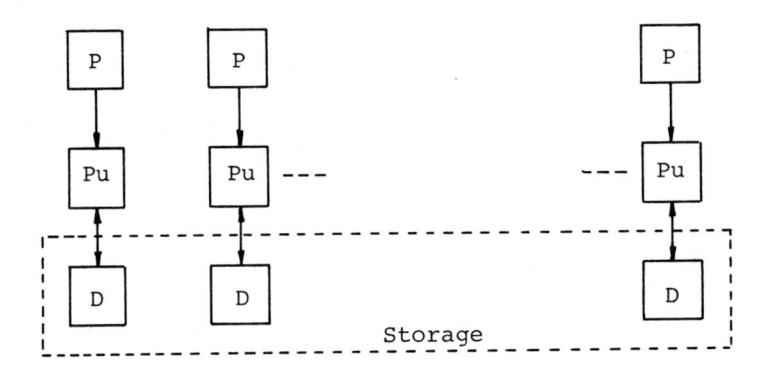

STATE-OF-THE-ART SYSTEMS FOR PICTORIAL INFORMATION PROCESSING

B. Kruse
IMTEC, Image Technology AB, Box 5047, S-58005
Linköping, Sweden

STATE-OF-THE-ART SYSTEMS FOR PICTORIAL INFORMATION PROCESSING

Recent years have shown a renewed interest in computer architectures for image processing. Several processors and complete image processing systems have emerged throughout the world. Although the field is still fast moving and there is not yet any degree of perspective available, it is hoped that the collection of systems to be described will give an indication of the state-of-the-art. In the list of references a variety of different machines and systems can be found of which only a few have actually been realized. The systems that we will present to some detail in the following sections are operational or near-operational. They have been chosen arbitrarily for illustrative purposes as exponents of different techniques.

Clip 4

Clip 4 has been developed from Clip 3 that we touched upon in the introductory remarks. The research efforts from Clip 3 and onwards have resulted in an operating parallel array processor system including television input/output: the Clip 4 system, described in Duff (1982). Fig. 13 shows the overall architecture with the processor array in the middle.

The Clip 4 system is intended more as a research vehicle than as a production image processor. Therefore there has been no efforts made to make input/output efficient. As can be seen in fig. 13 the array is loaded serially from the shift register storage. Loading and unloading could of course be done in parallel if the shiftregisters were incorporated in

Figure 13 Clip 4 system architecture. (From Duff (1982)).

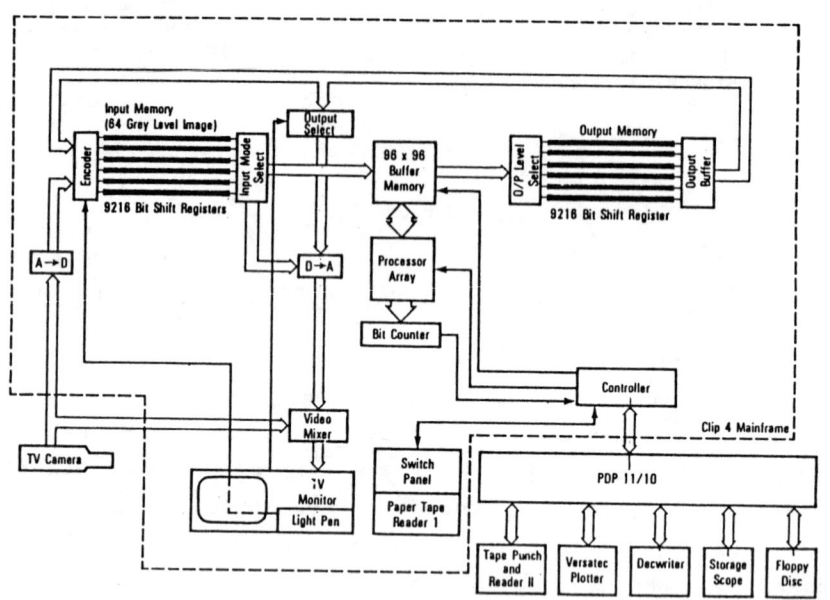

the array and could be operated simultaneously to the array. However, this would have increased the already large chip area.

The algorithm for area computation, as previously described, is not very effective. The operation belongs to a class which is very important: counting of events in the image. The events may often be computed by parallel operations but when it comes to counting their number the inefficiency is apparent. Consequently, the architecture is enriched by a "line-parallel" bit counter (fig. 13) attached to one side of the array. Shifting the array contents line by line into the bit-counter results in an accumulation of the number of events. The number can be transferred to the host-computer for control and measurement purposes. In principle execution time is still proportional to N, the side of the array. However, the proportionality factor is much smaller than in the previous case.

The individual processing element of the Clip 4 array is shown in fig. 14. It includes an input circuit, a boolean processor, three bit-registers, a memory and miscellaneous gates. The inputs interconnect the module to its nearest neighbors in the array. This can be done both for an orthogonal and for a hexagonal lattice structure. Programmable control singals determine which inputs are to be fed to the input OR-gate. Hence, the output of the neighboring elements are fed in parallel to the unit and do not have to be accessed one after the other. The registers A and B of fig. 14 are general purpose registers while C is used mainly in arithmetic operations to save the carry. The boolean processor can perform any two combinatorial functions of the two signals A and P depending on the setting of the control signals.

One of the boolean processor outputs is fed to the surrounding processing element while the other is input to the memory D. The signal path bypassing the boolean processor is fed directly to the output OR-gate which permits rapid asynchronous propagation from input to output.

A more detailed and comprehensive description of the Clip 4 system is given in Duff (1982). The system has been operational since the spring of 1980.

DAP and MPP

The discristed array processor (DAP) of ICL Ltd., Flanders et al. (1977), and the massively parallel processor

Figure 14 Clip 4 processing element. (From Duff (1982)).

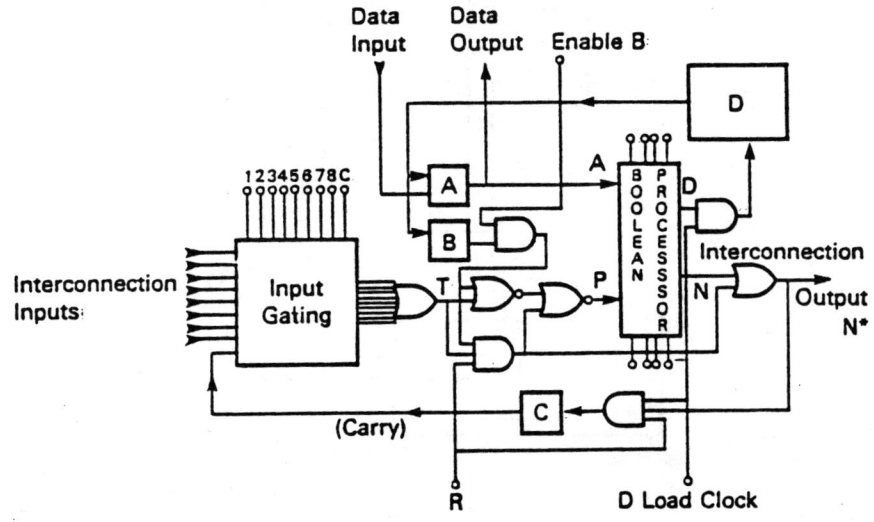

(MPP), Strong (1982), conceived by the Goddard Space Flight Center are basically of the same structure as the Clip 4 processor array. However, there has been given more attention to memory size and bit-serial arithmetic in these machines.

The DAP is not primarily intended for picture processing but for demanding computational problems in general. The array has been embedded in the main memory of a host computer on an interrupt basis, enabling the host to reference data or load and unload an entire data array simultaneously to the operation of the processor. Therefore DAP could also be called an intelligent memory with distributed logic.

External to the DAP array there are a set of registers that communicate with the array both vertically and horizontally. Consequently information stored in the array can be accessed word-wise or bit-wise depending on the problem. Fig. 15 shows schematically how datawords stored in the array rows can be accessed by the column. This technique has also been successfully used in the Staran computer, Potter (1982), for speeding up pattern matching and search problems.

In MPP the processing element has been supplied with an internal variable-length shiftregister that speeds up bit-

Figure 15 Column and Row Highways in DAP. (After Flanders et al. (1977)).

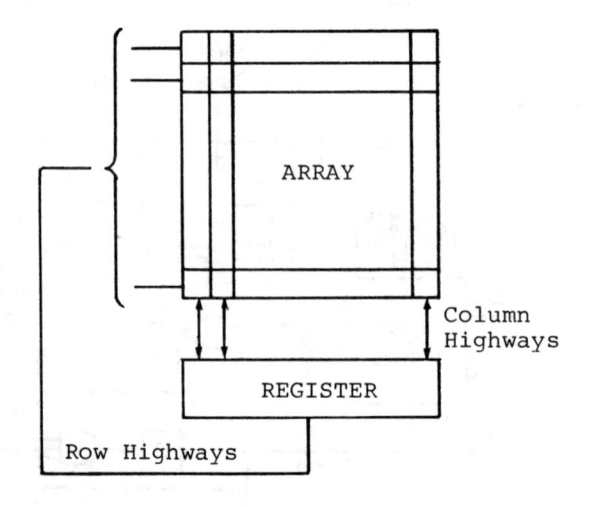

serial computations. See fig. 16. As in DAP, the memory is
external to the processing element. It is built from commer-
cially available memory chips and does not compete for chip
area with the logic as in Clip 4. The loading of a bit-plane
into the processor array is done by shifting data column-wise
by means of an independently operated shiftregister. Data is
presented to the array from the "west", one column at a time
for each shift. The number of cycles for the full operation
is equal to the number of modules along the side of the array.
However, since it is performed independently to other activi-
ties the number of cycles lost may be as low as one, that is
the cycle in which the shiftregister content is moved to
memory in parallel over the entire array.

The processing power of the SIMD architectures that
we have described is beyond what has been achieved in other
existing architectures. The DAP which has been operational for
several years is superior to "ordinary super computers" and
the MPP promises an even higher throughput when it comes into
operation in the fall of 1982. The number of primitive operations such as addition and multiplication is on the order of
one billion operations per second.

Figure 16   MPP Processing element. (From Strong (1982)).

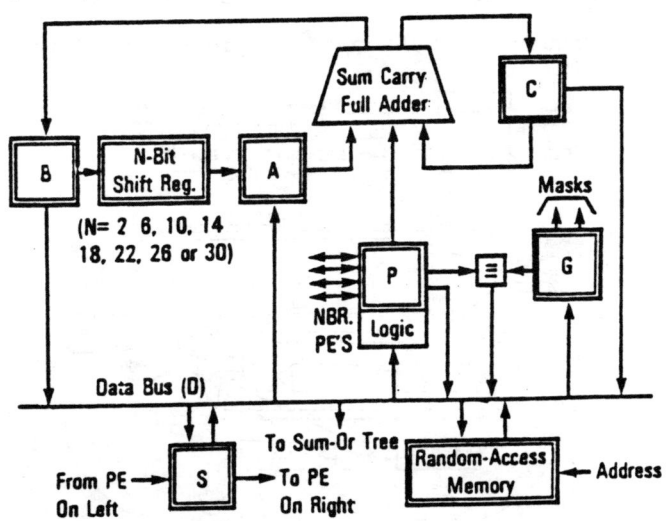

Cytocomputer

A typical exponent for the pipe-line approach is the Cytocomputer described by Sternberg (1979). Contrary to the array architecture, where the processing units have a two-dimensional interconnection pattern, the units in a pipe-line are stringed together in a linear unilateral structure, fig. 17. The operation of each unit is individual and determined once for each task so that the entire program is executed as the data flows downstream. Following an initial delay to fill the pipe-line, images can be processed in synchronism with the scanning of an image.

In the Cytocomputer there are two kinds of transforms that can be performed by the pipe-lined units, the so-called "silhouette"- and "umbra"-transforms. It is beyond the scope of this paper to go into a detailed description of the operations. In essence they are identical to the structuring element transformations of mathematical morphology, Serra (1980).

Contrary to the previously described architectures, in which any program could be executed, the programs that can be run in the cytocomputer must be able to be decomposed into primitive operations of the said type. An operation, such as the extraction of a single connected component for example is difficult if not impossible to perform efficiently. On the

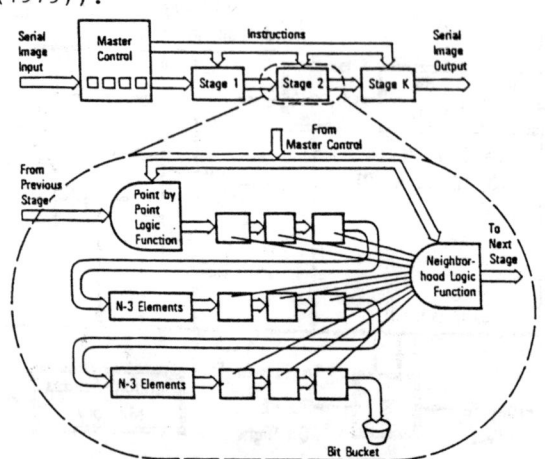

Figure 17 Cytocomputer pipe-line. (From Sternberg (1979)).

other hand global operations of counting nature are easily added at the end of the pipe-line. The cytocomputer has been operational since 1979.

DIP-1, FLIP, ZMOB and PM4

The architecture of many systems derives its speed from a moderate number of fairly powerful modules that can be reconfigured according to different problems. Because of their low number they have to be more versatile than the one-bit modules of the processor arrays. Typical operations are addition, multiplication on eight-bit data and in one case also floating point operations. The main problem in this type of architecture is to implement the interconnection network. The use of a large number of processors (hundreds) prohibits random interconnection or that a time-multiplexed network is used.

The processors are interconnected so that the dataflow in the system is mapped after the actual problem. Independent calculations are performed in concurrently operating processors whereas steps in a chain of calculations are pipelined. The overall speed is determined by the cycle of the most time-consuming operation.

In DIP-1, Gerritsen (1977), and FLIP, Luetjen (1980), the modules are few, seven in DIP-1 and 16 in FLIP. Despite that, the interconnection network is non-trivial. Each of the 16 FLIP processors, for example, delivers its 8-bit data output to all other processors' two 8-bit input ports and the data exchange processor PEP. See fig. 18.

The modules of FLIP are identical whereas in DIP-1 there are five different categories. In addition to arithmetic processors there are table-look-up functions for general data mapping and for binary 3 x 3 neighborhood processing. Two modules buffer neighborhoods in order to lower the requirements on the memory bus.

The number of modules in ZMOB, Rieger et al. (1980), and PM^4, Briggs et al. (1982) is on the order of hundreds. Special consideration as to the interconnection of the processors is therefore of vital importance. In ZMOB the pro-

Figure 18 Flip multi-processor architecture.
(From Luetjen (1980)).

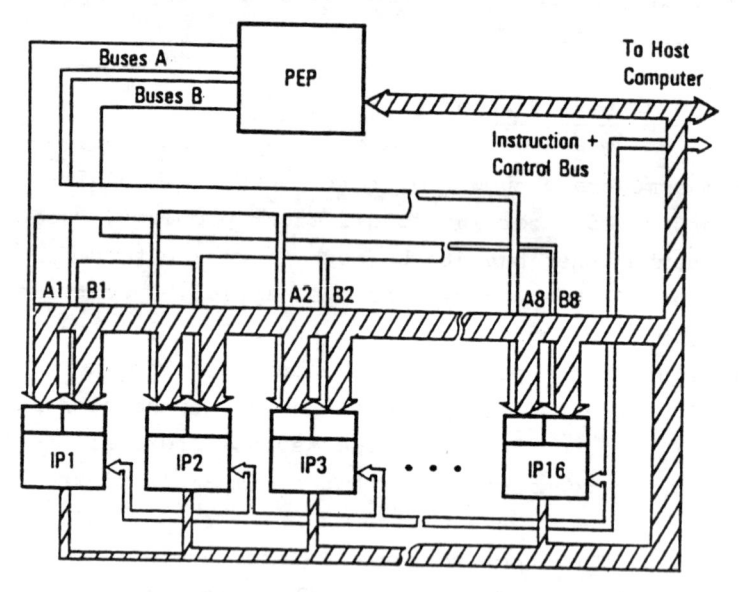

-cessors are built from ordinary micro-processors with their own memory for program and data. They share a common bus that communicates data in a "conveyor-belt" fashion. The 256 modules are positioned along the bus each with an interface that intercepts the data as it is moved along the bus at high speed. The micro-processors operate at a much lower speed with the effect that the transfer rate is in accordance with their processing capacity.

The architecture for PM^4, fig. 19, was designed to allow for concurrently operating processor groups executing SIMD processes. Hence, the system is a MIMD structure. The data processing is mainly performed in the Processor-Memory units (PMU) controlled by another set of so called Vector Control Units (VCU). The PMU's are connected to a three-level memory hierarchy and the Inter-Processor Communications Network (IPCN) that control the permutation of data within a SIMD process. Due to the large number of PMU's the IPCN will not accommodate arbitrary interconnection of the modules even through the chosen delta network, fig. 20, is less costly than

State-of-the-art systems for pictorial information processing 433

Figure 19 PM^4 architecture. (From Briggs et al. (1982)).

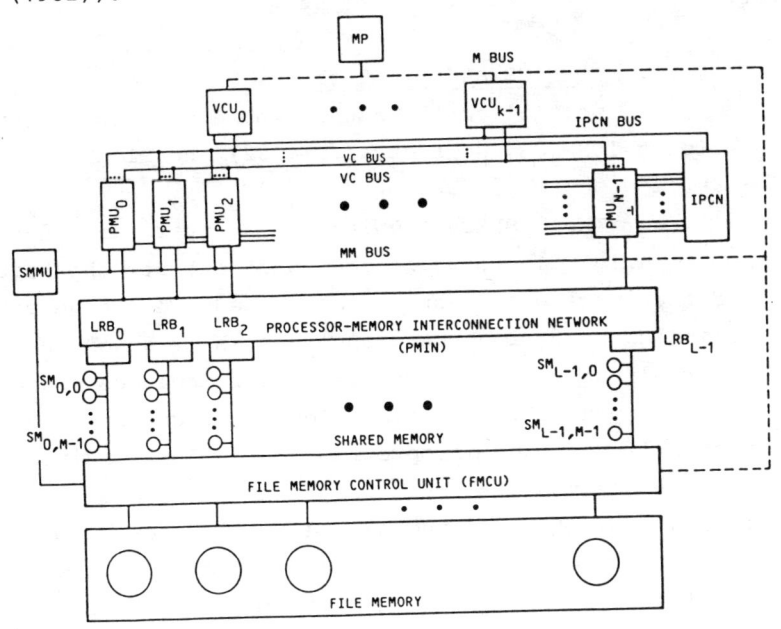

Figure 20 Delta interconnection network. (From Briggs et al. (1982)).

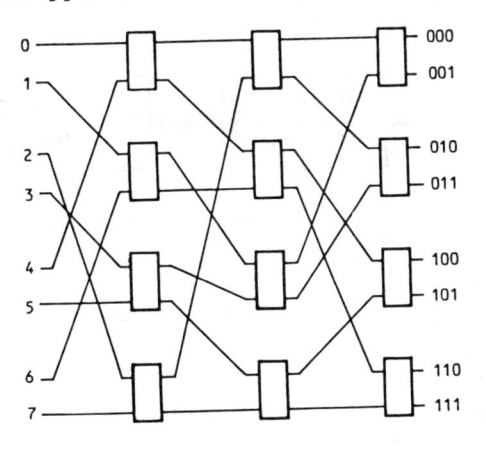

a full crossbar switch. For further details the reader is referred to Briggs et al. (1982).

TOSPICS and PICAP II

We will conclude the exposé of architectures by describing TOSPICS, Mori et al. (1982) and PICAP II, Kruse et al. (1982).

The TOSPICS architecture centers around the image memory, fig. 21, to which the input/output, the host computer and the image processor are connected. The image processor communicates over a DMA-channel which is an extension of its internal bus, fig. 22. The processor includes seven special purpose modules for typical picture processing operations controlled by one microprogram controller. Individual operation of the modules is therefore not possible. However, concurrent operation on the same data flow can be achieved.

Special purpose hardware has been developed for the modules. For example in the two-dimensional convolution processor a fixed network of multipliers and adders are used to accumulate eight products per cycle, fig. 23. A two-dimensional convolution with an eight by eight filter matrix can be computed in eight cycles.

Figure 21 Tospics system architecture. (From Mori et al. (1982)).

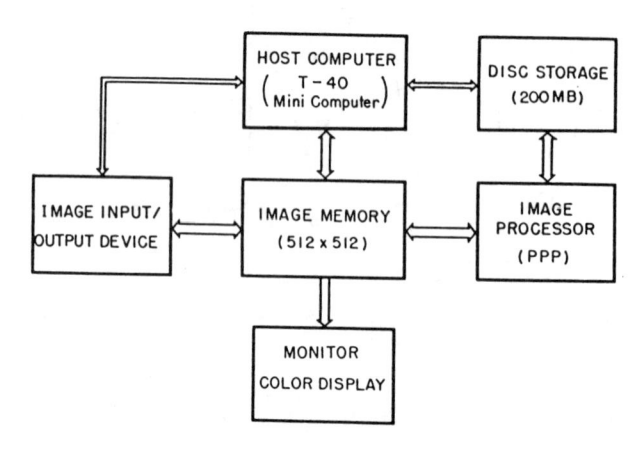

State-of-the-art systems for pictorial information processing 435

Figure 22 Tospics image processor. (From Mori et al. (1982)).

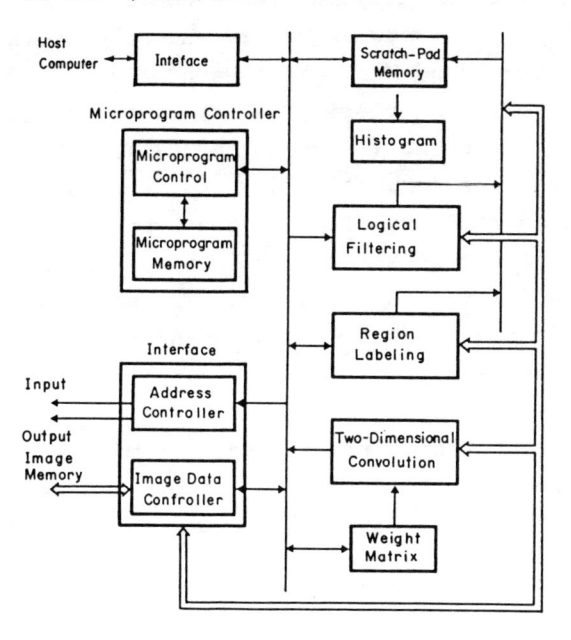

Figure 23 Tospics convolution network. (From Mori et al. (1982)).

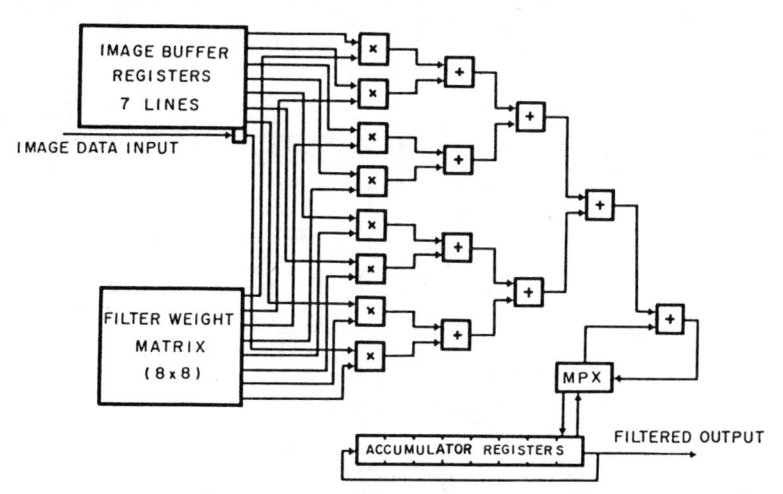

The overall PICAP architecture is of type MIMD. The processors run independently sharing a common memory, fig. 24. Therefore the datapath between memory and processors has to be fast. The time-shared bus, Danielsson (1980), for which the processors contend on a priority basis, has a bandwidth of 40 Mbyte/s as compared to 4 Mbyte/s in TOSPICS. To match the speed of the bus, the 4 Mbyte memory is interleaved four ways. The memory has an ordinary linear address space allowing for arbitrary picture sizes. Memory can be accessed from any processor or channel attached to the bus. The host computer communicates over a DMA-channel with a two-dimensional address mechanism that enables reference to arbitrary two-dimensional subfields of an image. The concurrent execution of several processors represent one form of parallelism in PICAP.

To increase speed further the processors themselves make use of parallelism in various forms. The filter processor,

Figure 24   PICAP system architecture

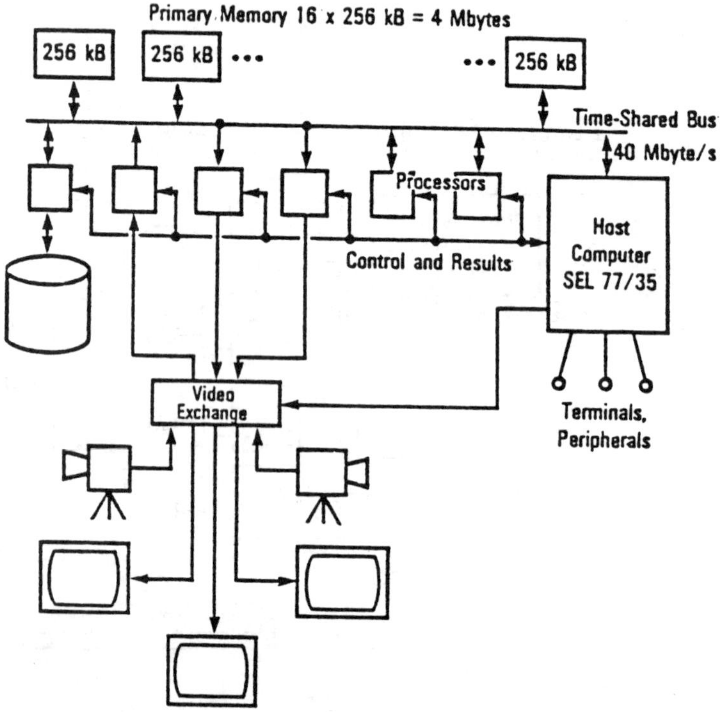

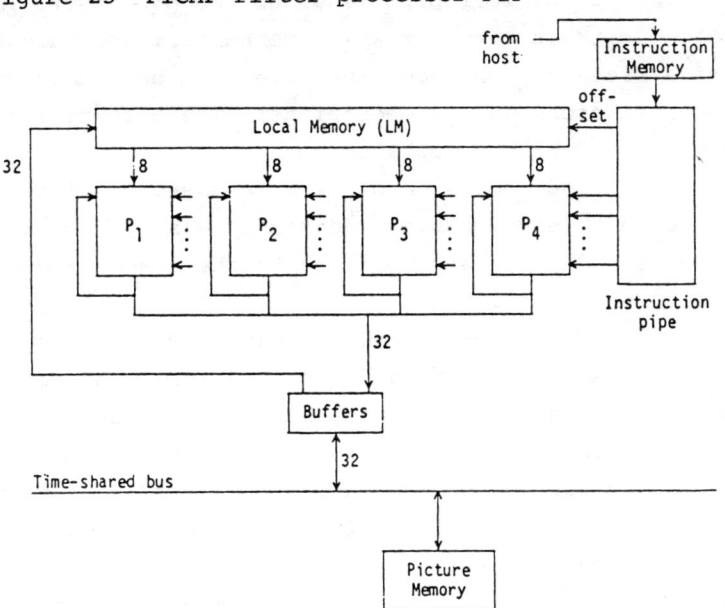

Figure 25 PICAP filter processor FIP

FIP, in fig. 25 may have concurrently operating pipe-lines controlled by a common microprogram and a 32-bit CPU that sets up the data-paths and memory communication. FIP includes a local memory (LM), the high speed cache, that enables access to all neighborhoods along a line in the image, fig. 26. The cache addressing is relative to a baseregister and four adjacent picture elements are accessed simultaneously

Figure 26 Buffering of image data in the local memory

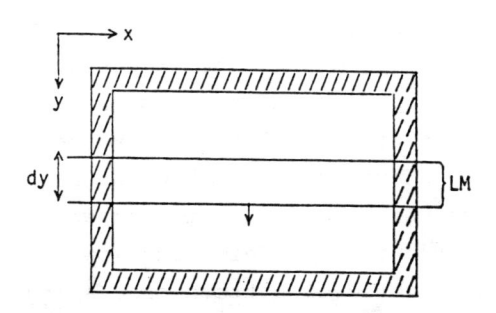

and fed to the four pipe-lines of the processor. The address space fo LM is circular which means that regardless of the image size only one new line at a time has to be brought from the main store (unary operations). The base-register scans the image during computation.

The address mechanism is basically linear but two- and even three-dimensional data may easily be accessed. For example a 3-D operator working in the space-time domain is easily executed in FIP.

The pipe-lines of FIP are divided in two, one for linear operation, typically convolution, and one for logical operation. Auxilliary memory is provided for storing intermediate results that have to be reinserted into the pipe-lines for further processing. This means that any type of compound operator using different neighborhoods, can be accomodated.

The basic PICAP system has been operational since June 1980 and new facilities are currently added to the system.

The future

The systems that have been brought forward in this paper are by no means the only ones. The exemplified architectures have been chosen mainly because they are familiar to the author and because they give an indication of the state-of-the-art in the field.

To prophesize the development of a young art is always difficult since the rapid evolution of technology will probably bring new un-foreseen possibilities for parallel processing that may change the trends completely. However, there are some fundamental remarks that probably will be valid regardless of technological advances.

In fig. 27 we have tried to show the applicability of different architectures. A solution to the general image understanding problem requires capacity for low-level neighborhood processing to high-level pattern processing such as identification of objects in moving scenes. The input image is usually represented as a regular array in the spatial domain and this representation is often kept through the first few preprocessing stages. After segmentation the representation is of diminishing interest and the final result is perhaps represented as a verbal description of the scene. The regularity of the data in the low-level processing end should favour processor array architectures and pipe-lines, whereas in the opposite end multi-processor bus-oriented data-base architectures probably will be necessary. The trend for such architectures will probably become stronger as soon as the low-level processing methods have matured.

Figure 27 Applicability of architectures

Task:	Low-level processing	— High level processing
Data:	Regular structures	— Non-regular structures
Architecture:	Array or pipe-line machines	— Multiprocessor, bus-oriented data-base machines

Antonsson, D. & Danielsson, P.E. & Gudmundsson, B. & Hedblom,
    T. & Kruse, B. & Linge, A. & Lord, P & Ohlson, T.
    PICAPII- A System Approach to Image Processing. IEEE
    Comp. Soc. Workshop on Comp. Architecture for
    Pattern Analysis and Image Database Management, Nov.
    1981.

Briggs, F.A. & Fu, K.S. & Hwang, K. & Patel, J.H. A Shared
    Resource Multiple Microprocessor System for Pattern
    Recognition and Image Processing. In Special
    Computer Architectures for Pattern Processing. Eds.
    Fu, K.S. & Ichikawa, I. CRC Press 1982.

Cordella, L. & Duff, M.J.B. & Levialdi, S. Comparing
    Sequential and Parallel Proccessing of Pictures.
    Third Intern. Conf. on Pattern Recognition 1976,
    pp703-707.

Danielsson, P.E. The Time-shared Bus - The Key to Efficient
    Image Processing. Fifth Intern. Conf. on Pattern
    Recognition, Miami 1980.

Danielsson, P.E. & Levialdi, S. Computer Architectures for
    Pictorial Information Systems, Computers, Vol.14,
    No.11, 1981.

Duff, M.J.B. & Watson, D.M. & Deutsch, E.S. A Parallel
    Computer for Array Processing. Proc. IFIP Congress
    74, 1974.

Duff, M.J.B. CLIP4: A Large Scale Integrated Circuit Array
    Parallel Processor. Third Intern. Conf. on Pattern
    Recognition, 1976, pp728-733.

Duff, M.J.B. CLIP4. In Special Computer Architectures for
    Pattern Processing, Eds. Fu,K.S. & Ichikawa,T., CRC
    Press 1982.

Duff, M.J.B. & Levialdi, S. (eds), Languages and Architectures
    for Image Processing, Academic Press, 1981.

Flanders, P.M. FORTRAN Extensions for a Highly Parallel
    Processor. Infotech State of the Art Report on
    Supercomputers, Vol. 2, 1979.

Flynn, M.J. Some Computer Organizations and their
    Effectiveness. IEEE Trans. Comp., 948,1972.

Fu, K.S. & Ichikawa, I. (Eds.) Special Computer Architectures
    for Pattern Processing, CRC Press 1982.

Gerritzen, F.A. & Aardema, L.G. Design and Use of a Fast
    Flexible and Dynamically Microprogrammable
    Pipe-lined Image Processor. First Scandinavian Conf.
    on Image Analysis, Linköping, Sweden, Jan. 1980.

Golay, M.J.E. Hexagonal Parallel Pattern Transformations. IEEE Trans. Comp., Vol. C-18, pp733-740,1969.

Granlund, G. GOP, A Fast and Flexible Processor for Image Analysis, Proc. fith Intern. Conf. on Pattern recognition, pp489-492, IEEE Comp. Soc., 1980.

Gray, B.S. Local Properties of Binary Images in Two Dimensions. IEEE Trans. on Comp., Vol. C-20, pp551-561,1971.

Kruse, B. A Parallel Picture Processing Machine. IEEE Trans. on Comp., Vol. C-22, No. 12, pp1075-1087, Dec. 1973.

Kruse, B. The PICAP Picture Processing Laboratory. Third Intern. Conf. on Pattern Recognition 1976, pp875-881.

Kruse, B. & Gudmundsson, B. & Antonsson, D. FIP- the PICAPII Filter Processor. Proc. Fifth Intern. Conf. on Pattern Recognition, Miami, Dec. 1980.

Kruse, B. & Danielsson, P.E. & Gudmundsson, B. From PICAPI to PICAPII. In Special Computer Architectures for Pattern Processing. Eds. Fu, K.S. & Ichikawa, I. CRC press 1982.

Kulpa, Z. & Bolc, L. (Eds.) Digital Image Processing Systems, Springer-Verlag 1981.

Luetjen, K. & Gemmar, P. & Ischen, H. FLIP: A Flexible Multi-processor System for Image Processing. Proc. Fifth Intern. Conf. on Pattern Recognition, Miami, Dec. 1980.

McCormick, B.H. The Illinois Pattern Recognition Computer. IEEE Trans. on Electronic Comp., Vol. EC-12, pp791-813,1963.

Mori, K. & Kidode, M. & Shinoda, H. & Asada, H. Design of a Local Parallel Processor for Image Processing. In Special Computer Architectures for Pattern Processing. Eds. Fu, K.S. & Ichikawa, I. CRC Press 1982.

Onoe, M. & Preston,JR, K. & Rosenfeld, A. Real Time/Parallel Image Processing, Plenum , 1981.

Potter, J.L. Pattern Processing on Staran. In Special Computer Architectures for Pattern Processing. Eds. Fu,K.S.& Ichikawa,I. CRC Press 1982.

Preston, JR., K. & Duff, M.J.B. & Levialdi, S. & Norgren, P.E. & Toriwaki, J. Basics of Cellular Logic with Some Applications in Medical Image Processing. Proc. IEEE Vol. 67, No. 5, May 1979.

Preston, JR., K. & Uhr, L. (Eds.) Multicomputers and Image Processing, Academic Press 1982.

Reeves, A.P. A Systematically Designed Binary Array Processor. IEEE Trans. on Comp., Vol. C-29, No. 24, April 1980.

Reddaway, S.F. The DAP Approach. Infotech State of the Art Computer Architectures for Pictorial Information Systems. Vol. 12 pp309-329, 1979.

Rieger, C. & Bane, J. & Trigg, R. ZMOB: A Highly Parallel Multi- processor. Proc. of the Workshop on Pict. Data Descrip. and Management, IEEE Comp. Soc. pp298-304, 1980.

Rosenfeld, A. & Pfalz, J.L., Sequential Operations in Digital Picture Processing, J. Ass. Computer. Mach., Vol. 13, No. 4, pp471-494, Oct. 1966.

Rosenfeld, A. Connectivity in Digital Pictures, J. Ass. Comp. Mach., Vol. 17, No. 1, pp146-160, Jan 1970

Rosenfeld, A. Cellular Architectures: From Automata to Hardware. In Multicomputers and Image Processing, Eds. Preston, JR., K. & Uhr, L., Academic Press 1982.

Serra, J. Image Analysis and Mathematical Morphology. Academic Press, 1980.

Siegel, H.J. PASM: A Reconfigurable Multimicrocomputer System for Image Processing. In Languages and Architectures for Image Processing, Eds. Duff, M.J.D. & Levialdi, S. Academic Press 1981.

Sternberg, S. Parallel Architectures for Image Processing. Third Intern. IEEE COMPSAC, Chicago, 1979.

Strong, J.P. Basic Image Processing Algorithms on the Massively Parallel Processor. In Multicomputers and Image Processing, Eds. Preston, JR., K. & Uhr, L. Academic Press 1982.

Unger, S.H. A Computer Oriented towards Spatial Problems. Proc. IRE, Vol. 46, pp1744-1750, 1958.

THE GOP IMAGE COMPUTER

G.H. Granlund and J. Arvidsson
Linkoeping University, Valla, 581 83 Linkoeping, Sweden

Abstract. The GOP image computer consists of an algorithmic structure as well as a hardware structure. In the algorithmic structure is employed a new type of information representation, which allows a high degree of data compression. It includes a class of context-sensitive symmetry operations, which have proved very powerful for processing of image information. These features together allow processing in a hierarchical structure, which has proved necessary for effective processing. The hardware structure allows implementation of this class of algorithms, as well as most other image processing operations suggested. The processor has a two-stage architecture, where the first stage has a power corresponding to one billion operations per second, providing a considerable data reduction. The second part allows a flexible combination of partial results at floating point accuracy.

INTRODUCTION

There are two basic problems in computer vision. One is the lack of a fundamental understanding of the problem of how to describe and operate upon image information. The other problem is that images contain vast amounts of information, which will put high demands upon storage capacity and processing power.

Traditionally various disparate methods, such as edge and line detection, line description, texture analysis, description of regions etc, have been employed for the solution of problems in image analysis. See Niemann (1981). These methods have used different modes of representation, which have made it difficult to relate different types of information. Such a communication between different information representations is commonly regarded as necessary, as no one-level operation can be expected to unambiguously interpret complex events.

In order to manage the large amounts of data involved, different types of fast special purpose image processors have been developed. See Danielsson & Levialdi (1981). Many of these processors have been oriented towards the use of logical operations upon binary images.

In order to provide a solution to these problems, GOP (General Operator Processor), a special purpose computer, has been developed for processing of image information and other types of structural information. With the GOP processor it is possible to analyze and interpret images with a high structural complexity.

OVERALL HIERARCHICAL STRUCTURE

Hierarchical structures have been proposed for the processing of structural information. See Tanimoto & Pavlidis (1975), and Rosenfeld (1980). So far it has proved difficult to devise effective hierarchical structures for use in information processing. A major problem has been that of communication between levels in the hierarchy. If information from a higher level is to communicate with and make sense at a lower level in the function of context, there are very strong restrictions upon the representation of information as well as upon the operations that separate different levels. The operations will among other things have to act as interfaces between different levels of the hierarchy in bottom-up as well as in top-down direction.

The GOP Image Processor implies an algorithmic structure which allows operations within hierarchical structures. See Granlund & Knutsson (1982). Due to limitations in space it will only be possible to give a qualitative discussion of the more fundamental properties. Certain aspects of the presentation will be in terms of 2-dimensional signals, but the principles are applicable to signals of any dimensionality.

In the processing of image information, the hierarchical structure may appear as in figure 1. The original image appears at the lowest level. An operation with a particular size neighborhood produces a transformed image at the next level. Another operation uses this transformed image to produce a higher level transform, and there is a sequence of bottom-up operations. There is also a top-down flow of information which enables information from a higher level to control operations at a lower level.

On each neighborhood a set of operation kernels are applied, each one measuring a particular feature of the event within that neighborhood. As an example at low levels in an image hierarchy, the operation can measure the edge content in four different directions. As edges of all orientations cannot be expected within a particular, small neighborhood, we can infer the mutual exclusion of edges of perpendicular orientation. This leads to a 2-component vector representation, where perpendicular edges will be represented by vectors of opposing directions.

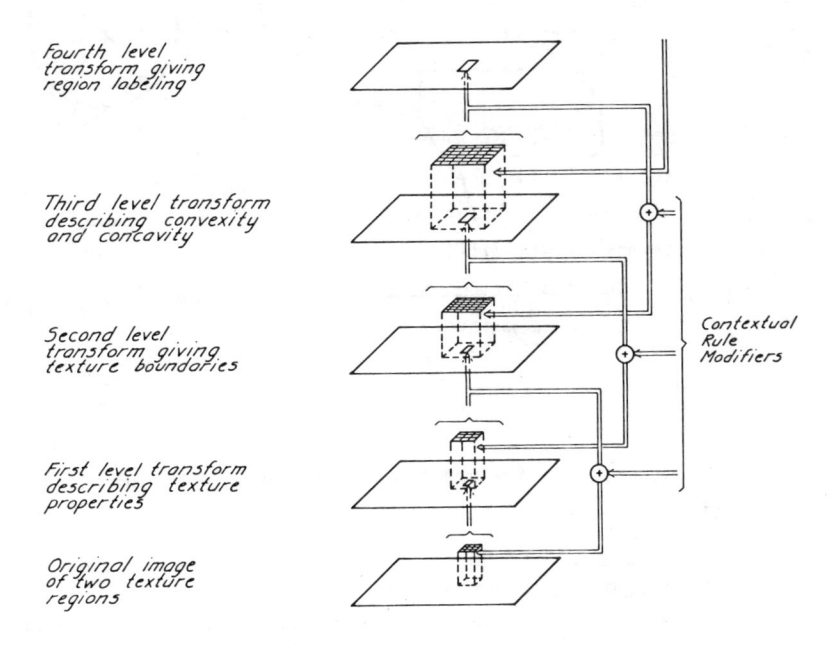

Figure 1 Hierarchical structure relating information representations and operations

This interpretation at low levels can be extended to a general principle for information representation at all levels. Features derived from a particular neighborhood are mapped into a 2-dimensional space and arranged in a circular fashion, where conflicting or incompatible features are located at opposing positions, while relatively independent features are at right angles. The feature measurements are weighted together to form a vector describing the event within the neighborhood. The direction of the vector indicates the dominant class membership of the event, and the magnitude gives the certainty of this choice. See figure 2. This information representation is consistently used through all levels of the hierarchy. However, the meaning of the information varies with the level of the hierarchy. We may somewhat simplified say that information is represented in terms of relative similarity - dissimilarity.

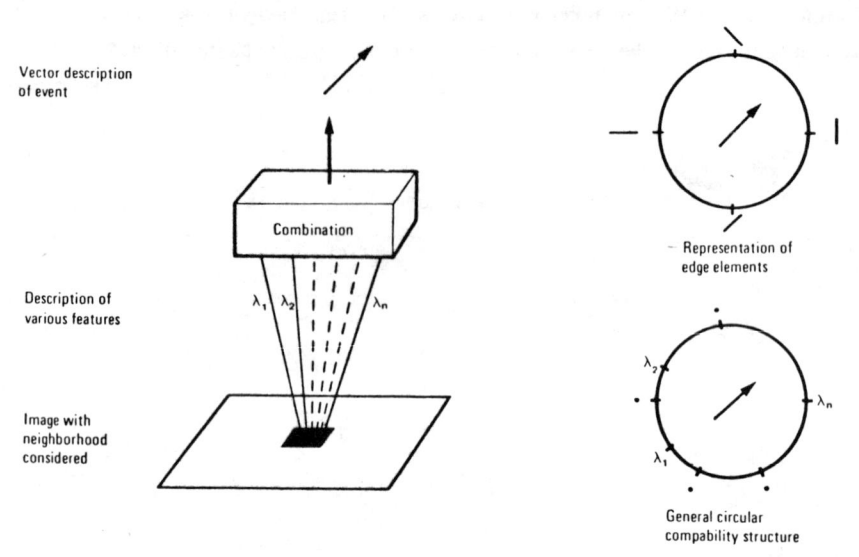

Figure 2 Illustration of feature generation and information representation

STRUCTURE OF OPERATIONS

The general structure of an operation is indicated in figure 3.

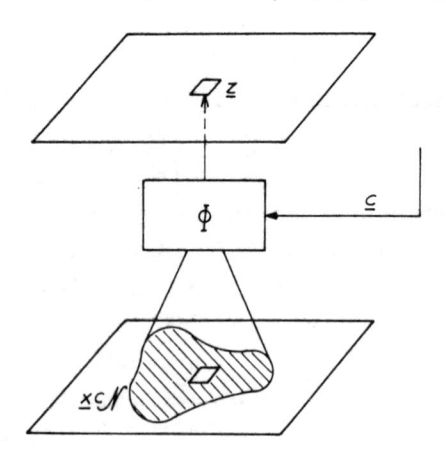

Figure 3 General structure of operation

An output vector $\underline{z}$ is computed as a function of the data elements $\underline{x}$ within the neighborhood N and the context vector $\underline{c}$.

$$\underline{z} = \Phi\,(\underline{x} \subset N,\ \underline{c})$$

The data elements, $\underline{x} \subset N$ within the neighborhood N, the output data $\underline{z}$ and the context descriptor $\underline{c}$ are all assumed to be vectors. In the ensuing discussion we will in particular assume that they are two-dimensional vectors, which does not imply any fundamental limitations of the generality of the approach.

A more detailed formulation of an operation is given in figure 4. The computation of $\underline{z}$, the state description vector, is performed in two steps.

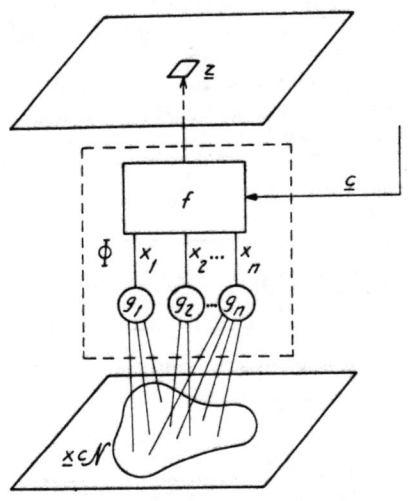

Figure 4 More detailed structure of operation

In the computation of $\underline{z}$, the state within the neighborhood N is described by a number of scalar measurements or features.

$X_k \qquad k = 1, \ldots, n$

Every feature X_k describes a particular aspect of the state within the neighborhood N, e.g. the gradient in a particular direction. There are two ways to formulate X_k:

1 $X_k = g_k (\underline{x} \subset N_{rs}, \underline{c})$

2 $X_k = g_k (\tilde{x}_r, \tilde{\underline{x}}_s, \underline{\Delta}_{rs}, \underline{c})$

In the first formulation X_k is viewed as a function of image elements $\underline{x}$ within a subregion N_{rs}, part of the entire neighborhood N. In the second formulation X_k is a function of two complex valued variables $\tilde{\underline{x}}_r$ and $\tilde{\underline{x}}_s$, which are each estimates describing events within the neighborhoods N_r and N_s, where $N_r * N_s = N_{rs}$. See figure 5. The vector $\underline{\Delta}_{rs}$ is the spatial relative vector between events $\tilde{\underline{x}}_r$ and $\tilde{\underline{x}}_s$; $\underline{\Delta}_{rs} = \Delta_{rs} \exp j\varphi$.

In general the second formulation will be used. An event represented by an estimate $\tilde{\underline{x}}_r$ is assumed to be a condensed representation of the data within a subneighborhood or region $N_r \subset N$. See figure 6 for an intuitive representation of a neighborhood with subregions.

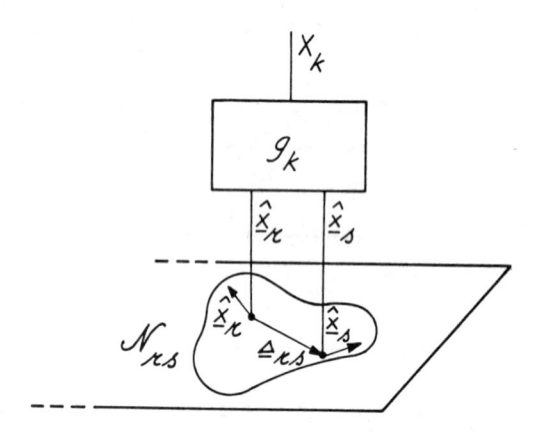

Figure 5 Features derived as functions of events

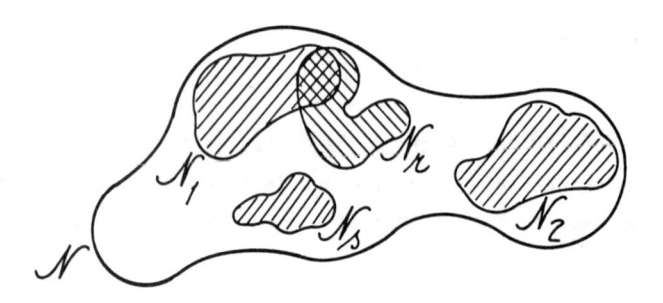

Figure 6 Intuitive representation of a neighborhood with subneighborhoods

It is assumed that the estimate $\tilde{x}_r$ is a weighted sum of the data $\underline{x}_h$ within the region N_r.

$$\tilde{\underline{x}}_r = \sum_{N_r} \underline{w}_{rh} \cdot \underline{x}_h \qquad \underline{x}_h \subset N_r$$

The estimates $\tilde{\underline{x}}_r$ may well not be explicity available during a particular computation, but only implicit in the computation of the features X_k. The estimates $\tilde{x}_r$ have sampling properties dependent upon the size of the neighborhood and of the weighting coefficients $\underline{w}_{rh}$. The implications of these sampling properties have been discussed elsewhere. See Knutsson et al (1981).

INFORMATION REPRESENTATION AND CONTEXT DEPENDENCY

As seen in the preceding section the first part of an operation produces a set of n features X_k which generally all will be nonzero for the particular neighborhood considered. A priori it would be necessary to represent all the components as a description of the state within the neighborhood.

The general principle with regard to information representation is that the original n features produced, X_k, are condensed into only 2, which means that the output is represented as a complex valued variable $\underline{z}$. We will already now suggest the notion that arg ($\underline{z}$) points out a dominant class membership of the state within the neighborhood considered. The magnitude of $\underline{z}$

$$z = |\underline{z}|$$

can be viewed as a deterministic magnitude, a probability, or some form of confidence or certainty.

The main reason for the choice of dimensionality 2 for the output vector is that this is the lowest dimensionality that allows a continous representation of class membership. This does not imply that the description of an image or an image transform is limited to two variables, but that the descriptors are represented by sets of two variables.

This reduction of dimensionality is based upon two assumptions:

1. The assumption of internal compatibility restrictions between features given a particular contextual situation
2. The assumption that only certain aspects of the information within a particular neighborhood are relevant in a particular contextual situation

The first assumption is built upon the observation that certain features are not likely to occur simultaneously within a particular neighborhood given a particular contextual situation. As a typical example we can mention the case at low levels where the simultaneous occurrence of lines and edges at perpendicular orientations is a rather unlikely event. Consequently, two features which a priori could be independent can be mapped into a single variable with opposite signs, when such a restriction is valid.

The second assumption is based upon the belief that an interpretation of an image can only involve certain aspects of the image at a given time. The full discussion of the reasons for these fundamental restrictions is outside the scope of this paper. The assumption is consequently made that in all but trivial situations context has to be invoked to restrict the information propagated to the next level.

An important question is: How can contextual information be represented and in what way does it restrict the information propagated? In this presentation we will have to restrict the treatment of this question to a formal one. It is postulated that the effect of context is to produce a transformation matrix $\underline{C}$ as a function of the context vector $\underline{c}$.

$$\underline{C} = [f(\underline{c})]$$

where $\underline{C}$ is an n x 2 matrix
The output vector $\underline{z}$ is then defined

$$\underline{z} = \underline{C}\,\underline{X} = [f(\underline{c})]\,\underline{X}$$

where $\underline{X}$ is the feature vector or a function thereof.

It is apparent that the expression implies a mapping from n to 2-dimensions of the feature vector $\underline{X}$. This can be visualized as the context vector $\underline{c}$ defining a plane in n-dimensional space upon which the feature vector $\underline{X}$ is projected to produce the 2-dimensional output vector $\underline{z}$.

The general strategy for selection and implementation of the context vector $\underline{c}$ is outside the scope of this paper and still an open question. Strategies that have proved useful have been selection of features that support dominant trends around the neighborhood, and selection among feature sets which at a higher level have the ordinary relationship of compatibility and incompatibility. These strategies have been used for image enhancement and coding as has been described elsewhere. See Knutsson et al (1981) and Wilson et al (1982).

CLASSES OF SYMMETRY OPERATIONS

We will now take a look at the operations producing the feature components X_k. The output vector $\underline{z}$ from an operation can also be viewed as a vector sum

$$\underline{z} = \sum_k X_k \exp j\psi_k$$

where the magnitude function is

$$X_k = g(\underline{\hat{x}}_r, \underline{\hat{x}}_s, \underline{\Delta}_{rs}, \underline{c})$$

and the argument function is

$$\psi_k = f(\theta_r, \theta_s, \arg \Delta_{rs}, \underline{c})$$

where

$$\underline{\hat{x}}_r = \hat{x}_r \exp j\theta_r \quad \text{and} \quad \underline{\hat{x}}_s = \hat{x}_s \exp j\theta_s$$

One standard form of the magnitude function is

$$X_k = \hat{x}_r \cdot \hat{x}_s \cdot g_\theta(\theta_r, \theta_s, \Delta_{rs}, \underline{c})$$

We can see that the magnitudes of the individual events form a product with the remaining part of the function. One view of the magnitudes $\hat{x}_r$ and $\hat{x}_s$ is that they represent a probability of existence of the events. Consequently $\hat{x}_r \cdot \hat{x}_s$ would imply the joint probability of the simultaneous events. A feature is a function of two events in a data space of some dimensionality; typically dimensionality two in the case of image information.

Going back to the implementation of different varieties of operator functions it is hypothesized that we are interested in particular symmetric relationships between the events considered:

1. It appears that wherever we have symmetry around us in nature, it indicates structure as opposed to randomness
2. The recognition or existence of symmetry implies that a data compression can be made
3. Symmetry operations can be viewed as producing reciprocal reference pointers between hierarchically ordered data structures
4. Symmetry operations have an inherent robustness. As indicated earlier the operations are of type differencing operations. This would generally cause sensitivity to noise and instability. The symmetry properties expressed imply the assumption of anisotropy which allows operations to be differencing in one dimension while they are integrating in another dimension
5. The symmetry operations proposed imply continuous functions of the input data, a property that appears necessary in a hierarchical system to allow communication between levels

As one example of symmetry operation we will give the following:

$$\underline{z} = \Sigma\, X_k \cdot \exp j\phi_k$$

where disregarding context

$$X_k = g(\underline{\hat{x}}_{kr}, \underline{\hat{x}}_{ks}) = \hat{x}_{kr} \cdot \hat{x}_{ks} \left|\sin \frac{\theta_{kr} - \theta_{ks}}{2}\right| \text{ and } \phi_k = 2\varphi_k$$

We observe that the magnitude of a particular feature X_k is proportional to the product of the event magnitudes $\tilde{x}_{kr}$ and $\tilde{x}_{ks}$ and dependent upon the difference between the class memberships expressed by θ_r and θ_s. Used at low levels in image processing it provides the dominant orientation and magnitude of a line or edge going through the neighborhood.

The symmetry operations define a hierarchy of information primitives. In the case of 2-dimensional signals the image primitives may imply lines and edges, texture, curvature and divergence, convexity and concavity, partial closure, regions, combinations of regions, and so on.

THE HARDWARE PROCESSOR

Image processing operations generally involve two phases of computation. First convolutional types of operations, in which a set of kernels operate on the image. Secondly, certain non linear functions operate on the data obtained in the first phase. These two types of computation require two different architectures to be performed efficiently, suggesting a structure according to figure 7. A two stage structure has been adopted in the GOP processor.

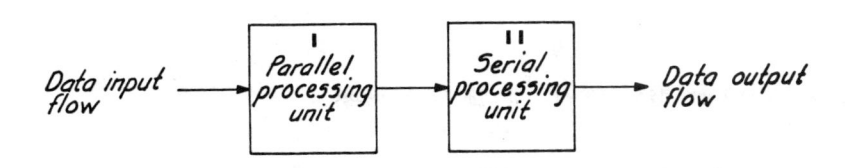

Figure 7 Simplified block diagram of system architecture

Part I of the processor is a reconfigurable pipelined parallel processor, where data from the image segment memory and weights from the kernel memory are combined in four parallel pipelines. Part II of the processor has an entirely different architecture. After processing in part I, the amount of information is reduced considerably. Now a high degree of flexibility is required to combine intermediary results derived by part I. These combinations are usually highly non linear operations, determined from one point to the other by some particular transform of the image to process. Part II is consequently a serial, special-purpose floating point processor. A block-diagram of the processor is shown in Fig 8.

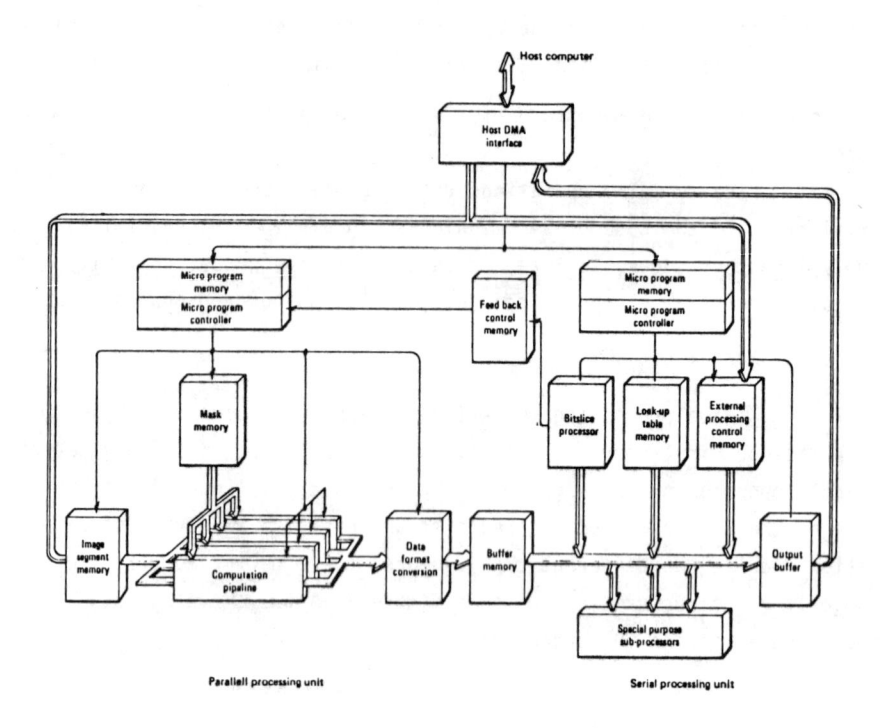

Figure 8   Block diagram of GOP Image Processor

The image segment memory has a capacity of 64K bytes extendable to 32M bytes. It can be restructured to fit the current processing situation, such that any number of input images of any size can be involved in processing simultaneously and operators with a size up to 128x128 pixels can be used. Software in the host computer determines the allowable length of the image segment, which depends upon the number of images and the kernel size. Usually the image segment has a width equal to the size of a neighborhood. Data is moved from the host to the processor one line at a time. There it substitutes the oldest line in a "rolling" fashion.

The kernel data is stored in a memory of size 4x4K words of 24 bits. This storage can be restructured in a number of ways. The normal configuration is that the kernel memory is divided into four sections, each section providing one pipeline with weight coefficients. In parallel with the kernel data memory is another memory of 2x4K words of 24 bits. This later memory contains coordinate displacement values used together with scanning pointer to form the addresses of the selected image points. This means that points can be picked arbitrarily within the 64K byte image segment memory to form a neighborhood subregion of up to 4096 points sampled in any order or arrangement. This allows e.g. kernels of different sizes to be used on different input image planes.

The kernel memory can be organized to contain up to 4096 different kernels with any distribution of size within the limits of the kernel memory itself. These kernels can be freely combined in up to 4096 different kernel sets. Which one of these kernel sets to use can be determined by part II, e.g. in response to image data intended for control of the processing.

The main elements in the pipelines are 8 12-bit ALU:s, 8 12-bit parallel multipliers, 16 16-bit parallel multipliers, 8 4K word x 12-bit look-up tables, 8 24-bit parallel scalers, 8 48-bit parallel scalers, 8 16-bit rounding adders, 8 48-bit accumulator-adders and 96 data-routing multiplexers. In order to allow fast computation, most of the processing in part I of the processor uses fixed point arithmetic. However, great care has been taken not to cause errors due to overflow. Consequently, at the end of a pipeline there is a dynamic range of 48 bit.

Before data enters part II of the processor it is converted to either of three types: 16 bit fixed point, 24 bit fixed point or 24 bit floating point. The data type mode can be selected with respect to demands on accuracy and speed in the computation in part II. The communication from part I of the processor to part II is done over a dual two-port memory of 2x2x1K words of 24 bits. Part I can write into one half of this memory at the same time as part II reads from the other half.

The central units in part II are a program control unit and several arithmetic units. A fast program memory of size 64K bytes gives a cycle time of 150 nsec for the processor. Data and addresses are transferred between the units on five 24 bit buses. The stack and scratch memory is organized as a 2x24 bit two-port memory of 64K bytes (expandable), and it allows two operand reads and one operand store in one basic machine cycle. A memory of 4K words is available for external processing control. In this memory, lines from up to 8 images can be stored to control the processing point by point. The program control unit, which is designed for efficient execution of high-level language control structures, has arithmetic resources for calculation of three operand addresses during one machine cycle.

On the buses are attached a number of arithmetic processing units. They perform common floating and fixed point operations such as addition, multiplication, division, function approximation, parallel scaling and shifting, data type conversion, etc. The computations within the processor can be performed using 16/24 bit fixed point and 24 bit floating point representation in any desired mix. Most fixed point operations are executed in one machine cycle. Floating point addition, multiplication and division require three cycles. The floating point units can, however, be utilized in a pipelined mode in array operations to produce one new result each machine cycle. The function estimation unit allows high speed evaluation of common trancendental functions. A logarithm is for example computed in less than 1 μsec.

Part II controls part I regarding what operations to perform but does not interfere during computations set up for a neighborhood. However, the configuration of the pipelines can be changed after the computation and an entirely different configuration can be set up instantaneously for a different type of computation on the same (or different) neighborhood. This allows maximal flexibility in conjunction with high speed. In normal processing the pipelines remain in the same mode for the whole image, and part I and II run simultaneously at maximum speed with data exchanged over the twin buffer.

SOFTWARE

A processor with the flexibility of GOP requires an extensive software system to provide easy control of the power of the system. The software system has three levels. At the highest level is a language I4PL (Interactive Incremental Interpretative Image Processing Language). This language is developed around a core of LISP, which provides a great flexibility in its ability to handle complex structures of operations. The language gives a simple interactive way to generate procedures through combination of earlier coded operations together with an automatic variation of parameters.

The medium level language is PASCAL, in which the operations are defined. This makes it possible for the advanced user to define his own operations. He can select the appropriate pipe line procedures for part I, and he can in a flexible way define how the intermediary results from part I are to be combined in part II. He also expresses the context dependency of operations from the control images.

The lowest level languages are assembly languages for GOP parts I and II. The average user of the system will never have to be concerned about this level.

This structure and the implementation of the languages has been chosen in order to make the software system transportable between different computers. However, an orientation has been made towards DEC:s computers PDP-11 and VAX-11, as these are the most often occurring ones in image processing systems. A typical configuration of an image processing system using the GOP processor is given in figure 9.

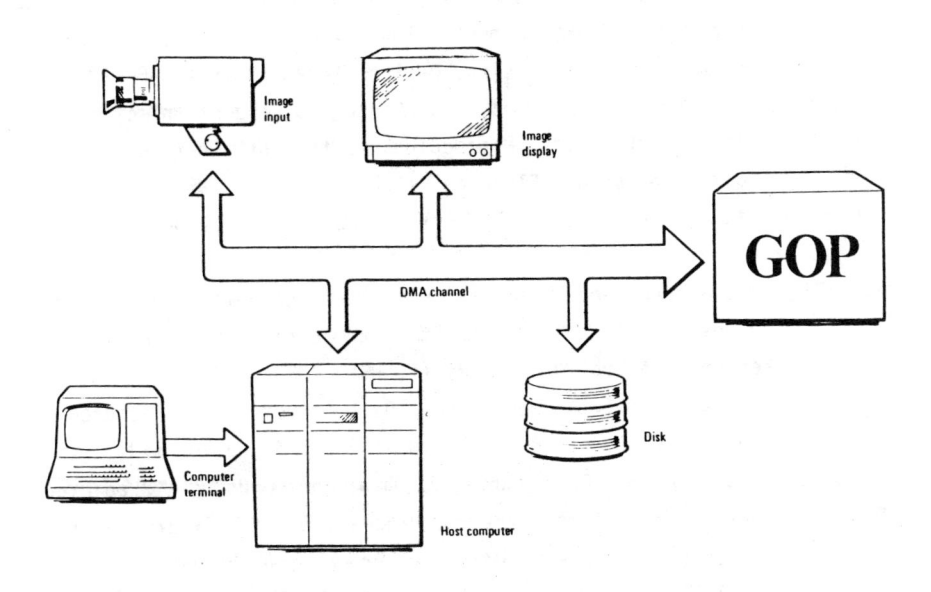

Figure 9   Typical configuration of image processing system

CONCLUDING REMARKS

The first prototype of the GOP processor, which was completed in 1979 has been used for various problems in image analysis, enhancement and coding. See Knutsson et al (1981) and Wilson et al (1982).

ACKNOWLEDGEMENT

This work was supported by the Swedish National Board for Technical Development. The authors also wants to express their appreciation of the enthusiastic work done by the GOP group.

REFERENCES

Danielsson, P-E., Levialdi, S. Computer Architectures for Pictorial Information Systems. Computer Magazine, Nov 1981, pp 53-67

Granlund, G.H., Knutsson, H. Hierarchical Processing of Structural Information in Artificial Intelligence. Proceedings of IEEE Int. Conf. on Acoustics, Speech and Signal Processing, Paris, France, May 1982.

Knutsson, H., Wilson, R., Granlund, G.H. Anisotropic Filtering Operations for Image Enhancement and Their Relation to the Visual System. Proceedings of IEEE Computer Society Conf. on Pattern Recognition and Image Processing, Dallas, Texas, August 3-5 1981

Knutsson, H., Wilson, R., Granlund, G.H. Content Dependent Anisotropic Filtering of Images. Proceedings of the Int. Conf. on Digital Signal Processing, Florence, Italy, Sept 2-5 1981.

Niemann, H. Pattern Analysis. Springer Verlag Berlin-Heidelberg-New York, 1981

Rosenfeld, A. Quadtrees and Pyramids for Pattern Recognition and Image Processing. Proceedings of 5th Int. Joint Conf. On Pattern Recognition, pp 802-811, Miami, USA, 1980

Tanimoto, S., Pavlidis, T. A Hierarchical Data Structure for Picture Processing. Computer Graph. Image Proc. 4, pp 104-119, 1975

Wilson, R., Knutsson, H., Granlund, G.H. Image Coding Using a Predictor Controlled by Image Content. Proceedings of IEEE Int. Conf. on Acoustics, Speech and Signal Processing, Paris, France, May 1982.

LANGUAGES FOR IMAGE PROCESSING

S. LEVIALDI
Institute for Information Sciences, University of Bari, Italy

1. Introduction

Two essential features of a language, any language, are expression and communication and must be kept in mind when designing new formal languages for computer programming. In the case of natural languages, the first feature allows to extrinsecate thoughts in a way which is natural to the speaker whilst the second feature implies the presence of a listener(s) who should understand the message unambiguously or, more plausibly, giving a meaning which is "similar" to the one given by the speaker. The two main goals at which these features are aiming are naturality and precision (in meaning): a language will be very powerful if these goals are reached for a wide number of expressions used in areas of communication; technical, legal, medical, etc.

Words, in a natural language, are created by sound similarity (like "buzz" for the sound of a fly), by metaphor (like "acute" for intelligence, taken from acoustics), by composition like NASA which stands for that organization of which NASA are the initials, or by derivation like joule as an energy unit from the name of a scientist.

One of the main purposes of a word is its understandability which is, in itself, a problem since each word has a meaning(1) which largely depends on time, on the speaker, on the listener, on the purposes of the communication, etc. Since a definition of the meaning of a word is very difficult, generally people refer to a "significant field" which has fuzzy borders; such borders will move in accordance to the original function that was assigned to the word and to the specific function which that word is performing at a given instance.

These problems, which are prominent in the study of natural languages also appear in the area of formal languages and should be remembered when designing new constructs for performing operations on data, in our case, on the pixels of a digital image.

One first point is whether image processing, as a separate field, really needs special purpose languages in order to comply with the tasks which are required in a natural and precise manner. Let us now give the two possible answers to this question: namely there is no real need for a special language (we have enough languages at present) and yes, it would be nice if a specially designed language would be available for image processing.

The first answer is given by people who feel that the effort spent in programming machines for image processing by means of Fortran is so high that it would be really uneconomical to reprogram in a new language. Libraries of subroutines, perhaps written in Assembly language of each of the machines which is used for processing images, do exist and are generally Fortran-callable. Provided a good documentation exists on the meaning of the parameters and on their usage within the software environment where the programs may run, there is no real need for writing new code in another language, they claim. Furthermore a Committee for the standardization of image processing software (funded by the National Science Foundation) has arrived to the same conclusion, namely that it is better to impose some standards on existing and future subroutines rather than design a new language for image processing.

The second answer is provided by people who feel that the new computer architectures(2) which are beeing designed and built will allow the programmer to run the algorithms in a way which is more efficient especially if some form of parallelism is provided . Moreover, a greater transparency of the program, if it is written in a modern language (like

Pascal, for instance) greatly helps in understanding a program written by another person and also helps in modifying an existing program which was written a long time ago. In short: to enable parallel execution of an algorithm and readability of the program. Fortran does not appear as a suitable language.

2. Image operations

Let us now see which are the main operations which are generally performed on images and which should be considered when studying a language: either an old one or a new one. One approach, taken in (3) is to divide these operations as follows: 1) utilities, 2) representation, 3) arithmetical operations, 4) geometrical transformations, 5) enhancement, 6) analysis and 7) classification. In 1) we have all input-output operations, management of special memory areas for the images, etc., in 2) all operations required for video display, printer or plotter output; in 3) the usual arithmetical operations on real variables, integers or boolean also taking into account the possibility of compressing more pixels in the same word. In 4) we have those operations required for registration, scale changes, rotation, perspective generation, etc; 5) cointains algorithms for improving the image quality by enhancement ad restoration, digital filtering, etc; 6) includes segmentation operations, histogramming, statistics of the image, means, medians, etc. and finally 7) has the operations required for classification (statistical classifiers with or without supervisor).

There are nearly 50 languages (operating) for image processing and it is estimated that the man effort to produce these languages is of about. 5 mancentury, for all these languages there is at least one of the operations mentioned in the above classification, which is included. This means that, in general, no language contains all the seven types of operations.

Perhaps it is interesting to note that there are about 2700 formal languages although most of them are dialects of the main ones and some of the them are rarely used.

One possible way to subdivide languages for image processing is with respect to their level (high level like Fortran) or low level like CAP4 (for a machine called Clip 4); compiled languages (like Algol) on interpreted (like APL), languages which do not contain the operating system (like Basic) and those which do contain it (like Glol). Still other possible features for a classification are whether the language contemplates parallel expressions (like GLYPNIR) or not, whether the constructs are of symbolic nature (like Glol, A=B to assign the contents of B to A, where both A and B are arrays) or whether the constructs are parametric like in SUPRPIC where COPY (A,B) is written to execute the same operation.

3. UPL: a Universal Processing Language

There are many ways in which a language may be described and it would be nice if a language for image processing could contain all the nice features we have been discussing, let us now call this perfect language UPL (Universal Processing Language) and summarize the properties of this language. We aim at a _natural_ language meaning that it should be simple to express image processing operations without the need to build up complex expressions; at a _transparent_ language which could allow the immediate understanding of a program without the need to include long detailed comments; it should also be _transportable_ so that it might run on any machine (!). Moreover it would be important for it to be _extendible_ so that for any new construct nothing should be redifined, it should also provide expressions for _parallel_ operations on images with clean control structures. Finally, _interactivity_ should also be allowed for obtaining intermediate results which could be displayed and in order prove correctness of programs, UPL should be formally defined (in the Backus-Naur form). Lastly, it should also be efficient using all the system resources without overloading the memory so as to obtain a high performance speed. It is clearly seen that all these features, some of which are of contrasting nature, are impossible to obtain-yet it is important to specify them so as to have a broad overview of the problems that should be faced when designing a language for image processing.

4. Some languages for image processing

4.1 An interactive command language, SUSIE (4)

It is written in PDP12 assembly language and has two types of instruction-commands, depending on whether the instruction is immediately executed or slightly delayed. All commands work on a 128x128 image having a memory occupation of 8k words of 12 bits; another part of the memory, also of 8k words is reserved for a second image, the first one contains image A (current image) the second part of the memory contains image B (alternative image).

Within the immediate-commands group we have NE (negate) which complements the values of image A, TH32 (thereshold) which imposes a thereshold with value 32 on image A, THX1 which is similar to the above command only that the value of the threshold is the one contained in register X1. Another command AVX1 will compute the mean of values in A and place such value in register X1; 6XLP will execute 6 times a local mean operation on an established subset of elements in A.

Within the delayed-commands group there are macroinstructions which may be defined by means of DM XY = (3 x LP ; GR ; TH4) which means: execute 3 times a local mean, execute a gradient and then put a threshold on 4 whenever a command such as XY is given.

The memory map shows 16k words for A and B, 8k words for SUSIE, 6k words for the library of programs and 2k words for the buffer area; by means of a special command one may look at A or B.

The operator GR (gradient) substitutes for every value of B (b_{ij}) the result obtained from the mean of the absolute value of the differences between a vertical pair of elements added to a horizontal pair of elements centered on a_{ij}:

$$b_{ij} \leftarrow 1/2 \; (\; |a_{i+1,j} - a_{ij}| + |a_{i,j+1} - a_{i,j}| \;)$$

4.2 A high level language which allows parallelism, GLYPNIR(5).

This language was developed in order to program smoothly and efficiently ILLIAC IV which had 64 processors that were interconnected horizontally and vertically. Each one had 2k words of 64 bits in their local memory and one central processing unit which made all the 64 processors execute the same instruction or leave a subset of them idle.

The memory units in this machine could be of three different types: a word, a superword (sword) and a slice. The first one could be addressed by two parameters α and β : α indicates which processor and β indicates the address within the memory of that processor. The sword, made of 64 words of 64 bits has only one address which holds for all the processors; finally the slice has 64 words but each word has a different address within the memory of each processor.

There are two kinds of variables in GLYPNIR: CU which occupy words or word vectors and PE which occupy swords or sword vectors, for example:

    CU INTEGER C1

    CU ALPHA B2

    CU REAL VECTOR Z [100]

and

    PE REAL X,Y

    PE ALPHA B

    PE REAL VECTOR Z [100]

If X and Y are swords then

X ← X + Y

means the sum of 64 words of X with 64 words of Y, the result will be in X.

The superword vector Z is of 64 words of 100 words, for boolean variables which will have 64 values of true/false, the expression will be TRUE only if all 64 values are true, FALSE otherwise. Nevertheless there are other possibilities as will be shown later on.

A special construct, SOME will allow TRUE to hold if "some" values, out of the 64 possible ones, are true and, by the same token, the construct EVERY will produce FALSE if for some processor, the result is false.

GLYPNIR is a structured language, it belongs to the family of Algol-like languages, has dynamic allocation of memory and contains all the control structures of Algol with some specific differences due to the multiprocessor structure of the machine on which GLYPNIR runs.

<BE> is a boolean expression with 64 values of true/false and <S> is a construct (statement), then the control structures which are in GLYPNIR may be listed as follows:

```
IF < BE > THEN < S >
IF < BE > THEN < S1 > ELSE < S2 >
FOR ALL < BE > DO < S >
LOOP    I ← I1, I2, I3   DO < S >
DO < S > UNTIL < BE >
WHILE < BE > DO < S >
GO   TO < LABEL >
```

Let us now look at an example,

IF X < 0   THEN   X = -X   ELSE   X = 0

for the data given below, the successive line produces the results of the statement.

X_0	X_1	X_2	-------------	X_{63}
-2	3	-1		1
2	0	1		0

Note that it is not possible to insert a GO TO statement within an IF, THEN, ELSE since this would imply that each processor would follow a different execution path in the program and this is not allowed by the machine architecture.

This means that if one processor executes the GO TO statement all the others must do the same.

```
        IF      X < O       THEN BEGIN
                X ← -X
                GO TO       HERE
                END
        ELSE    BEGIN
                X ← O
                GO TO       THERE
                END
```

which means: if x < O is true for all the processors then x ← -x

if x < O false (not true for all processors) then x ← O

A more elegant way of writing the same program is the following one:

```
        FOR  ALL  < BE >  DO  < S >
```

which is perfectly equivalent to

```
        IF  < BE >  THEN  < S >
```

In both cases if<BE> is false (all the values must be false) then the next instruction is executed: the ELSE part of the program.

The iterations are performed by LOOP, THRU and FOR until a counter reaches O or by DO, WHILE until a boolean expression is satisfied. If I is an Integer CU variable, then:

THRU I DO < S > means execute < S > I times.

In the following example, S will be executed a different number of times according to the processor which is involved.

FOR I ← I1 STEP I2 UNTIL I3 DO < S >

I is the variable having initial value I1, final value I3 with step I2, but all these Is are superwords (swords) and therefore each processor will be driven by the component of the sword belonging to its memory: its own value of I1, I2 and I3.

Let us now consider an algorithm for the square root extraction of 64 different numbers (one for each processor).

One obvious way would be to use a subfunction (SQRT(C)), the other would be to write an instruction that implements the square root algorithm:

algorithm: $X_j' = 1/2\ (X_j + C_j/X_j)$

termination test:

$$\left| \frac{((X_j)^2 - C_j)}{C_j} \right| < \varepsilon$$

WHILE ABS (XxX - C)/C > EPSILON DO X ← (X + C/X)x.5

which is natural and readable, remembering that 64 processors are acting simultaneously on different data until the convergence of the last one of them.

In closing with this language, GLYPNIR is efficient for ILLIAC IV and allows the parallel access in memory of the data, the parallel execution of the 64 processors so helping the programmer by means of the swords and slices that he may use in parallel algorithms as the basic memory bricks of data. This is achieved without hiding the machine architecture which remains a pioneering effort in the area of new computer architectures for image processing.

4.3 A high level language for multispectral images(6)

A Telemecanique minicomputer T1600 with 32k words of 16 bits is used in managing image files and processing images for remote sensing applications. The high level language for image manipulation was oriented to this machine, it has been formally defined and has the following declarations for variables, particularly suited for image processing.

VAR

IM1 : IMAGE (a, b, c, d); where a is the number of rows,
b is the number of columns,
c is the number of pixels per word,
d is the image channel (its wavelength)

VAR1 : ENTIER:

FVAR;

CONST

DEUX = 2;

FCONST

The control structures are the classic ones:

SI < BE > ALORS < S1 > SINON < S2 > FSSI;

TQ < BE > FAIRE < S > FTQ;

CAS < S > FINCAS;

Some subsets of images may be selected by using a data structure called FENETRE (10, 10, 50, 2, 2) which means: pick up a window having the topmost left corner at coordinates 10, 10 (from the top-left corner) with 50 x 50 pixels, with a packing factor of 2 pixels per word from channel 2.

The contents of two images may be added (as in Basic):

C: = A + B

This language is implemented in assembly code (P1600) and

allows recursive procedures, it is of interactive nature and enables the expansion of already written macroprocedures.

4.4. A Language which allows parallelism, PIXAL(7).

An Algol-based language aiming at two possible implementations: one on a sequential machine (an HP minicomputer) the other on a SIMD machine (CLIP family(8) of machines) was defined formally and implented on the HP Assembly language.

The particular data structures of PIXAL are MASK and FRAME which allow the definition of particular templates with given size, coefficients and pivoting element like

MASK M1 $[1:3, 1:3]$ ON $[1,1]$ OF (0,0,0,1,2,1,0,0,0)

wich stands for a square structure of 3x3 elements which, when read by rows, has the values indicated by the parameters contained in the brackets of OF and is centered on the topmost element at the left (coordinates 1,1).

Whenever the pivoting element (on which the window will be centered) corresponds to the geometrical center of the window there will be no need to specify ON $[\ ,\ ]$ (default option).

The FRAME structure is used to extract submatrices from the image like

FRAME F2 $[1:3,1:3]$ (3x3 submatrix from the image on which the FRAME will be applied).

Two separate ways for expressing parallelism are considered in PIXAL: a global one (the predicate will be TRUE only if it holds for all the elements of the image) and a local one (the predicate is considered on each element and may be TRUE or FALSE locally).

As an example of the global case we may have:

A:= 0   all elements of A are put to 0

A:= A2 + 1   all elements of A2 will be incremented by 1 and
put in A

IF   A=0   THEN GO TO HERE

ELSE   A:= 1   which means execute the instruction at the address HERE only if all the elements of A are equal to 0, otherwise put all the elements of A to 1.

In the local parallelism a special control structure PAR-PAREND was defined which delimits the part of the program which expresses such parallelism. For example

PAR

   IF   C=1   THEN   C:=0

                   ELSE   C:=1

PAREND

stands for a test on each element of image C, if the element is 1 then its new value will be 0, otherwise it will be 1; if image C is a binary image this program will complement it.

A termination test for an image processing algorithm may require to check whether the image is "empty", i.e. is full of 0s.

For example

IF   B=0   THEN GO TO   STOP

          ELSE

             PAR

                IF   B $\neq$ 0   THEN B:= B + 1

             PAREND

STOP

This program, for a 3 by 3 submatrix, in one step would produce the following result

Languages for image processing

	B			B'	
1	0	1	2	0	2
0	3	2	0	4	3
1	2	1	2	3	2

PIXAL also contains internal functions like SUM (B) which adds all the elements of B, COMPARE (M,A) which will output 1 if the contents of a mask M and those of a local neighborhood of an element of A are equal, it will output 0 otherwise.

Two other useful functions are OVERLAP (F,A) and OVERWEIGH (M,A) which allow the extraction of some elements of A (contained in the frame F) for every element of A and the multiplication of some elements of A by the coefficients of the mask M, for all the elements of A.

As a final example, let us consider a filter for extracting vertical lines present in a grey level image by means of a simple mask.

```
PROGRAM "VERTFILT"
BEGIN
GREY ARRAY G [1:100, 1:100] ;
INTEGER MASK MV [1:3,1:3] OF  (0,1,0,0,2,0,0,1,0);
    PAR
        G:= SUM(OVERWEIGH(MV,G))
    PAREND
END
```

The filter will act on all the elements of G, one at a time, producing a new G with enhanced vertical lines.

As it can be seen from these examples, PIXAL is particularly transparent for expressing local operations for image processing and, being a superset of Algol has all the facilities for sequential computation. Unfortunately this project is still on a completion phase and therefore no testing of the compiler has been performed except on the control structures and functions which have been shown in this paper.

4.5 Image Processing C (9)

It is a high level language which may be considered as an adaptation for image processing of a language called C.(10) This language was implemented as a program in the PDP/11/34 driving subroutines in a machine called CLIP4 (11) (a truly SIMD machine with 96x96 processors having 35 bits of memory per processor, the interconnection pattern allows 4-,8- or 6-neighobors per element at will). The C compiler was chosen because it provided good bit-manipulation facilities, it is a modern block-structured language and it was easy to produce programs in separately compiled modules.

The IPC language is the expansion of the C language including a large subroutine library and few new constructs.

The data type that was added required the address(es) of the data in the memory of CLIP and some information about the data (unsigned, 2's complement, etc.). The precision was not fixed in order to take maximum advantage of the CLIP memory, which is addressable in bits since the machine is an array of binary processors. IPC has subroutines to perform dynamic memory allocation and also to swap images to and from secondary memory.

The subroutine library enables to perform arithmetic, boolean, I/O operations as well as local convolutions on images. For example, "mult(im1,im1)" will multiply the data in CLIP associated with im1 with the data associated with im2 putting the result in another area of memory (automatically allocated) and returning (in IPC in the PDP-11) an image structure noting the address, precision, sign, etc of the result.

An assignment operator enables to copy the data in the image on its right-hand side to the image on its left-hand side; the results of appliying a subroutine are returned in a separate image so that nested function calls are permitted.

The assignment operator is "$=" and, as an example, may be

used as follows

          cube ∅= mult (im, mult(im,im));

In order to facilitate program development and algorithm testing, a "command" subroutine is provided which gives a prompt at the user's terminal so that he may display images interactively or print copies on a hard copy device. The images may be called by name, by data associated to them, the user can request absolute memory addresses.

In order to fully appreciate the nature of IPC, we enclose here an example program(9) which extracts the connectivity number (i.e. the number of 8-connected objects - the number of 4-connected holes). In order to do this, two masks of 2x2 elements are used to compute the number of points (binary elements) which satisfy the specified neighboring conditions:

          connectivity number = n(1 0) - n(1)
                                   (0 0)   (1 0)

"n( )" means the number of points where the given mask matches the neighbor values.

```
/*
 * Example of IPC - find connectivity number of binary image
 *                  ( = (number of objects) - (number of holes) )
 *
#include        <clip.h>              /* defines images etc. */
main( )
{
        image   pic;                  /* declare variables */
        masktype        mask1, mask2;
        int     n1, n2, con_no;
        openclip();                   /* initializes CLIP4 */
        setname(pic);                 /* initializes image variables */
```

```
set8mask( X X X
          X 1 0
          X 0 0 , &mask1);      /* define first mask */


set8mask( X 1 X
          1 0 X
          X X X , &mask2);      /* define second mask */
pic $= inpics(1);               /* take binary picture from
                                   camera */
n1 = volume(mark(pic,mask1));
              /* find no. of points where first mask fits */
n2 = volume(mark(pic,mask2));
              /* find no. of points where second mask fits */
con_no = n1 - n2;
printf("Connectivity number = %d\n",con_no);
```

"Mark" is a subroutine which returns an image which is one at the points where the mask fits, and zero elsewhere. "volume" returns the sum of all the pixels.

It is interesting to note that this provides a quick way of counting objects, because it only takes one CLIP operation to fill in the holes in an object.

IPC is transparent and efficient, at least for using on a CLIP4 machine: it is therefore not very portable. There are no easy ways to program sequential algorithms on it since it has been directed mainly towards parallel processing when using a SIMD machine such as CLIP4. Serial algorithms will be expressed in the standard C language and will run on the PDP computer.

The authors suggest another language for interactive image processing on CLIP4; a language called POPX which is an extension of a previous language designed for artificial intelligence (POP-2). The main difference with IPC is that POPX is interpreted (whilst IPC was compiled). Some operators act directly on images, like "im1 + im2 / im3" or statements like

: varsim pic;   ! declare variables !
: inpic(6)>>pic;   ! take in 6 bit picture from camera!
: volume(pic) / (96.0 - 96.0)=>43.20! now find average grey level!

":" is the prompt by the computer, ">>" assigns the value of the (image) expression on the left-hand side to the right-hand side, and "=>" prints the value of an expression.

For example, an algorithm for edge extraction which is based on the substitution of each value of a pixel by the maximum value in its local neighborhood and is coded under "^^", will be programmed in POPX as follows:

: varsim edges;
: ^^pic - pic >> edges;   !large values give strong edges!
: edges > 2 >> edges;   !now threshold to give binary image!

The comparison between "edges" and 2 returns a binary image which is one where the condition is true.

In closing this part which specifically refers to high level languages for the CLIP4 machine, we may refer here the conclusions of the authors of these two languages (extensions of C and POP-2 respectively).

The use of these languages is strongly limited by the small memory (35 bits) available to each processor since parallel operations often require partial copies of intermediate results and therefore extra memory. IPC is used more than POPX because large programs are executed at a higher speed in a compiled language; the interactive facilities of IPC (which are of a reduced nature) suffice for most paractical purposes; compilation time is short compared with the advantages in writing and editing programs.

5. Conclusion

High level languages for image processing will certainly become useful tools for designing new algorithms for image analysis and for testing a variety of architectures, perhaps by simulation. There are still a number of open questions as to the constructs which be added to existing modern languages (of the Pascal family) in order to simplify image data handling, to specify parallel control structures and to provide all the necessary commands for interactive use of the system and for image retrieval from a pictorial data base. As can be seen all these requirements add up to a considerable amount of complexity and of conflictuality especially regarding efficiency on a specific machine since the language, being high level, should be machine independent and therefore not optimized to any particular machine.

REFERENCES

(1) B. Bréal, "Essai de sémantique", 5a ed. Paris, 1921.

(2) P.E. Danielsson, S. Levialdi, "Computer Architectures for Pictorial Information Systems", IEEE Computer, November 1981, 53-67.

(3) J.K. Preston, Jr., "Image manipulative languages, a preliminary survey", Workshop on High Level Languages for Image Processing, Windsor, 1979.

(4) B. Batchelor, "Interactive image analysis as a prototyping tool for industrial inspection", Workshop on High Level Languages for Image Processing, Windsor, 1979.

(5) D.H. Lawrie, T. Layman, D. Baer, J.M. Randal, "GLYPNIR: a programming language for ILLIAC IV", Com. ACM, vol. 18, 3, 1975.

(6) F.C. Argilas, S. Castan, A.M. Gleizes, "Un langage specialise dans le traitement d'images", 2° Congress Reconaissance des Formes et Intelligence Artificiel, Toulouse 1979.

(7) S. Levialdi, A. Maggiolo-Schettini, M. Napoli, G. Tortora, G. Uccella, "On the design and implementation of PIXAL, a language for image processing" in Languages and Architectures for Image Processing, edits. M.J.B. Duff and S. Levialdi, Academic Press, London, 1981, 89-98.

(8) M.J.B. Duff, "The Cellular Logic Array Processor", Comp. Journal, 20, 1977.

(9) G.P. Otto, D.E. Reynolds, "High Level Languages for CLIP4", 1981, Report, Image Processing Group, Dep. of Physics and Astronomy, University College London.

(10) B.W. Kernighan, D.M. Ritchie, "The C Programming Language", Prentice-Hall, Englewood Cliffs, New Jersey, 1978.

(11) T.J. Fountain, "CLIP4 Hardware Manual", Report 81/1, Image Processing Group, Dep. of Physics and Astronomy, UCL (1981).

(12) A. Maggiolo-Schettini, "Comparing some high-level languages for image processing" in Languages and Architectures for Image Processing, edits. M.J.B. Duff and S. Levialdi, Academic Press, London, 1981, 157-164.

ACHIEVING PORTABILITY IN IMAGE PROCESSING SOFTWARE PACKAGES

S. W. Krusemark
Department of Electrical Engineering, Virginia Polytechnic
Institute and State University, Blacksburg, VA 24061 USA

R. M. Haralick
Department of Electrical Engineering, Virginia Polytechnic
Institute and State University, Blacksburg, VA 24061 USA

Abstract. The first key to portability is in the use of a kernel of routines that interface to the peculiar operating system of each machine. The kernel provides sophisticated but standard operating system services required by the image processing software. It makes the operating system of each computer appear identical and, nicest of all, when carefully designed it does not pose a difficult implementation problem. Above this interface, all image processing applications programs can be machine-independent, written in a structured language such as RATFOR, without sacrificing power or ease of use on any machine.

1.0 INTRODUCTION

Transportability is achieved by writing image processing applications programs that obtain all operating system services from the standardized operating system interface. Moving to a new machine then requires only the implementation of the kernel supporting the standard calls.

1.1 Footnotes

Some of the research that went in to the kernel was funded by NSF grants to VPI&SU, University of Maryland, and Rensselaer Polytechnic Institute. Detailed technical discussion of the overall set of conventions described here can be found in Hamlet and Haralick (1981) or Hamlet and Rosenfeld (1979) and detailed discussions of the kernel conventions can be found in Krusemark and Haralick (1981) or Guerrieri (1981).

1.2 Operating system interface

Operating systems all support the capabilities for file operations, memory allocation, process control, input/output and interrupt handling. Machine independence can be achieved by obtaining these services only through prespecified subroutine calls that remain the same from computer implementation to computer implementation. This collection of entry points constitutes a kernel that makes all operating systems look the same to the image processing package.

Three factors shape the definition of the kernel:

1. The services must be powerful enough and easy to use.
2. The kernel must be capable of being built around any existing operating system.
3. The implementation of the kernel for a new system must be easy for a local systems expert.

The first two factors tend to increase the size of the kernel; the last limits its size.

In an operating system whose services are nearly the same as those of the interface, the kernel would be only a calling-sequence converter, transforming the subroutine calls into monitor calls with some additional error checking added. When the system provides very different services, however, the code may be as large as a few thousand lines of code. Section 2.0 discusses the operating system interface and section 3.0 discusses the image I/O primitives that sit on top of the operating system interface.

1.3 Command and process interface

An image processing package must communicate with its users, then carry out the tasks they specify. The jobs of scanning commands and interacting with a user are unlike the processing that takes place after the command is decoded. This job can be accomplished in a command processing module.

In the command-processing module, all interaction with the user results in a standardized description of a processing request. Any routine may later use this information without concern for details of format. Furthermore, the information is checked once and for all at the beginning. For example, a file may be required to exist, have a certain format, etc. These matters can be interactively straightened out with the user before the information is stored for the next module. The processing module that then deals with the standardized request can be conventionally linked to the command routines, or overlaid.

Unfortunately, the general overlay organization is an unsatisfactory mechanism because the implementation capabilities differ widely in different computers. This is a fundamental transportability problem. Our approach to the overlay problem is to make a simple kind of overlay a standardized operating system service called "program exchange," in which the executing code calls for itself to be replaced by another ready-to-execute program module. This simple overlay mechanism is supported by every modern operating system.

1.4 Image I/O primitives

The image I/O routines can be entirely written in a high level language like RATFOR. They format the image data appropriately, and call upon the kernel I/O routines to accomplish the data transfers to and from the image file. The format for an image must be general, allowing images to be any practical size and any number of bands. Logical records can be subimage blocks of arbitrary number of rows and columns. The most common format would have logical records correspond to one image row. Since image data often does not have high precision, the data can be packed. This bit packing occurs inside the image I/O routines and is completely transparent to the programmer. Section 3.0 discusses the details of the image I/O routines.

1.5 RATFOR

RATFOR is a structured FORTRAN language that has the constucts: IF THEN ELSE, REPEAT UNTIL, WHILE, and blocks of code. Thus an IF statement can execute more than one statement such as:

```
IF( A > B )
    [
    A = B
    B = C + B
    ]
ELSE
    [
    B = A
    A = C + A
    ]
```

Another example is the DO loop:

```
DO J = 1, 20
    ARR(J) = 0
```

to zero out an array of 20 positions. Note that unless otherwise blocked ([]) the object of a DO, IF, or WHILE is a single statement.

RATFOR permits the use of symbolic names instead of numeric quantities. These symbolic names are also helpful mnemonics for the programmer. An example is the use of .OLD instead of the integer 1 as the parameter in an open call to indicate that an old file is to be opened. The RATFOR preprocessor translates .OLD into a 1 so that the output is FORTRAN compatible. The use of the symbolic names instead of the numeric value is accomplished by the RATFOR DEFINE statement. Thus .OLD is translated to the numeric 1. A CHARACTER statement can be implemented by translating CHARACTER to INTEGER, for example. A table of such DEFINEs can be built that define system wide constants and by using the INCLUDE statement in RATFOR, these DEFINEs can be put in every file on the system. All uses of such symbolic names in this paper start with the period for clarity.

The INCLUDE statement in RATFOR is a mechanism to merge one file into another. This, for example, allows a single COMMON to be defined in an INCLUDE file. If a change is necessary the INCLUDE file is changed and the system is recompiled. Without the INCLUDE one would have to find ALL occurrences of the common and change them (hopefully finding them all) and then rebuild the system. The INCLUDE file can be used for tables of DEFINEs so that machine constants can be changed quickly, for commons, and even sometimes for code.

2.0 OPERATING SYSTEM INTERFACE CONVENTIONS

The operating system services provided by the kernel include program exchange, random sequential file handling, memory allocation, and user interrupt handling. As discussed in section 2.1, we suggest each kernel subroutine be an integer function subprogram whose value indicates a completion condition code. This facilitates error handling. Section 2.2 discusses program exchange, section 2.3 discusses file operations, section 2.4 discusses memory allocation subroutines and section 2.5 discusses interrupt handling.

2.1 Error handling

When experienced users make use of well-tested software, error conditions arise only occasionally. But while the users are learning, or the software is under development, most processing is error handling. In providing error returns, a set of subroutines should make it easy for the calling program to deal with complex error situations, yet at the same time the overhead should be low. Simple situations should not require the caller to make use of all aspects of the error mechanism. For software development, there is another important factor: it must be easy to quarantee that _every_ error is detected, even those that cannot occur once the software is working properly.

To take care of error processing, each kernel interface routine is an INTEGER FUNCTION. If there are no errors, the routine returns a non-negative number (whose value may have significance as a part of normal processing). However, if there are errors to report, the routine delivers a negative value indicating the type of error.

In the sections that follow, this error mechanism is described in full only for the first routine (OSCHAN in section 2.2).

2.2 Program exchange

At any time during execution, a program can terminate by calling for its successor. The routine accomplishing this is:

INTEGER FUNCTION OSCHAN(PROGRAM)

where PROGRAM is a string of characters identifying the new program. The only potential failures for OSCHAN involve the nonexistence of PROGRAM.

The error handling works the following way. Let us suppose that the code -3 is assigned to the file error that the named file does not exist. (Perhaps -1 means that there was a read error, -2 is end of file, and so on.) The caller may anticipate that the new program may not exist, so that

```
      IEV = OSCHAN(PRG)
      IF (IEV == -3)  # DOESN'T EXIST
         WRITE(6, 1)
    1 FORMAT(' NO SUCH PROGRAM')
```

is a call in which the anticipated error is explicitly processed.

All routines in the kernel treat errors in this way, but the details will be suppressed in the discussions to follow. This example is typical of the presentation in other ways: the code is written in RATFOR, and little attention is paid to FORTRAN conventions about variable names, in the interests of clarity.

2.2.1 <u>Parameter passing for program exchange</u>. When passing control from one program to another parameters are needed to allow information to be passed as well as control. These parameters could be passed by the user in a program dependent way but since many machines do have fast parameter passing capability and a transportable version of these routines can be written, the routines OSSEND and OSRECV were created. The two routines to send and receive information through a program change are:

        OSSEND (PARAMETERS,LENGTH)
        OSRECV (PARAMETER,READLENGTH,BUFFERLENGTH)

The send and receive are capable of repeated use to allow several sets of parameters to be sent. This makes passing several arrays of numbers, file names, etc. possible without the need to recopy them into a single rather long array. An example is to send two arrays with the new program name in PRG

        OSSEND (ARRAY1,50)
        OSSEND (ARRAY2,5)
        OSCHAN (PRG)

The receive end looks slightly different because the send and receive operate like a push-down stack (first-in-last-out) so:

    OSRECV (ARRAY2,LEN2,5)
    OSRECV (ARRAY1,LEN1,50)

(Error checking on returned length can be done.)

The push-down stack concept allows layers of code similar to subroutine calls and returns. Each exchanged program reads in only the last series of arrays sent by OSSEND.

2.3 File operations

Because file operations differ greatly from system to system, it is necessary to duplicate most of the operations performed by the local operating system (and the input-output control system, if there is one) in the interface. Fortunately, most of the coding can be done in RATFOR, and much of it carries over from machine to machine.

Files can be random access files or sequential files. The operations which programs need to perform on such files are SETUP, OPEN, READ/WRITE, and CLOSE. The first, second, and last could be done by the transfer operations, but they are needed for certain special actions, and they always provide useful error control. Furthermore, an OPEN operation can be used to verify the existence of a file, and to acquire its present characteristics, even if no transfers are contemplated. Similarly, a CLOSE operation can perform an action like deleting the file.

The complete file name is stored in an array (one character per word), and passed to a standard interface routine: OSINFD. This routine moves the name into an array called FILE. The array FILE, is called a file descriptor, and is then passed to each system interface routine. The routine looks like:

OSINFD(FILENAME,FILE)

Space is provided in the array FILE for the routines to assign and maintain some kind of internal description invisible to the user. The array parameter containing the file descriptor is designated FILE in each of the following routines to be described.

A new random file descriptor must be initialized with the correct size information before the open. This is the job of OSPINF whose calling sequence is:

OSPINF(FILE,ATTRIBUTE,VALUE)

This routine puts the (ATTRIBUTE,VALUE) pair into the file descriptor array FILE. The .MODE (record mode) attribute determines whether the file is treated as an integer, real, etc., file. The .LREC (record length) attribute determines the length of each record. The .NREC (number of records) attribute determines for random files the number of records in the file. These are the most commonly used options. The other attributes will not be discussed due to space limitations. An example is to set number of records:

OSPINF(FILE, .NREC, 60)

The OPEN of a random file is done by a call to OSOPNR:

OSOPNR(FILE, TYPE)

where TYPE can be .NEW or .OLD. TYPE is .NEW to create a new file and .OLD to open an old file. To open a new file, the file descriptor must have the required number of records and record length. To determine the number of records and record lengths of an old file, the routine OSGINF is available. OSGINF is the reverse of OSPINF.

No two calls to OSORNR (without an intervening OSCLOS) may name the same file, with one exception: using different arrays for the name, a file may be opened once .OLD and once .NEW at the same time. The intent is to allow a copy-and-update operation that does not happen in place.

All files must be closed since unclosed files may not necessarily exist after a program exchange. OSCLOS provides the close service:

OSCLOS(FILE,OPTION)

The action of the close depends on the TYPE of open and the OPTION:

OPTION	TYPE	ACTION
.DELETE	.OLD	The existing file is deleted. Any open NEW files are not touched.
.DELETE	NEW	The new file is deleted. This may leave an old copy (which may or may not be also open at the time).
.KEEP	.OLD	The existing file is closed.
.KEEP	.NEW	This is where any problems that will occur can occur. The disk is checked for a pre-existing file. If one exists, it is deleted or otherwise removed from consideration. The NEW file is closed so that the file replaces the pre-existing file. If a pre-existing file does not exist, then the new file is closed such that it appears with the correct names etc.

The actual data transfer to or from open files is done with:

OSRDR(FILE,RECORDNUMBER,BUFFER,READLENTH,WAIT)
OSWTR(FILE,RECORDNUMBER,BUFFER,WRITELENGTH,WAIT)

The routines read from and write to the file having FILE for its file descriptor. In the BUFFER array is the data. The mode (INTEGER, REAL, etc.) is determined by the mode specification of the last call to OSPINF. Since the read and write lengths can be other than a full record, the RECORDNUMBER is the starting record number (the first record is one). If the length is greater than a single record, then multiple records are transferred. If the length is shorter than a full record, the extra data written is garbage on a WRITE and on READ it is ignored.

If the WAIT parameter is .WAIT, return does not occur until the operation is complete. If the wait parameter is .NOWAIT, then the data transfer is started but control returns to the calling program immediately allowing the program to continue until the data is actually needed, at which time

OSWAIT(FILE,WAIT)

is called. The WAIT can be used again if the program wants to only check if the operation is done.

The last operation specific to the random files is the ability to change the number of records on a random file.

OSGROW(FILE, NREC)

allows the NREC (number of records) to be changed on a random file. There are two ways to implement this depending on the machine: 1) to copy the file, necessary on machines that cannot grow files, and 2) to simply let the file get bigger. If the code is implemented using OSGROW rather than the machine dependent capability of making a file can change its size then transportability is enhanced.

Sequential input/output is more complicated because the kernel handles files and devices such as terminals, printers, magnetic tapes, etc. The first call is to OSINFD which creates the file descriptor. The second call is to OSPINF and sets the characteristics of the operation. Then the file is opened by a call to:

OSOPNS(FILE,TYPE)

The variable TYPE can take the value .INPUT or. OUTPUT.

The same as specified for the Random files can happen for a .INPUT and .OUTPUT opened with the same name. The operation of .INPUT is like .OLD and .OUTPUT is like .NEW when the close is done. Devices are specified by special reserved file names such at TT for terminal and PRINT for line printer.

The reads are done by:

OSRDS(FILE,BUFFER,ACTUALLENGTH,BUFFERSIZE)

Since in many cases the calling program will not know the actual length until after the read is done, the amount of space available is specified by BUFFERSIZE.

The write is done by:

OSWTS(FILE,BUFFER,LENGTH)

Since the data length is known, this is the simplest of the routines.

2.4 Memory allocation

Dynamic memory allocation can be difficult because it involves manipulation of addresses. To solve this in a transportable system, the mainline code for each image processing operation becomes a subroutine:

SUBROUTINE OSMAIN(DYNARRAY)

The programmer writes this routine as if it were the main program, but is provided with an argument at the outset. The parameter is an array that may be passed about among the routines of the package, and which has been set up by the calling side of OSMAIN so that it can change in size up to some set internal maximum. Initially, DYNARRAY will have one element, but following a call to

OSALOC(SIZE)

it will be as if it were dynamically altered to

INTEGER DYNARRAY(SIZE)

OSALOC changes the size of the dynamic array DYNARRAY that was passed to the user via OSMAIN. The user requests the number of integer words (SIZE) that is needed and the routine OSALOC checks to see if this is available. If it is, a value of .OK is returned, otherwise a negative value is returned. This array is intended to be the main user work area.

This dynamic allocation can be implemented in two different ways. The first is very simple and easy to do: simply allocate a very large array and pass it to OSMAIN. Then OSALOC simply checks if the request is for too much memory. The second is to set the base of the array at the end of the user memory space. Then OSALOC extends user memory by the requested amount. Thus the user only pays for what is used.

2.5 Interrupt handling

As an example of the use of an interrupt service, imagine the problem of terminating unwanted output. Most systems have some means of alerting a running program to a user "attention" typed in, implemented as an interrupt. If the user invokes this during output, the interface main program will take control and then call OSINTR. OSINTR sets a flag which the print routine tests before each line, and returns in the "dismiss" mode. Printing terminates as soon as that flag is tested and found to be present. After printing is stopped the package program proceeds normally.

The same initial system main program (described in the last section) can set up interrupt processing. It can set up the system interrupt to branch to a particular location. Then, should an interrupt occurr, it can call the user supplied routine:

OSINTR(TYPE)

TYPE describes the problem that caused the interrupt so that the processing in the package can be intelligent. By providing two kinds of returns from OSINTR, the package can "dismiss" the interrupt normally and continue (perhaps after taking corrective action); or, it can specify a "restart" in which the main program again calls subprogram OSMAIN, and thus cuts short the interrupted code forever. If OSMAIN is written to test a global flag, it can know whether it is starting or restarting.

3.0 IMAGE I/O ROUTINES

The image I/O primitives take care of file I/O for image data files. They permit image data to be handled at one logical level higher than the I/O provided by the kernel. The image I/O routines can be written as entirely portable code and include an open, a read, a write, and a close. Haralick (1977) discusses the image access protocol conventions described here. The open (RDKINL) is passed an array containing values some of which must be initialized for opening a new file and others of which do not have to be. This array is called the IDENT array. On opening an old image file the image files IDENT array is returned. This array is accessable to the user and is passed to the read and write routines (RREAD and RWRITE).

3.1 IDENT array

The IDENT array must have certain specified parameters defined in order to be legal for opening a new image file. The IDENT array is always zeroed first and so a 'not set' value is zero. The values that must be set are the logical size of the image, and its mode (integer, real, double integer, or double precision). If it is an integer image then either the number of discrete levels of grey shades, the minimum/maximum grey-tone values, or the number of bits to hold the grey-tone shades must also be set. Other optional parameters include the subimage block size (if different from an image row), the number of bands in the image, and the number of symbolic bands. If not set these optional parameters default to reasonable values.

3.2 Error handling

Error handling for the image I/O primitives as well as for all machine independent code uses an event variable IEV and alternate return %XXXX mechanism. The argument %XXXX is an alternate return as defined by FORTRAN, the XXXX standing for a statement number such as 1234. When no errors are generated in the routine a normal return is taken and processing continues at the statement following the call. If, however, an error occurrs the IEV event variable is set and the alternate return is taken and processing continues at the statement with the label XXXX given in the call.

3.3 Random file open

The subroutine RDKINL (Random Disk Initialize) performs the random file open for both new and old files.

CALL RDKINL(FD, IDENT, OLDNEW, IEV, %XXXX)

The IDENT is the identification array as described earlier. The argument OLDNEW has one of the two values denoted by the symbolic definitions .NEW and .OLD, for new and old files, respectively, to be opened.

On a new file the values in the IDENT array are checked for consistency and unspecified values are initialized. Then a random file of the correct size is created and the IDENT array is written to the first record of the file.

On an old file the routine first checks for its existence on the disk. If it does not exist an error is generated and the alternate return is taken. If it does exist it is opened and the first record is read into the IDENT array.

3.4 Random file read/write

The routine RREAD takes the data on the image and copies it to the buffer and the routine RWRITE copies the data in the buffer to the image, unpacking and packing where needed.

CALL RREAD

(FD,BUF,BND,BLKNO,IDENT,WAIT,IEV,%XXXX)

CALL RWRITE

(FD,BUF,BND,BLKNO,IDENT,WAIT,IEV,%XXXX)

Since image access is random, the two arguments BND and BLKNO are needed to get the correct band and block number from the SIF file. Blocks are numbered starting in the upper left corner and proceed down the left side to the bottom of the image. The word block corresponds to a subimage or logical record.

The buffer (BUF) should be dimensioned for the correct size of one block. Only one block can be obtained at a time. However, if a value of zero (0) is specified for the band number argument (BND) the block specified is returned for all bands. Normal use is to specify the band number and block number thus returning only one block from one band. Also, the most common shape of a block is a row of the logical image. The argument WAIT has a value indicating whether or not the I/O must complete before returning.

3.5 File close

CLOSE compliments the open, in this case RDKINL.

CALL CLOSE(FD)

All files must be closed. The subroutine CLOSE can be used to close any file descriptor. (It is not an error to close an already closed file, but all files must be closed before exiting or they may cease to exist as expected.)

3.6 Use and example of image I/O routines

The following example will help to clarify the order and use of the image I/O routines.

The RATFOR INCLUDE, MACA1, defines the standard system-wide set of symbolic definitions is the first line of "code". The file descriptors FDI, FDO are for input and output name information. The data BUF is passed from above so that the calling program needs to make it large enough to do what needs doing.

The code first opens the input file with RDKINL, sets up the output image sizes, and opens the output also with RDKINL.

The next section is the actual algorithm which in this example simply copies the input to the output.

After the algorithm is finished for all bands and all lines then the files must be closed, and finally a return to the calling program is made.

Errors are checked such as the check that the number of points per line is less than or equal to the buffer size. Errors once detected are dealt with at the bottom of the routine in the manner shown. This starts the alternate return chain.

```
  INCLUDE MACA1
#
  SUBROUTINE NUMBCH( FDI, FDO, BND, NBND, LBUF,
                THRSH, IEV, * )
#
  REAL THRSH
  CHARACTER FDI( .FDLENGTH ), FDO( .FDLENGTH )
  INTEGER BUF( LBUF ), IDENT( .IDLENGTH )
  INTEGER OBND, IBND
  INTEGER JDENT( .IDLENGTH ), BND( NBND )
#
  CALL RDKINL( FDI, IDENT, .OLD, IEV, %9000 )
#        open input file and temp file
#
  IF( IDENT( .IDNPPL ) > LBUF ) GOTO 9010
#        check if too large an image
#        set up output IDENT record
#
```

```
    DO I = 1, .IDLENGTH
       JDENT(I) = 0
 #
  JDENT( .IDNPPL ) = IDENT( .IDNPPL )
 #         output number of points per line
 #         same as input file
 #
  JDENT( .IDNLINS ) = IDENT( .IDNLINS )
 #         output number of lines = input
 #
  JDENT( .IDNBITS ) = 8 # number of bits is 8 bits
 #         ( 0 to 255 )
  JDENT(.IDNBNDS ) = NBND
 #
 #         default to 1 band, row format
 #         image INTEGER image.
 #
  CALL RDKINL( FDO, JDENT, .NEW, IEV, %9000 )
 #
 #         This section of code is the 'real'
 #         image processing.
 #
 #         This example is simply a copy from
 #         input to output
 #
  NLIN = IDENT( 7 )
 #
  DO OBND = 1, NBND
     $(
     IBND = BND( OBND )
     DO LIN = 1, NLIN
         $(
         CALL RREAD( FDI, BUF, IBND, LIN, IDENT,
                    .WAIT, IEV, %9000 )
         CALL RWRITE( FDO, BUF, OBND, LIN, JDENT,
                    .WAIT, IEV, %9000 )
         $)
     $)
 #
  CALL CLOSE( FDI )        # close input file
  CALL CLOSE( FDO )        # close output file
 #
 #
  RETURN
 #
  9000 CONTINUE
 #
 #         Error in lower routine.
 #         IEV already set.
  GOTO 9999
  9010 CONTINUE
 #
  IEV = -3013    # buffer smaller than data to go
                   in it.
```

```
#
 9999 CONTINUE
      CALL CLOSE( FDI )    # even on error must close
                                all files
      CALL CLOSE( FDO )    # input and output
#
      RETURN 1             # take alternate return
#
      END
```

4.0 OMITTED CONVENTIONS

Since the kernel is a complicated concept it is difficult to fully explain in the small amount of space available in this paper. The full technical discussion is available in Krusemark and Haralick (1981). Also left out of the discussion in this paper is the concept that devices such as terminals and printers are basically a form of sequential file. The undiscussed sequential I/O primitives OSRDS and OSWTS allow terminal graphics as well as the normal mode of terminal interaction.

One thing that some software packages do not do, but which is important to do is a form of history keeping. This record keeping should be maintained in the standard image file so that any and all steps that a particular image has been through can be recorded. Because of space limitations we can not describe here the conventions we suggest for these routines.

5.0 GIPSY

There is a system called GIPSY (General Image Processing System) at VPI & SU that uses these routines and although it has been in existence a short while we have already transported over 100,000 lines of code from an IBM 370/VM running CMS to a VAX 11/780 running VMS in about one man week. It took less than seven man weeks for a programmer initially unfamiliar with the VAX 11/780 to write the kernel. GIPSY currently has over 200,000 lines of code.

6.0 REFERENCES

Guerrieri, E., Software/O.S. Interface Kernel User Manual, Preliminary Version, Techincal Report, IPL, Rensselaer Polytechnic Institute, March 1981.

Hamlet, R.G. and R.M. Haralick, Transportable "Package" Software, Software Practice and Experience, to appear.

Hamlet, R.G., and A. Rosenfeld, Transportable Image-Processing Software, Proc. Nat. Computer Conf., Vol 48, AFIPS Press, June 1979, pp 267-272.

R.M. Haralick, Image Access Protocol for Image Processing Software, IEEE Transactions on Software Engineering, SE-3 (1977), pp. 190-192.

Krusemark, S.K. and R. M. Haralick, Operating System Interface, Technical Report, SDA 81-1, VPI&SU, April 1981.